00326717

PNEUMATIC HANDBOOK

8th Edition

PNEUMATIC HANDBOOK

8th Edition

By Antony Barber
C.Eng., M.Sc., F.I.Mech.E., M.R.Ae.S.

U.K.	Elsevier Science Ltd, The Boulevard, Langford Lane, Kidlington, Oxford OX5 1GB, U.K.
U.S.A.	Elsevier Science Inc., 665 Avenue of The Americas, New York 10010, U.S.A.
JAPAN	Elsevier Science Japan, Tsunashima Building Annex, 3-20-12 Yushima, Bunkyo-ku, Tokyo 113, Japan

Eighth Edition 1997

Library of Congress Cataloging-in-Publication Data
Barber, Antony
    Pneumatic handbook/by Antony Barber—8th ed.
      p. cm.
    Includes index
    ISBN 1-85617-249-X
    1. Compressed air—Handbooks, manuals, etc.   2. Pneumatic machinery—Handbooks, manuals, etc.
I. Title.
TJ985.P54   1997
621.5'1–dc21   97-33006 CIP

British Library Cataloguing in Publication Data

A catalogue record for this book is available from the British Library

ISBN 1 85617 249 X

Printed in Great Britain by Galliard (Printers) Ltd, Great Yarmouth

Published by

Elsevier Advanced Technology
The Boulevard, Langford Lane, Kidlington, Oxford OX5 1GB, UK
Tel: +44 (0) 1865 833842
Fax: +44 (0) 1865 843971

Preface

Elsevier Science present the latest edition of this comprehensive work on all aspects of compressed air. There have been many advances in hydraulic and electric technology in recent years, yet compressed air still keeps its place as an important medium for power generation, transmission, and control. Its versatility makes it the ideal choice for many applications for which no other form of power is as suitable. As in previous editions, the PNEUMATIC HANDBOOK covers the whole range of pneumatic engineering in one volume. It starts with the basic principles and properties of air and includes the theory of air compression, the principles of air treatment and the design of compressors, tools, machinery, actuators, sensors and valves. Low pressure and vacuum applications are also covered.

The importance of efficiency in the generation and transmission of compressed air is emphasised throughout and useful guidance is given on the design of factory installations to achieve this. The author has made a survey of pneumatic equipment available in the UK, Europe and elsewhere in the world; the latest significant developments, wherever produced, are described. Legislation introduced by the European Community by way of Technical Directives in respect of safety in the use of compressed air, pressure vessel design, noise control and environmental control is discussed at length. British, American and ISO Standards are an important source of design material and reference is made to them throughout the book. Included is a comprehensive list of Standards, Codes of Practice and Technical Directives.

The book is written with needs of the practical engineer in mind, and only sufficient theory is included for a proper understanding of the principles involved.

The publishers are grateful for the assistance of manufacturers who have supplied information and for the trade associations CETOP, Pneurop, British Compressed Air Society (BCAS) and the Compressed Air and Gas Institute (CAGI). Acknowledgement is given in the text for material where it has been used.

Antony Barber
antony.barber@btinternet.com

Acknowledgements

Acknowledgements are due to the following for contributions to particular chapters. When an illustration of a proprietory product is used in the text, further acknowledgement is given where it occurs.

Atlas Copco	Desoutter
British Compressed Air Society (BCAS)	Flair
British Fluid Power Association (BFPA)	Institution of Mechanical Engineers
British Standards Association (BSI)	International Standards Organisation
CETOP	Pneurop
CompAir Broomwade	ultrafilter international
Compair Holman	University of Bath
CompAir Maxam	Willpower Breathing Air

The union of three Worldclass brands

The union of three Worldclass brands makes Parker Pneumatic *the* force in pneumatic control and automation, providing modern and innovative products for virtually any application; anywhere.

First choice for pneumatics

By providing our customers with first class products, coupled with speedy and reliable delivery, we are fast becoming the first choice for users of pneumatics anywhere in the world . We'll make sure you find the products you need, regardless of whether they are

in conjuction with our customers to meet the strictest operational requirements; whether it's high-temperature, salt spray or

food hygiene requirements etc., we've got what it takes.

Premier pneumatic partnership

Working in close partnership with the U.K.'s leading pneumatics distribution network, enables us to provide not only premier customer service, but unrivalled local off the shelf deliveries from a range of products second to none.

Schrader Bellows

standard off the shelf components or custom built to suit the needs of your application.

Solutions to your problems is our business

With the combined expertise and application knowledge resulting from this union, we are able to offer a wide number of 'industry specific' products designed

Atlas Automation

Parker Hannifin plc

Pneumatic Division
Walkmill Lane • Bridgtown
Cannock • Staffs • WS11 3LR
Tel • 01543 456000 • Fax • 01543 456161

Contents

SECTION 1

Basic Principles

NOMENCLATURE AND UNITS

TERMS AND DEFINITIONS

PROPERTIES OF AIR AND GASES

GAS LAWS AND THERMODYNAMICS

NOMENCLATURE AND UNITS

Nomenclature

Unless otherwise specified in the text, the following symbols are used consistently throughout this book:

A = cross sectional area (m²)

c_p = specific heat at constant pressure (J/kg K)

c_v = specific heat at constant volume (J/kg K)

C = flow coefficient for orifices and nozzles = discharge coefficient corrected for velocity of flow = $C_d/(1-\beta^4)$

C_d = discharge coefficient for orifices

D = internal diameter of pipe (m)

d = internal diameter of pipe (mm)

f = Darcy–Weisbach friction factor

g = gravitational constant (9.81 m/s²)

K = coefficient of heat transmission W/(m² K)

k = thermal conductivity W/(m K)

L = length of pipe (m)

M = mass (kg)

N_u = Nusselt number = $Ql/k\Delta\theta$ — see text

N = rotational speed (r/min)

n = rotational speed (r/s)

P = absolute pressure (N/m²)

P' = gauge pressure (N/m²)

p = absolute pressure (bar)

p' = gauge pressure (bar)

P_r = Prandtl number = $(Cp\mu)/k$ — see text

Q = heat (J)

q = flow rate (m³/s)

R_o = universal gas constant = 8.314 J/(K mol) = 8314 J/(K kmol)

R = individual gas constant (J/kg K) = R_o/M where M = molecular weight of the gas

R_e = Reynolds number = $\rho vl/\mu$ — see text

S = stroke length (m)

t = time (s)

T = absolute temperature (K) = temperature in °C + 273

V = volume (m^3)

V_s = specific volume (m^3 /kg)

v = velocity (m/s)

w = mass of gas (kg)

W = work done (kW)

Z = compressibility factor

Greek letters

β = ratio of small to large diameter in orifices and pipes

γ = ratio of specific heats = c_p /c_v

Δ = signifies an increment in a unit

ε = absolute roughness of pipes (mm)

ν = kinematic viscosity (m^2/s)

μ = dynamic viscosity (Pa s)

ρ = fluid density (kg/m^3)

ϕ = relative humidity

η = efficiency

η_a = isentropic (adiabatic) efficiency

η_v = volumetric efficiency

ψ = ratio of air consumed to air delivered by compressor

Subscripts for diameters and areas

$_1$ defines smaller

$_2$ defines larger

Subscripts for fluid flow, pressures, temperatures and volumes

$_0$ defines base or reservoir conditions

$_1$ defines initial, inlet or upstream

$_2$ defines final, outlet or downstream

Other subscripts are defined as they occur

Note on units

SI units are used consistently throughout this handbook. Certain conventions within the SI structure are used generally throughout the European compressed air industry. These are as follows:

Compressor capacity:	the traditional unit for compressor capacity is m³/min, which most manufacturers prefer to quote. Litre/s (l/s) is however the recommended unit (1 l/s = 0.06 m³/min).
Specific power consumption:	kW h/m³ or J/l.
Pressure:	bar is preferred usage for air pressure, although Pascal is theoretically better (1 bar = 10⁵Pa). MPa is also found.
Length:	metre for dimensions greater than 1 metre, millimetre for dimensions less than 1 metre.
Mass:	kilogram or tonne.
Torque:	Newton metre (N m).
Rotational speed:	rev/min is widely used although rev/s is the preferred unit.
Moisture content:	kg/m³ or g/m³.

Conversion factors

For the convenience of those readers more familiar with Imperial units, the following conversion factors may be used.

The values quoted are adequate for most engineering purposes, *ie* accurate to at least 3 significant figures:

Acceleration	1 ft/s²	= 0.305 m/s²
Area	1 ft²	= 0.093 m²
Density	1 lb/ft³	= 16.02 kg/m³
Energy, work, heat	1 ft lbf	= 1.356 J
	1 Btu	= 1055.06 J
Entropy	1 Btu/°F	= 1899 J/K
Force	1 lbf	= 4.448 N
	1 tonf	= 9.964 kN
Length	1 in	= 0.0254 m
	1 ft	= 0.3048 m
Mass	1 lb	= 0.4536 kg
	1 ton	= 1.016 tonne
Moment of inertia	1 lb ft²	= 0.04214 kg m²
Power	1 hp	= 0.746 kW
Pressure	1 lbf/in²	= 0.06895 bar = 6895 Pa
Specific capacity	1 (ft³/min)/hp	= 0.03797 (m³/min)/kW
Specific energy	1 Btu/ft³	= 0.01035 kWh/m³
Temperature	1 °F	= 9/5 °C + 32
Torque	1 lbf ft	= 1.356 N m

Velocity	1 ft/s	= 0.305 m/s
Volume	1 ft^3	= 0.0283 m^3
Volume flow	1 ft^3/s	= 0.0283 m^3/s
Viscosity (dynamic)	1 lb s/ft^2	= 47.88 Pa s
	(n.b. 1 centipoise = 10^{-3} Pa s)	
Viscosity (kinematic)	1 ft^2/s	= 0.0929 m^2/s
	(n.b. 1 centistoke = 10^{-6} m^2/s)	

TERMS AND DEFINITIONS

Terms and definitions used in the Compressor Industry (see also BS 1571 and BS 5791):

Adiabatic compression
One in which the compression takes place with no heat transfer across its enclosure.
(See isentropic.)

Aftercooling
Removal of heat from the air after compression is completed.

Clearance volume
The volume remaining inside a compression space at the end of the compression stroke.

Compressibility factor
Ratio of the actual volume of a compressed gas to its volume calculated in accordance with the ideal gas laws.

Compression ratio
Ratio between discharge and final pressure.

Compressor
A machine which draws in air or gas at one pressure and delivers it at a higher pressure.

Compressor, booster
In which the inlet gas has already been compressed.

Compressor, displacement
In which the pressure rise is achieved by allowing successive volumes of air to be inspired and exhausted from an enclosed space by displacement of a moving member, *eg* piston, screw and sliding vane.

Compressor, dynamic
In which the pressure rise is achieved by converting kinetic energy into pressure energy in its passage through the machine, *eg* radial and axial fixed vane.

Compressor, multistage
In which the increase in pressure takes place in two or more stages.

Compressor capacity
The volumetric flow rate of the gas as compressed and delivered, referred to the conditions of temperature and pressure at the inlet. For air this is referred to as Free

Air Delivered (FAD). When referred to standard conditions, 1.013 bar and 15°C, this is known as standard capacity. BS 5791 suggests that this expression should be replaced by "volume rate of flow".

Condensate	Liquid formed by condensation of water vapour.
Displacement	Actual volume swept by the first (or only) stage of a displacement compressor.
Efficiency, adiabatic	Ratio of the adiabatic power consumption to the shaft input power.
Efficiency, isothermal	Ratio of the isothermal power to the shaft input power.
Efficiency, volumetric	The ratio of the compressor capacity to its displacement.
Filter	A device for removing contaminants from a fluid.
Free air	Atmospheric condition at the inlet point of a compressor.
Free Air Delivered	See above under Compressor capacity.
Ideal compression	Conditions appertaining to the isentropic compression of an ideal gas.
Intercooling	The cooling of compressed air (or gas) between compression stages.
Isentropic process	A thermodynamic process which takes place at constant entropy. ($\Delta Q/T$ = constant.) A reversible adiabatic process is also isentropic, so the two terms are often used interchangeably in compressor technology. Isentropic is the preferred usage when describing a compression cycle.
Isothermal process	A thermodynamic process which takes place at constant temperature. Isothermal compression is not realisable in practice, since it would have to take place very slowly. However it is theoretically the most efficient and is therefore used as the basis of comparison.
Packaged compressor	A complete powered compressor unit ready to operate.
Polytropic process	Compression or expansion of a gas where the relationship PV^n = constant applies.
Pressure ratio	Ratio between absolute discharge pressure and absolute inlet pressure.
Pressure regulator	A valve or similar device for reducing line pressure to a lower (constant) pressure.

Pulsation damper	A chamber designed to eliminate pulsations present in air flow.
Relative clearance volume	Ratio of clearance volume to volume swept by a compressing element.
Relative humidity	Ratio of water vapour actually present in air to the saturated water vapour content, at a given temperature and pressure.
Separator	A device for removing liquids (*eg* water and oil) from compressed air.
Specific capacity	Compressor capacity per unit of input power.
Specific energy	Input power per unit of compressor capacity.
Stage pressure ratio	Compression ratio for any particular stage in a multistage compressor.
Surge limit	The flow limit below which operation of a dynamic compressor becomes unstable.
Temperature, absolute	Temperature above absolute zero in Kelvin.
Temperature, ambient	Temperature of the environment.
Temperature, inlet	Total temperature at the standard inlet point of the compressor.
Temperature, discharge	Total temperature at the standard discharge point of the compressor.
Temperature, total	Temperature at the stagnation point of an airstream.
Vacuum	Difference between atmospheric and a lower system pressure.
Vacuum pump	A machine for generating a pressure below atmospheric.

PROPERTIES OF AIR AND GASES

Clean, dry air is a mechanical mixture of approximately 78% by volume of nitrogen and 21% oxygen, the remaining 1% being made up of minor quantities of other gases (see Table 1). The composition of air remains substantially the same between sea level and an altitude of about 20 kilometres, but its density decreases with increasing altitude, and varies with pressure and temperature. At sea level, at a pressure of 1 bar and a temperature of 15°C, the density of air is 1.209 kg/m^3. Thus 1 kg of air has a volume of 0.827 m^3.

At a specific temperature and pressure, the number of molecules in a unit volume is constant for any gas or mixture of gases. For a temperature of 0°C and a pressure of one atmosphere, this number is 2.705 x 10^{19} molecules per cubic centimetre.

At standard temperature and pressure, the mean velocity of gas molecules is of the order of 500 m/s with a mean free path between intermolecular collision, for air, of the order of 3 x 10^{-16} millimetres. The rate of collision under such conditions is responsible for the pressure exerted by the air (or any gas) on a surface immersed in it, or on walls containing the gas. The pressure therefore of any gas depends on its mass density, the number of molecules present and their mean velocity.

The effect of a change in temperature is to modify the value of the mean velocity. Hence in the absence of any other change, the resultant pressure will vary with the temperature. Similarly, any change in volume will effectively modify the mass present and again affect

TABLE 1 – Typical composition of dry air (ISO 2533)

Components	Mass %	Vol %
Oxygen O_2	23.14	20.947
Nitrogen N_2	75.52	78.084
Argon Ar	1.288	0.934
Carbon dioxide CO_2	0.048	0.031 4
Hydrogen H_2	0.000 003	0.000 05
Neon Ne	0.001 27	0.001 818
Helium He	0.000 073	0.000 524
Krypton Kr	0.000 33	0.000 114
Xenon Xe	0.000 039	0.000 008 7

the pressure. Thus pressure, temperature and volume are interrelated. This relationship can be expressed in the following form for a perfect gas (which assumes that the molecules are perfectly elastic, are negligible in size compared with the length of the mean free path and exert no force on each other):

$$PV = wRT$$

Note that an alternative form of the equation is:

$$PV_s = RT$$

where V_s = specific volume in m^3/kg.

The former is to be preferred.

TABLE 2 – Vapour pressure of moist air
(pressure in millibars)

| Temp. | Relative humidity in percent | | | | | | | | | |
°C	10	20	30	40	50	60	70	80	90	100
−10	0	1	1	1	1	2	2	2	2	3
−5	0	1	1	2	2	2	3	3	4	4
0	1	1	2	2	3	4	4	5	5	6
5	1	2	2	3	4	5	6	7	7	8
10	1	2	4	5	6	7	9	10	11	12
15	2	3	5	7	9	10	12	14	15	17
20	2	5	7	9	12	14	16	19	21	23
25	3	6	10	13	16	19	22	25	29	32
30	4	8	13	17	21	25	30	34	38	42
35	6	11	17	22	28	34	39	45	51	56
40	7	15	22	30	37	44	52	59	66	74
45	10	19	29	38	48	57	67	77	86	96
50	12	25	37	49	62	74	86	99	111	123

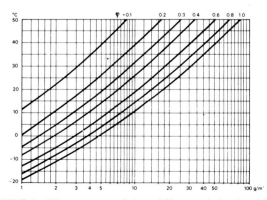

FIGURE 1 – Water content of air at different relative humidities.

The value of R in SI units is 8314 J/(K kmol), which for air is equivalent to 287 J/(kg K). For other gases see Table 4.

Atmospheric air

Atmospheric air normally contains water vapour, and the total pressure of the air is the sum of the partial pressures of the dry air and the water vapour (follows from Dalton's Law of Partial Pressures). The air is saturated when the partial pressure of the water vapour is equal to the saturation pressure of the water vapour at that temperature. The saturation pressure is dependent only upon the temperature (see also Table 2 and Figure 1).

When air is cooled at constant pressure, the *dew point* is reached when the partial pressure is equal to the saturation pressure. Any further cooling will then result in water separating by condensation (see also Table 3).

Water content of moist air

The water vapour content of moist air can be determined from the following equations derived from the gas law:

$$m_w = \frac{P_s V_t}{R_w T} \times 10^5$$

$$m_A = \frac{(P - \phi P_s)V}{R_A T} \times 10^5$$

where:

m_w	=	mass of water vapour in kg
m_A	=	mass of dry air in kg
ϕ	=	relative humidity
P_s	=	saturation pressure of water vapour at given temperature (see Table 3)
V_t	=	total volume of air/water vapour in m³
R_W	=	gas constant for water vapour = 461.3 J/(kg K)
R_A	=	gas constant of dry air = 287.1 J/(kg K)
T	=	absolute temperature of air/water vapour mixture (K)

The gas constant of the air/water vapour mixture can also be determined as:

$$R_m = \frac{(287.1 m_A) + (461.3 m_w)}{(m_A + m_w)}$$

From this the density of the air/water vapour mixture follows as:

$$\rho = \frac{P_s}{R_m T} \times 10^5$$

where ρ is the density in kg/m³

TABLE 3 – Saturation pressure P_s and density P_w of water vapour at saturation.
Values below 0°C refer to saturation above ice

t °C	P_s mbar	P_w g/m²		t °C	P_s mbar	P_w g/m²
-40	0.128	0.119		5	8.72	6.80
-38	0.161	0.148		6	9.35	7.26
-36	0.200	0.183		7	10.01	7.75
-34	0.249	0.225		8	10.72	8.27
-32	0.308	0.277		9	11.47	8.82
-30	0.380	0.339		10	12.27	9.40
-29	0.421	0.374		11	13.12	10.01
-28	0.467	0.413		12	14.02	10.66
-27	0.517	0.455		13	14.97	11.35
-26	0.572	0.502		14	15.98	12.07
-25	0.632	0.552		15	17.04	12.83
-24	0.699	0.608		16	18.17	13.63
-23	0.771	0.668		17	19.37	14.48
-22	0.850	0.734		18	20.63	15.37
-21	0.937	0.805		19	21.96	16.31
-20	1.03	0.884		20	23.37	17.30
-19	1.14	0.968		21	24.86	18.34
-18	1.25	1.06		22	26.43	19.43
-17	1.37	1.16		23	28.09	20.58
-16	1.51	1.27		24	29.83	21.78
-15	1.65	1.39		25	31.67	23.05
-14	1.81	1.52		26	33.61	24.38
-13	1.98	1.65		27	35.65	25.78
-12	2.17	1.80		28	37.80	27.24
-11	2.38	1.96		29	40.06	28.78
-10	2.60	2.14		30	42.43	30.38
-9	2.84	2.33		31	44.93	32.07
-8	3.10	2.53		32	47.55	33.83
-7	3.38	2.75		33	50.31	35.68
-6	3.69	2.99		34	53.20	37.61
-5	4.02	3.25		35	56.24	39.63
-4	4.37	3.52		36	59.42	41.75
-3	4.76	3.82		37	62.76	43.96
-2	5.17	4.14		38	66.26	46.26
-1	5.62	4.48		39	69.93	48.67
0	6.11	4.85		40	73.78	51.19
1	6.57	5.19		41	77.80	53.82
2	7.06	5.56		42	82.02	56.56
3	7.58	5.95		43	86.42	59.41
4	8.13	6.36		44	91.03	62.39

Condensation of water in compressed air

A common problem in compressed air is the calculation of the amount of water which condenses out when a compressor takes in air at one temperature and pressure and delivers it at another. A further equation, which derives from the above, may be used.

The mass of water which can be held in air at a temperature of T_2 K and a pressure of P_2 bar per 1000 litres (1 m³) at a temperature of T_2 K and a pressure of P_1 bar is given by

$$m_W = \rho_1 \; \frac{1 - \phi P_{S1}/P_1}{1 - P_{S2}/P_2} \; \times \; \frac{T_2}{T_1} \; \times \; \frac{P_1}{P_2}$$

P_{S1} is the vapour pressure at P_1 (from Table 3)
P_{S2} is the vapour pressure at P_2 (from Table 3)
ρ_1 is the density at P_1 (also from Table 3)

As an example a compressor takes in 1000 l/s of air at a relative humidity of 80% at 15°C and 1 bar and delivers it at 8 bar. Find the mass of water condensing out when the discharge air is 30°C.

Mass of water in 1000 litres of saturated air at 15°C from Table 3 = 12.83.

Mass of water at 80% relative humidity = 12.83 x 0.8 = 10.264.

Mass of water at 30°C and 8 bar can be found from the above equation:

$$30.38 \times \frac{1 - \dfrac{0.8 \; \times \; 0.01704}{1}}{1 - \dfrac{0.04243}{8}} \; \times \; \frac{303}{288} \; \times \; \frac{1}{8} = 3.962$$

Therefore the mass of water condensed out per 1000 litre = 10.264 - 3.962 = 6.302 g. The condensation rate for a compressor of 1000 l/s is 6.302 g/s.

Effects of compressibility

The properties of real gases differ from those of the ideal gas whose behaviour follows exactly the gas equation PV = wRT.

For most practical purposes this equation is adequate for engineering calculations. It is sometimes necessary to take into account the effect of compressibility near to the liquid state of the gas. This is conveniently done by replacing the gas equation by PV = ZwRT. In this expression, Z is known as the compressibility factor; it is the ratio of the actual volume of the gas to its value calculated in accordance with the ideal gas law.

Compressibility charts are available for a number of gases. These are derived experimentally and should be used if available. Alternatively a generalised chart can be used; such a chart is based on the fact that most gases behave in the same way when close to their

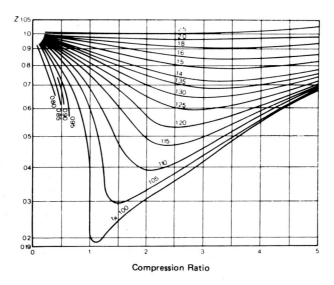

Compression Ratio

FIGURE 2 – Generalised compressibility factor Z.

critical temperature and pressure. Figure 2 shows such a chart. Another way of expressing this is by saying that nearly all gases behave the same when their reduced pressures and reduced temperatures are equal. The reduced pressure is the ratio of the actual pressure to the critical pressure and similarly the reduced temperature is the ratio of the actual temperature to the critical temperature.

When dealing with a pure gas the procedure is as follows:

a) Look up the critical pressure, P_{cr}, and critical temperature T_{cr}, from Table 4.

b) Calculate $P_R = P/P_{cr}$ and $T_R = T/T_{cr}$,   where P and T are the actual working pressure and temperature.

c) Read off the value of Z from Figure 2 .

When dealing with a mixture of gases such as natural gas, it is necessary first to calculate the *pseudo*-critical temperature and the *pseudo*-critical pressure of the mixture. If there are m gases in the mixture:

$$P_{RM} = \sum_1^m a_n P_{RN}$$

$$T_{RM} = \sum_1^m a_n T_{RN}$$

where P_{RM} and T_{RM} are the *pseudo*-critical pressure and *pseudo*-critical temperature of the mixture

TABLE 4 – Physical data for some gases

Gas	Formula	Molecular mass	Gas constant J/(kg K)	P_{cr} bar (abs)	T_{cr} °C	Normal boiling point °C	c_p kJ/(kg K)	$\gamma = c_p/c_v$
acetylene	C_2H_2	26.04	319.4	62	36	-83	1.659	1.26
air	—	28.96	287.1	38	-141	-194	1.004	1.40
ammonia	NH_2	17.03	488.2	113	132	-33	2.093	1.30
argon	Ar	39.94	208.2	49	-122	-186	0.522	1.67
benzene	C_6H_6	78.11	106.5	49	289	80	1.726	1.08
butane, N-	C_4H_{10}	58.12	143.1	38	152	-1	1.635	1.10
butane, iso-	C_4H_{10}	58.12	143.1	36	135	-12	1.620	1.10
butylene	C_4H_8	56.11	148.3	39	144	-7	1.531	1.11
carbon dioxide	CO_2	44.01	188.9	74	31	-78	0.833	1.30
carbon disulfide	CS_2	76.13	109.2	75	278	46	0.657	1.20
carbon monoxide	CO	28.01	296.6	35	-140	-191	1.039	1.40
carbon tetrachloride	CCl_4	153.84	54.0	44	283	77	0.833	1.18
chlorine	Cl_2	70.91	117.3	77	144	-34	0.498	1.35
cyan	$(CN)_2$	52.04	159.8	61	128	-21		1.26
dichloromethane	CH_4Cl_2	86.95	95.6	103	216	41		1.18
ethane	C_2H_6	30.07	276.5	49	32	-89	1.714	1.19
ethyl chloride	C_2H_5Cl	64.51	128.9	53	185	-12		1.13
ethylene (ethene)	C_2H_4	28.05	296.4	51	9	-104	1.515	1.24
Freon 12	CCl_2F_2	120.92	68.8	39	112	-30	0.611	1.13
Freon 13	$CClF_3$	104.46	79.6	39	29	-82	0.644	1.15
Freon 14	CF_4	88.01	94.5	37	-46	-128	0.694	1.16
Freon 22	$CHClF_2$	86.47	96.1	49	96	-41	0.646	1.20
helium	He	4.00	2077.2	2	-268	-269	5.208	1.66
hexane, N-	C_6H_{14}	86.17	96.5	30	234	69	1.617	1.08
hydrogen	H_2	2.02	4124.5	13	-240	-253	14.270	1.41
hydrogen bromide	HBr	80.92	102.8	82	90	-68		1.42
hydrogen chloride	HCl	36.46	228.0	83	52	-85	0.792	1.41
hydrogen cyanide	HCN	27.03	307.6	49	184	-89		1.27
hydrogen sulphide	H_2S	34.08	244.0	90	101	-59	0.996	1.33
hydrogen iodide	HI	127.93	65.0	—	151	-36		1.40
krypton	Kr	83.80	99.3	55	-64	-153	0.249	1.68
methane	CH_4	16.04	518.3	46	-83	-162	2.205	1.31
methyl ether	$(CH_3)_2O$	46.07	180.5	52	127	-23		1.11
methylene chloride	CH_2Cl_2	84.94	97.9	63	238	40	0.826	1.18
methyl chloride	CH_3Cl	50.49	164.7	67	143	-24	0.826	1.25
neon	Ne	20.18	412.0	27	-229	-246	1.032	1.66
nitrogen	N_2	28.02	296.8	34	-147	-196	1.038	1.40
nitric oxide	NO	30.01	277.1	65	-94	-152	0.990	1.39
nitrous oxide	N_2O	44.02	188.9	73	37	-89	0.874	1.30
oxygen	O_2	32.00	259.8	51	-118	-183	0.915	1.40
ozone	O_3	48.00	173.2	69	-5	-112		1.29
pentane, N-	C_5H_{12}	72.14	115.3	34	197	36	1.624	1.06
pentane, iso-	C_5H_{12}	72.14	115.3	33	187	28	1.601	1.06
propane	C_3H_8	44.09	188.6	43	97	-42	1.620	1.13
propylene (propene)	C_3H_6	42.08	197.6	44	92	-48	1.521	1.15
sulphur dioxide	SO_2	64.07	129.8	79	158	-10	0.607	1.25
water (superheated)	H_2O	18.02	461.4	221	374	100	1.860	*)
xenon	Xe	131.30	63.3	58	17	-108		1.66

*) for wet steam: $c_p/c_v \approx 1.035 + 0.1\ X$
where X is the steam content (X>0.4)
for superheated steam (up to 10 bar and 800°C):
$c_p/c_v \approx 1.332 + 0.112 \times 10^3 T$
where T is the temperature in °C

TABLE 5 – Attenuation of sound in air (dB/km)

f(kHz)	Relative humidity (%)									
	0	10	20	30	40	50	60	70	80	90
1	0	12	6	4	4	4	4	4	3	3
2	1	40	19	13	10	9	9	8	8	8
4	3	110	68	44	32	27	23	21	20	19
8	10	200	210	150	120	92	77	66	60	55
16	41	280	460	450	380	320	270	230	210	190
20	64	320	550	600	540	460	400	350	310	280

TABLE 6 – Standard atmosphere according to NASA

Altitude m	Pressure bar	Temperature °C	Density kg/m³
0	1.013	15.0	1.225
100	1.001	14.4	1.213
200	0.989	13.7	1.202
300	0.978	13.1	1.190
400	0.966	12.4	1.179
500	0.955	11.8	1.167
600	0.943	11.1	1.156
800	0.921	9.8	1.134
1000	0.899	8.5	1.112
1200	0.877	7.2	1.090
1400	0.856	5.9	1.060
1600	0.835	4.6	1.048
1800	0.815	3.3	1.027
2000	0.795	2.0	1.007
2200	0.775	0.7	0.986
2400	0.756	-0.6	0.966
2600	0.737	-1.9	0.947
2800	0.719	-3.2	0.928
3000	0.701	-4.5	0.909
3200	0.683	-5.8	0.891
3400	0.666	-7.1	0.872
3600	0.649	-8.4	0.854
3800	0.633	-9.7	0.837
4000	0.616	-11.0	0.819
5000	0.540	-17.5	0.736
6000	0.472	-24.0	0.660
7000	0.411	-30.5	0.590
8000	0.356	-37.0	0.525

a_n is the proportion of the nth component in the mixture

P_{RN} is the reduced pressure of the nth component

T_{RN} is the reduced temperature of the nth component

The value of Z is found from Figure 2 as before.

The factor Z can then be used to modify the gas equation as above.

The velocity and attenuation of sound in air

For any gas at moderate pressure, the velocity of sound is given by:

$$c = \sqrt{(\gamma RT/m)}$$

where m is the molecular weight. Thus to a first order, c is proportional to $\div T$ and is independent of the pressure. At low frequencies, the velocity of sound c_o, at 20°C in dry air is 331 m/s. The effect of water vapour on the velocity of sound is complex; as water is added to dry air the velocity at first decreases so that at 14% relative humidity, it is $0.995c_o$; above 30% relative humidity, the velocity increases linearly with increasing moisture content such that at 100% relative humidity it is $1.003c_o$.

The attenuation of sound in air varies with temperature, water vapour content and frequency. It is usually expressed in dB/km. Table 5 gives some typical values for the variation in attenuation with frequency and humidity at 20°C and a pressure of 1 atmosphere.

GAS LAWS AND THERMODYNAMICS

The gas laws are based on the behaviour of perfect gases or mixtures of perfect gases. They are:

Boyle's law – which states that the volume (V) of a gas, at constant temperature varies inversely with the pressure (P).

$$V_2/V_1 = P_1/P_2$$
$$\text{or}$$
$$P_1 \times V_1 = P_2 \times V_2 = \text{a constant}$$

Charles's law – which states that the volume of a gas, at constant pressure, varies directly with the absolute temperature (T).

$$V_1/V_2 = T_1/T_2$$
$$\text{or}$$
$$V_1/T_1 = V_2/T_2 = \text{a constant}$$

Amonton's law – which states that the pressure of a gas, at constant volume, varies directly with the absolute temperature.

$$P_1/P_2 = T_1/T_2$$
$$\text{or}$$
$$P_1/T_1 = P_2/T_2 = \text{a constant}$$

Dalton's law – which states that the total pressure of a mixture of gases is equal to the sum of the partial pressures of the gases present (partial pressure being the pressure that each gas would exert if it alone occupied the same volume as the mixture.

$$P = \sum_{1}^{m} P_n$$

Avogadro's hypothesis – which states that equal volumes of all gases under the same condition of pressure and temperature contain the same number of molecules. Since one mole of a substance contains the same number of molecules, the molar volume of all gases is the same. The number of molecules in one mole is 6.02257×10^{23}.

Amagat's law – which states that the volume of a mixture of gases is equal to the partial volumes which the constituent gases would occupy if each existed alone at the total pressure and temperature of the mixture.

Poisson's law – which states that for a process without any heat exchange with the surroundings, the relationship between pressure and volume follows the mathematical relationship:

$$P_1 V_1^\gamma = P_2 V_2^\gamma$$

where γ is the ratio of specific heat at constant pressure to specific heat at constant volume:
$\gamma = c_p/c_v$

The value of the exponent g varies with temperature and pressure (see Figure 1), but for pressures of less than 1 bar and all temperatures:

γ =   1.4 for air and di-atomic gases
  =   1.66 for mono-atomic gases
  =   1.3 for tri-atomic gases

General gas law – resulting from a combination of Boyle's and Charles's laws:

$$PV = wRT$$

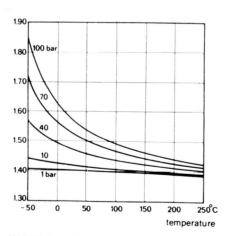

FIGURE 1 – The isentropic exponent for air.

Laws of thermodynamics

1st law – For any system of constant mass, the amount of work done on or by the system is equal to the amount of energy to or from the system.

2nd law – Energy exists at various temperature levels but is available for use only if it can pass from a higher to a lower temperature level.

3rd law – The entropy of a substance approaches zero as its temperature approaches zero.

Air compression

There are four theoretical ways of compressing air as shown in Figure 2. Of these only isothermal and isentropic need be considered as practical cycles for air compression. There is in addition a fifth cycle, polytropic, which is commonly used as a basis for performance estimation; it is intermediate between the isothermal and the isentropic curves. This takes into account the practical observation that real compression can take place neither at constant temperature nor without heat exchange to the surroundings. It is characterised by the value of the polytropic exponent, n, an empirical value which is not capable of theoretical determination, but which has generally accepted values for various kinds of compressors.

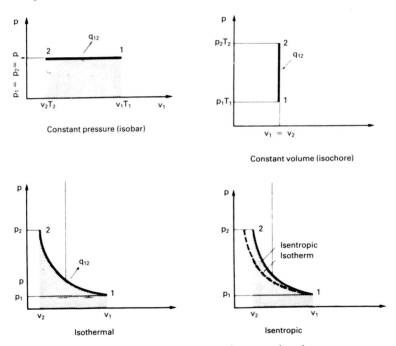

FIGURE 2 – Theoretical ways of compressing air.

Isothermal compression

$$P_1 V_1 = P_2 V_2 = \text{constant}$$

This is a hyperbola on the P-V diagram

The work done to compress *and to deliver* the gas is equal to the shaded area in Figure 2.

$$Q = \int_{V_1}^{V_2} PdV = \int_{V_1}^{V_2} \frac{wRT}{V} dV = -wRT\log_e(V_1/V_2)$$
$$= P_1V_1\log_e(P_2/P_1)$$

In practical units appropriate to a continuously working compressor:

$$W = 0.1\ p_1\ q_1\ \log e(P_2/P_1)$$

where p is in bar, q is the flow rate in l/s and W is the power in kW.

In this case, the total work also includes the amount of heat extracted to maintain the constant temperature.

Isentropic compression

No heat is extracted in this case and the governing equation is:

$$PV^\gamma = \text{constant}$$

The work done in compression and delivery is:

$$Q = \int_{P_1}^{P_2} VdP = wRT\left[\left(\frac{P_2}{P_1}\right)^{\frac{\gamma-1}{\gamma}} - 1\right]$$

$$= \frac{\gamma}{\gamma-1} P_1V_1\left[\left(\frac{P_2}{P_1}\right)^{\frac{\gamma-1}{\gamma}} - 1\right]$$

As previously stated, the work done in practical units quoted above is given by:

$$W = 0.1\ p_1q_1\left[\left(\frac{P_2}{P_1}\right)^{\frac{\gamma-1}{\gamma}} - 1\right]$$

Note that p_1 and q_1 refer to the inlet conditions of the compressor which is the usual way of quoting performance.

Polytropic compression

The equations appropriate to this type of compression are obtained by replacing γ by n, the polytropic exponent. This is more for practical purposes than a concept which has theoretical justification. n is usually obtained empirically or by reference to the performance of similar compressors.

In continuous steady-state conditions it is impossible to attain a cycle approaching isothermal. Most operations are closer to adiabatic. Isothermal represents the theoretical aim and so the efficiency of a unit is often calculated on that basis.

Isothermal efficiency, $\quad\quad\quad\quad\quad \eta_{is} \ = \ \dfrac{\text{isothermal power}}{\text{actual power}}$

Because a real compressor is usually much closer to isentropic than isothermal, isentropic efficiency is also used

Isentropic (or adiabatic) efficiency, $\quad \eta_a \ = \ \dfrac{\text{adiabatic power}}{\text{actual power}}$

It may sometimes happen in a very efficient compressor that this value exceeds unity. This occurs when exceptional measures have been taken to achieve internal cooling, as in some examples of oil injection in screw compressors.

SECTION 2

The Compressor

COMPRESSOR CLASSIFICATION
AND SELECTION

COMPRESSOR PERFORMANCE

COMPRESSOR INSTALLATION

MOBILE COMPRESSORS

ACCEPTANCE TESTS

COMPRESSOR NOISE REDUCTION

COMPRESSOR CONTROLS

AIR RECEIVERS

COMPRESSOR LUBRICATION

HEAT EXCHANGERS AND COOLERS

COMPRESSOR CLASSIFICATION AND SELECTION

Some of the various methods of compressing gas and air are shown in Figure 1. See also ISO 5390. Reciprocating compressors can be further classified as in Figures 2, 3 and 4 according to the arrangement of the cylinders.

Alternative forms of classification are possible, *eg* according to the motive power which drives the compressor – electric motor, diesel or petrol engine; the type of gas to be compressed; the quality of the compressed medium – the degree of freedom from moisture and oil (as for example might be required in a food processing plant or for underwater breathing supplies); the type of cooling – air or water and so on.

In choosing the correct compressor for a given installation, the following factors must be taken into account:

- Maximum, minimum and mean demand. If there is an intermittent requirement for air, but a large compressor set is needed to cater for peak requirements, the

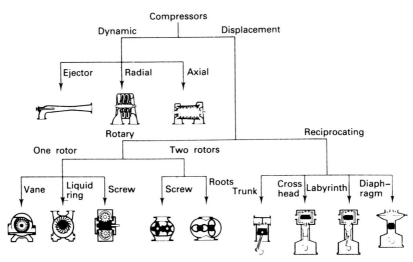

FIGURE 1 – Basic compressor types.

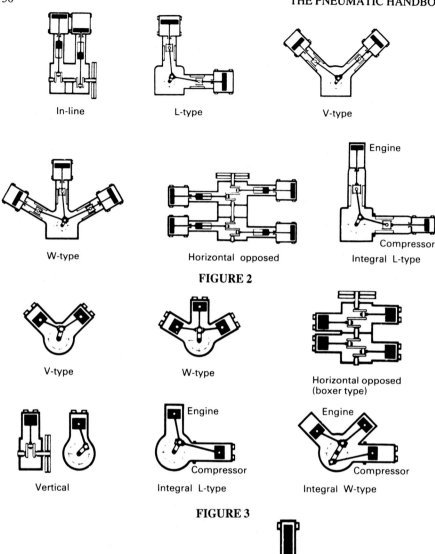

In-line

L-type

V-type

W-type

Horizontal opposed

Integral L-type

Engine

Compressor

FIGURE 2

V-type

W-type

Horizontal opposed
(boxer type)

Vertical

Integral L-type

Integral W-type

Engine

Compressor

Engine

Compressor

FIGURE 3

Horizontal with stepped piston
(four-stage)

Vertical with stepped piston
(two-stage)

FIGURE 4

installation can be very uneconomic to operate unless the control system is well designed. There may be a maximum placed on the power (electricity or gas) consumption. Depending upon the characteristics of the peak demand (*ie* whether it is long or short term), it may be appropriate to consider a large air receiver as an economic alternative to a larger compressor.

- Ambient conditions – temperature, altitude and humidity. At high altitude, because of the reduced air density, the compressor efficiency and capacity is reduced. High humidity can result in large quantities of water which have to be disposed of.

- Methods of cooling available – availability of cooling water or ambient temperature for air cooling. The possibility of using the waste heat for space heating or for process heating should be considered.

- Environmental factors – noise and vibration. Special foundations may be required, particularly with reciprocating compressors.

- Requirements for skilled maintenance personnel.

The actual means chosen for compressing the air is probably less important to the purchaser of a complete unit than is the purchase price and running costs. For small

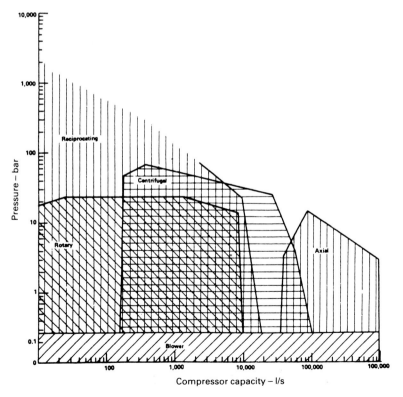

FIGURE 5 – Generalised performance envelopes for different types of compressors.

compressors used on construction sites or for intermittent stationary use, initial cost is the important parameter. For large, permanent, factory installations, running costs (particularly the off-load costs) are more important. The development of sophisticated methods of control allows almost any kind of compressor to be used efficiently. This means that the emphasis shifts from the compressor *per se* to the characteristics of the complete packaged installation. Figure 5 shows the broad map of application of the various types; there is much overlap. Table 1 is a broad description of characteristics. It should not be taken to give clear boundaries between the different types.

Reciprocating compressors

Reciprocating compressors are suitable for operating over a very wide range of speeds, with a practical maximum of about 10 bar delivery pressure from a single-stage unit, and up to 70 bar for a two-stage machine. Multi-stage reciprocating compressors may be built for special purposes capable of supplying delivery pressures up to and in excess of 700 bar. By carefully selecting the number of stages, the designer can also produce a machine which approaches the ideal or isothermal compression curve more closely than with any other type, with the possible exception of very large volume axial flow compressors (see the chapter on Compressor Performance). In the double-acting compressor the space on the other side of the cylinder is enclosed and so both sides are used for compression, giving two compression strokes for each revolution of the crank shaft. Alternatively, one side of the piston may be used for one stage and the other for the second stage in a two stage compressor. Individual cylinders may be also used for multi-stage compression, disposed in a number of arrangements.

Most reciprocating compressors are single-stage or two-stage, ranging from fractional horsepower units to very large machines with input requirements of the order of 2250 kW. The smaller compressors are usually single stage of a single or V-twin layout with air cooled cylinders and are powered by electric motors.

Intermediate sizes comprise a variety of different configurations. Vertical compressors may comprise one or more cylinders in line, 'V', 'W' and 'H' arrangements for multi-cylinder units, and also 'L' configuration which has both vertical and horizontal cylinders disposed about a common crankshaft. The angled arrangement of cylinders offers certain advantages, notably reduced bulk and weight and superior machine balance (since with careful design the primary forces can be accurately balanced). The 'L' configuration with a vertical low pressure cylinder and a horizontal high pressure cylinder is advantageous for larger machines, facilitating installation, assembly, maintenance and dismantling. Larger reciprocating compressors are commonly horizontal double acting, tandem or duplex. Basic two-stage units are shown in Figure 2. Variations include transferring the side thrust by means of a crosshead. When the space under the piston is used, it is essential to use a crosshead to convert the rotary action of the crankshaft into reciprocating motion in order to obtain a satisfactory seal of the piston rod where it passes through the compression chamber. The crosshead ensures that all the side thrust component from the crank shaft is taken by the crosshead guide and not transferred to the piston and cylinder. Single-acting reciprocating compressors are normally of the trunk piston type, whilst differential-piston double-acting compressors can be of either type – see Figure 3.

100 YEARS

# of innovation in compressed air technology.

CompAir
BroomWade

Over almost a century, CompAir BroomWade products have earned an enviable reputation for quality, efficiency and reliability. But we're still not satisfied.

A multi-million pound investment programme has transformed our plant into a world class manufacturing facility ensuring that our products meet your needs precisely - in performance, price and delivery.

There is a CompAir BroomWade compressor for every application from screw and reciprocating units to oil-free machines and bespoke packages. With ancillary equipment such as dryers, filters, receivers, energy recovery units and controls our comprehensive product range is supported by an unrivalled national distributor network.

Call us and benefit from a century of experience.

CompAir BroomWade Limited
Hughenden Avenue
High Wycombe
Bucks HP13 5SF
United Kingdom

Telephone +44 (0) 1494 605300
Facsimile +44 (0) 1494 462624

A Siebe Group Company

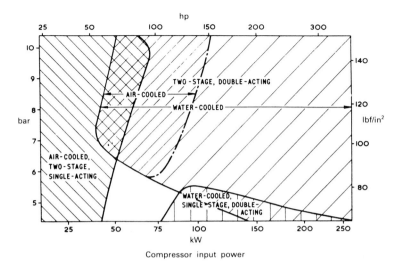

FIGURE 6 – Lubricated air compressors in industrial service – all piston types.
(CompAir Broom Wade)

Differential or stepped pistons may also be used as shown in Figure 4. Here one stage of compression takes place in the annular space between the shoulder of the piston and the corresponding shoulder in the cylinders.

Figures 2 to 4 also show designs where a compressor piston is connected to the same crank as the reciprocating engine which drives it. A common arrangement uses three cylinders of a four-cylinder block to form the engine, whilst the fourth cylinder forms the compressor cylinder. This is a convenient way of balancing the forces with a single cylinder compressor.

Figure 6 shows the pattern of reciprocating air compressors in industrial service, according to a recent survey. This represents most of the industrial units in service. For special purposes, as mentioned above, compressors are found well outside the range indicated. The versatility of reciprocating compressors means that they are the most common of all types. The disadvantages of reciprocating compressors are:

- They require special foundations to cater for the unbalanced inertial forces of the reciprocating pistons and connecting rods.

- The maintenance they need has to be done by skilled personnel.

- The inlet and delivery valves are prone to failure.

- The discontinuous flow of the compressed medium can cause vibrational resonance in the delivery passages and the distribution system.

In mobile compressors and in the medium range of stationary compressors, reciprocating compressors tend to have been superseded in recent years by screw and sliding vane designs in oil flooded and dry versions. They still maintain their dominance for large stationary (factory) applications

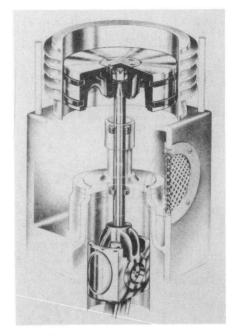

FIGURE 7 – Compressor supplying oil-free air. Note the PTFE piston rings and the rubber seals on the piston rod.

Reciprocating compressors for oil-free applications

For oil-free applications, special precautions have to be taken. Figure 7 shows how particular designs produce a high degree of cleanliness in the air. No lubrication is necessary in the upper cylinder because the pistons are fitted with PTFE rings; the combination of these rings, austenitic liners and a crosshead guide to minimise piston side loads extends operating life without complicating maintenance. To prevent oil entering the cylinder, a distance-piece incorporating rubber seals is fitted between each cylinder and crankcase. Pure air is produced straight from the compressor without requiring further separation equipment.

An alternative to PTFE piston rings is the use of a labyrinth seal on the piston. Fine grooves are cut in the cylinder wall and the piston skirt, and usually on the piston rod as well.

Valves for reciprocating compressors

One feature of a reciprocating compressor, which is unique to it, is the provision of check valves in the cylinder head. The delivery pressure is determined by the spring force behind the delivery valve and the pressure in the delivery passage . This is in contrast to a rotary compressor where the shape of the rotor casing alone controls the pressure. Inlet valves open when the cylinder pressure is lower than ambient, delivery valves open when the pressure in the cylinder equals the set pressure. The convenience of this is that the delivery pressure can be changed by modifying the valves only, rather than replacing the compressor casing.

Compressor valves are mostly of the plate type of design as illustrated in Figure 8.

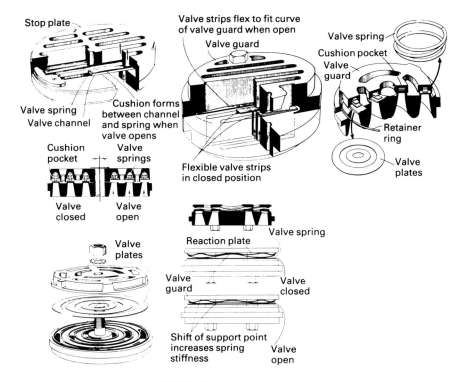

FIGURE 8 – Compressor valves.
(Mobil Oil Co.)

Detailed designs differ but they mostly consist of one or more flat ring-shaped plates backed by helical or flat springs. Valve design is a high art, and valves are best obtained from specialist manufacturers. Valves have to operate very quickly in response to the pressure difference, so the dynamic mass has to be kept low, there has to be a minimum of flow losses, and they must be capable of withstanding the impact forces at the end of the stroke.

Traditionally compressor valves have been made of steel or high grade iron, but there have recently been developed glass fibre reinforced plastics with a high degree of stability under the rapid impact conditions and high temperatures experienced in this duty. Polyetheretherketone (PEEK) has been successfully used for this purpose. This material retains good physical properties (strength and stability) up to 200 °C and has lower water absorption properties than other thermoplastics.

Rotary screw compressors

Figure 9 shows the distribution of screw units. It can be compared with Figure 6.

Undoubtedly the most common form of rotary positive displacement compressor is the helical screw compressor, developed primarily by the Swedish companies SRM and Atlas

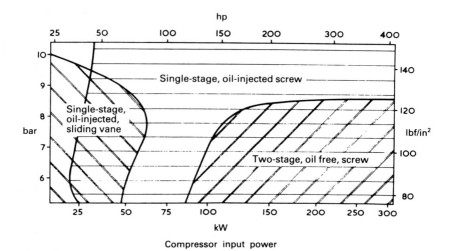

FIGURE 9 – Rotary type air compressors in industrial service.
(CompAir Broomwade)

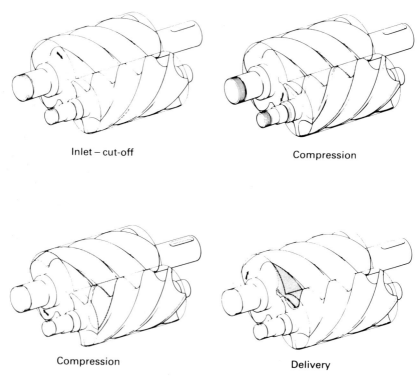

FIGURE 10 – Screw compressors compression cycle. *(CompAir Holman)*

Copco but now available from a number of manufacturers. The compressive cycle is illustrated in Figure 10.

Rotary screw compressors employ two intermeshing rotors with helical lobes. As the rotors revolve, the space between the unmeshing rotors increases, into which inlet air is admitted. On completion of the filling operation the inlet faces of the two rotors pass the inlet port and the air is sealed in the casing. With continued rotation the flute volume between the two rotors decreases, compressing the air. Compression increases with continued rotation until the built-in pressure ratio is reached and the air is discharged through the delivery port.

Screw compressors can be either oil-flooded or oil-free. Oil-flooded types rely on a substantial volume of oil injected into the compression space. This oil serves a number of purposes: it seals the clearance gaps between the screws and the casing; it lubricates the drive between the male and female screws; and it assists in extraction of the heat of compression.

Dry oil-free types cannot permit a drive to take place between the screws so there has to be external gearing. Some designs incorporate water injection. All-metal screws cannot be adequately lubricated by water so these types also require external gearing. There is an interesting new development in water-lubricated screws in which the screws themselves are made of a ceramic material fastened to a steel shaft. These screws have to be made very accurately in a special mould because only a limited amount of machining after manufacture is possible. They are finally run together in sacrificial bearings to achieve the final shape.

Whether the screws are water or oil-flooded, it will be apparent that in order to minimise leakage losses, the clearance between the screws has to be kept to a minimum at all rotational positions. The generation of the mathematical forms necessary to produce this minimum clearance and the subsequent manufacturing techniques are the secrets of efficient screw compressor technology.

A characteristic of rotary screws is that the power absorbed by the male rotor is about 95% of the total with the remaining 5% absorbed by the female rotor. The female rotors can be looked upon as being primarily rotary valves. Because of this power distribution between the rotors, it is customary to connect the male rotor (through gearing if necessary) to the prime mover. The screws are not designed as driving gears, but 5% is well within their capacity.

Oil-flooded and dry types each have their own advantages and disadvantages. Dry screw units can only have a limited pressure increase per stage (a maximum of about 3:1), otherwise the temperature increase would be excessive. Oil-flooded screws on the other hand can accommodate a normal pressure ratio of 8:1, and often as high as 13:1.

The total leakage flow path is larger than in a corresponding reciprocating compressor. In addition to the clearance between the rotors themselves and between the rotors and the casing, there is also the "blow hole", which is the triangular hole formed at the intersection of the two rotors and the casing cusp (Figure 11). It is not possible to have anything like a piston ring to seal the gaps. The leakage flow is the same whatever the operating frequency (it depends only on the pressure difference) so in order to keep the percentage loss through leakage to a minimum, the rotational frequency has to be high. This is true

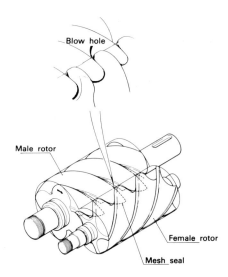

FIGURE 11 – Meshing of rotors showing the 'blow hole'. *(CompAir Holman)*

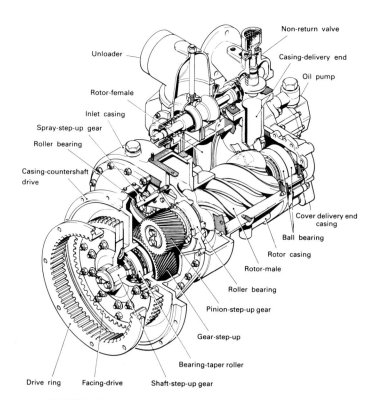

FIGURE 12 – Screw compressors air end. *(CompAir Holman)*

for both oil-flooded and dry types, but as there is no assistance from the oil in sealing the dry screw, its optimum frequency has to be much higher. The churning of the oil inside the compression chamber represents a power loss and a resultant fall in efficiency. There is an optimum operating speed at which such losses are balanced by losses through leakage. For oil-flooded units, the optimum linear peripheral speed of the screw is about 30 m/s and for dry screws about 100 m/s. For a typical oil-flooded air-end see Figure 12.

When driven from an internal combustion energy or an electric motor, a step-up gear drive (of about 3:1 in the smaller sizes of the oil-flooded types and about 10:1 in the dry types) is required. This is usually separate from the gearing provided to keep the two rotors apart. The latter has to be very carefully manufactured and set: the backlash in the timing gears has to be much less than that between the rotors, which itself has to be kept small to reduce leakage. A technique frequently adopted is to coat the rotors with a thin layer of plastic (PTFE or similar), so that on initial assembly the rotors are tight; after a careful running-in period the coating rubs off locally to form a good seal.

Much effort has been put into the careful design of the rotor profile. Originally rotors were part-circular in shape, but now much more mathematically complex shapes are favoured. Figure 13 shows a widely adopted shape. Note that the lobes are asymmetrical

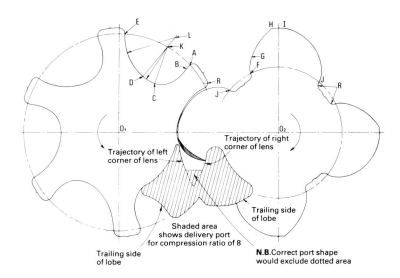

Lobes of the female rotor
B-A is the extension of the straight line O_1-B
B-C is generated by the point H of the male rotor
C-D is a circular arc with its centre K on the pitch circle
D-E is a circular arc with its centre L outside the pitch circle on the

Lobes of the male rotor
F-G is generated by A-B of the female rotor
G-H is generated by the point B of the female rotor
H-I is a circular arc with its centre on the pitch circle
I-J is generated by D-E of the female rotor

FIGURE 13 – Rotor geometry and delivery port shapes. *(CompAir Holman)*

in cross section. This ensures efficient compression with a limited leakage area. Many attempts have been made over the years to produce more efficient meshing profiles. The rewards likely to be obtained through the exercise of such ingenuity are likely to be small, although one of the main motives for producing alternative designs is an attempt to circumvent the strong existing patents.

One feature of the oil-flooded compressor is the necessity for removing the oil after compression. This is done by a combination of mechanical impingement and filtration. The residual oil left in the air after such treatment is very small (of the order of 5 parts per million) – good for most purposes but not for process equipment or breathing. See Figure 14 for a typical system. For further information see the chapter on Compressor Lubrication.

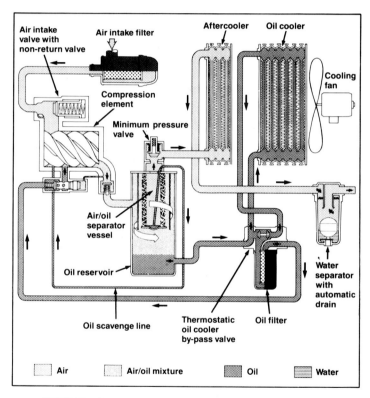

FIGURE 14 – Oil system for an oil-flooded screw compressor.
(Atlas Copco)

Diaphragm compressors

Diaphragm compressors are positive displacement, oil-free types. They can be driven mechanically or by hydraulic pressure. Mechanical diaphragm compressors are manufactured in smaller capacities than hydraulically actuated units and consequently overlapping of the two types occurs only over a limited range.

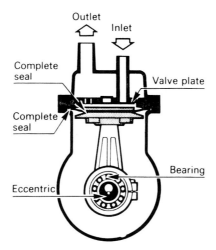

FIGURE 15 – Cross-section of a mechanically actuated diaphragm compressor.

FIGURE 16 – Diaphragm pump mounted on an electric motor. 17 litre/min at 2.8 bar.
(Charles Austen Pumps)

Mechanical units are less expensive to produce, are very compact and can be used on applications involving sub-atmospheric pressures (see Figures 15 and 16).

Hydraulically actuated units are able more easily to generate high pressure than are mechanical units as the latter are usually limited by considerations of bearing loading.

Apart from the fact that the compact form of a diaphragm unit enables it to be directly built into the electric motor or driver (or hydraulic pump in the case of an hydraulically operated machine, the main advantages offered by diaphragm compressors are:

- Complete gas tightness with only static sealing involved.
- Isolation of the gas being handled and thus the possibility of 100% oil-free compressed gas supply.

The main limitations of the type are low delivery rates (to a maximum of about 30 l/min in a single unit) and a limited compression ratio (to about 4.0). The latter limitation applies particularly to mechanically operated diaphragms where the diaphragm itself is normally of synthetic rubber. Hydraulically-operated diaphragm compressors may employ metallic diaphragms and be capable of generating much higher pressures. A further method of obtaining high pressure is to enclose the compressor and driver within a pressure vessel. The vessel is pressurised to a level equal to that present in the inlet. The same pressure is thus present on the underside of the diaphragm, reducing the pressure differential on the diaphragm when working.

Rotary sliding vane compressor

A well-tried, mechanically simple design, Figures 17 and 18. The rotor and the vanes are the only moving parts. The rotor is mounted offset in the casing; as can be seen, the

As rotor turns, gas is trapped in pockets formed by vanes

Gas is gradually compressed as pockets get smaller

Compressed gas is pushed out through discharge port

FIGURE 17 – Working principle of a rotary vane compressor.

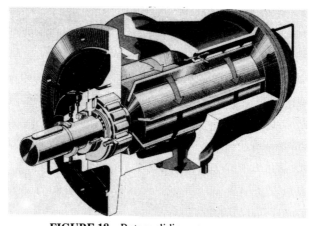

FIGURE 18 – Rotary sliding vane compressor.

compression is obtained through this eccentricity. Air is admitted when the crescent-shaped chamber is increasing in volume and delivered when the design pressure is reached. As in all other rotary positive displacement types, it has a built-in pressure ratio, dependent on the position of the delivery port.

The rotor is a simple cylinder, with longitudinal slots machined in it, and the vanes can be cut from plate material. These are available in three forms – lubricated, oil-free and oil-flooded.

Lubricated types inject an oil mist into the incoming air stream. Alternatively oil can be fed from a reservoir through the shaft or directly into the cylinders. Vanes are made of steel, in which case a floating ring (see below) is required, or of a synthetic fibre material (newer types). The use of synthetic fibre is advantageous in that the lower mass means a smaller centrifugal force.

Oil-free types, suitable for process applications require self-lubricating vanes made from carbon.

Oil-flooded types are similar to screw compressors in that large quantities of oil are injected into the compression space, which then has to be separated out and cooled. As with screw compressors, the oil lubricates, cools and seals. The vanes are made of aluminium, cast-iron or a synthetic fibre material.

The rotational frequency has to be low, otherwise the centrifugal forces on the vanes would cause frictional failure, so a direct motor drive is customary. Wear of the surface of the vanes that contacts the stator can be a problem, but because they are free in their slots, such wear is self-compensatory. Wear on the side of the vanes or on their ends causes a fall in efficiency, and they should be replaced when such wear exceeds the manufacturer's recommendations.

Medium or large capacity vane compressors running at high speeds normally incorporate a method of avoiding contact between the vanes and casing wall in order to reduce wear and minimize frictional losses. One method is to fit restraining or floating rings over the vanes, the internal diameter of these rings being slightly less than the cylinder bore. The difference is calculated to preserve a minimum clearance between the vane tips and the casing walls under operating conditions, see Figure 19.

The rings actually rotate with the vanes but the peripheral speed of each vane tip varies

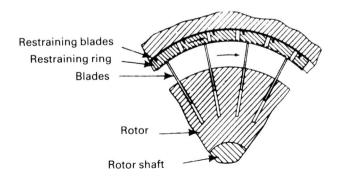

FIGURE 19 – Vane compressor with sliding rings.

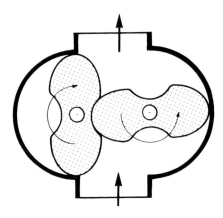

FIGURE 20 – Principle of the Roots blower.

with the degree of extension. The rings rotate at constant speed. There is some relative motion between vane and rings, but the contact between rotor and rings is essentially a rolling motion.

Low pressure blowers

These are meshing rotors suitable for low pressure ratios only. There is no built-in compression ratio, any increase in pressure being achieved by compressing against the back pressure in the system. Such a device is inefficient, its real advantage lying in the simplicity of manufacture. Figures 20 and 21 show an example of this type – the Roots blower.

Liquid ring compressors

In this design the rotor runs in a ring of liquid which seals and cools the compressed gas.

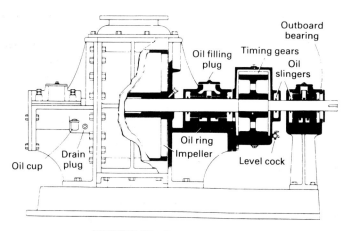

FIGURE 21 – Roots blower drive.

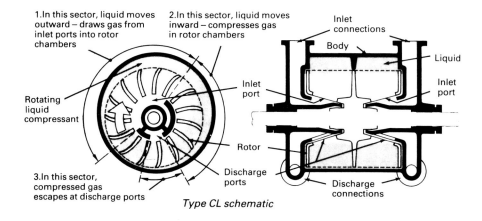

Type CL schematic

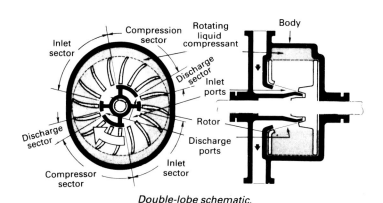

Double-lobe schematic.

FIGURE 22 – Type CL schematic (top) and double-lobe schematic. *(Nash Engineering)*

Two types are available – the single-acting, circular stator, and the double-acting, oval stator. Both are illustrated in Figure 22 . They both have a built-in pressure ratio, but the double-lobe body is more suitable for higher pressures (up to 7 bar). A variety of liquids can be used; water is a common one, but non-corrosive liquids can also be used according to application. The presence of a large quantity of liquid means that churning losses are high, but cooling is very good, so isothermal conditions are approached at the expense of mechanical efficiency. These units are also useful as vacuum pumps.

Rotary tooth compressors

One type is illustrated in Figures 23 and 24. The rotors are parallel in section and are externally meshed through timing gears; there is no possibility of one tooth driving the other as in rotary screws. They are suitable for producing oil-free air. They are available as low pressure single-stage blowers up to 2.5 bar and as two stage compressors up to 7

1.Suction: air at inlet pressure enters compression chamber. Outlet ports sealed off by female rotor hub.

2.Compression starts: in- and outlet ports closed off. Space between both rotors and housing becomes progressively smaller. Volume is reduced, pressure increases.

3.End of compression: the entrapped air is compressed to its maximum. Suction starts as inlet ports are opened.

4.Delivery: recess in female hub uncovers outlet ports and compressed air flows out.

☐ Inlet port   ☐ Intake air

■ Outlet port   ▨ Compressed air

FIGURE 23 *(Atlas Copco)*

FIGURE 24 *(Atlas Copco)*

bar. They are claimed to be suitable for pressures up to 10 bar. The rotors are of comparatively simple construction in that they do not require helical machining so they can be made from materials that would otherwise be difficult to manufacture, such as stainless steel.

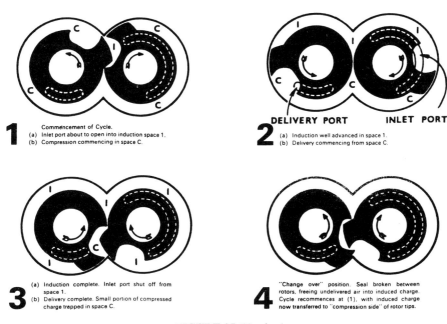

1 Commencement of Cycle.
(a) Inlet port about to open into induction space 1.
(b) Compression commencing in space C.

DELIVERY PORT   INLET PORT

2 (a) Induction well advanced in space 1.
(b) Delivery commencing from space C.

3 (a) Induction complete. Inlet port shut off from space 1.
(b) Delivery complete. Small portion of compressed charge trapped in space C.

4 "Change over" position. Seal broken between rotors, freeing undelivered air into induced charge. Cycle recommences at (1), with induced charge now transferred to "compression side" of rotor tips.

FIGURE 25 *(Northey)*

The inlet and delivery ports are situated in the end plates of the compressor housing with no inlet or delivery valves. There is very little end thrust on the rotating elements, which ensures low bearing loads. With the tooth form shown, there can be no drive between the rotors so there has to be external gearing.

Because of the straight tooth design, and the axial inlet and delivery ports, axial loads on the rings are eliminated. The compressor illustrated has rotor speeds between 3000 and 6000 r/min. Note that in this design, portions of the teeth intersect.

An alternative design has no intersecting part (Figure 25), which makes it closer to a Roots blower. This compressor element can be direct coupled to an electric motor with a rotary speed to 1500 r/min and below.

Turbocompressors

Turbocompressors (or dynamic compressors) convert the mechanical energy of the rotating shaft into kinetic energy of the air as it passes through the unit. The kinetic energy is converted into pressure energy, partially in the impeller and partially in the delivery passage or volute. These compressors are classified according to the path that the air follows in its passage through the machine – in a centrifugal compressor the flow is radial; in an axial unit it is axial; there is also a mixed type in which the flow is both radial and axial; see Figure 26. Turbocompressors are very compact units, producing large volumes of air in a small space. They are mainly found in process industries which require large volumes of oil-free air at a relatively constant rate.

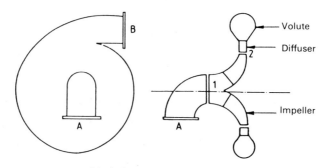

Principal parts of a turbocompressor.

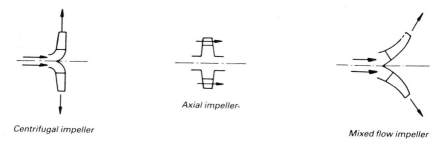

Centrifugal impeller

Axial impeller

Mixed flow impeller

FIGURE 26

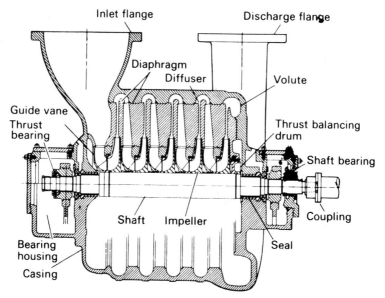

FIGURE 27 – Sectional view of a typical five-stage horizontally-split centrifugal compressor.

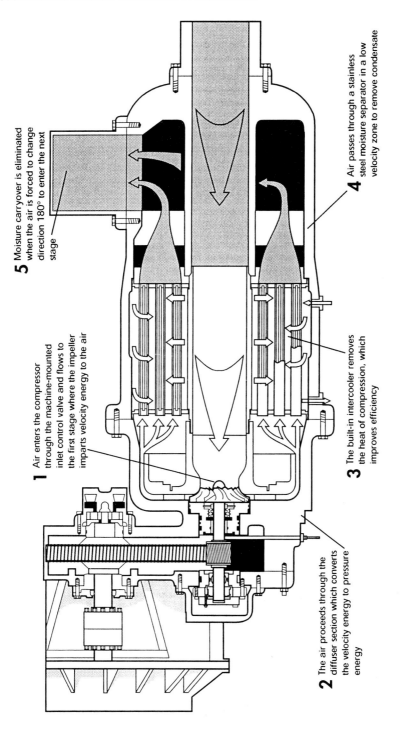

5 Moisture carryover is eliminated when the air is forced to change direction 180° to enter the next stage

4 Air passes through a stainless steel moisture separator in a low velocity zone to remove condensate

1 Air enters the compressor through the machine-mounted inlet control valve and flows to the first stage where the impeller imparts velocity energy to the air

3 The built-in intercooler removes the heat of compression, which improves efficiency

2 The air proceeds through the diffuser section which converts the velocity energy to pressure energy

FIGURE 28 – Single -stage centrifugal air end.
(Ingersoll Rand)

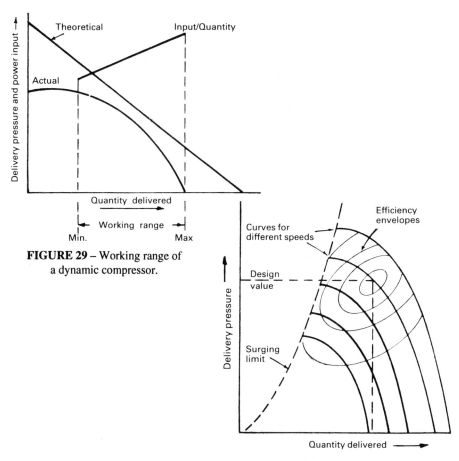

FIGURE 29 – Working range of
a dynamic compressor.

FIGURE 30 – Efficiency curves of a
dynamic compressor.

As in other types, compression can take place in stages. The pressure ratio which can be reached in a single stage is limited by the maximum peripheral speed of the impeller. Centrifugal machines normally generate a pressure ratio per stage of 1.5 to 2.0, although with improving technology pressure ratios are increasing. A pressure ratio of 6 can obtained in a 2-stage design. In an axial compressor the compression per stage is limited to about 1.3; they may have up to 20 stages. Centrifugal units may incorporate intercooling between stages, axial units rarely do. Single-stage centrifugal units are frequently used as blowers and exhausters.

Shaft speeds are high – 20 000 r/min is common in industrial units. 100 000 r/min is found in aircraft engines. At these speeds rolling bearings are unsuitable; plain or tilting pad bearings are necessary. Figure 28 shows a modern single-stage centrifugal unit. It can be seen that the actual compressor forms only a small part of the air end; the major part of the machine is in the step-up gear and the cooler.

Comparison of behaviour of centrifugal and axial compressors

Each type has its own characteristics: axial compressors are suitable for large flow rates only (20 to 100 m /s); centrifugal units are available supplying from 0.1 to 50 m /s. Axial compressors have a rather better thermodynamic efficiency, but only at the design point; they are susceptible to changes in throughput and pressure. Centrifugal units have good part-load characteristics; the off-load power absorbed is as low as 3%; this makes them particularly suitable for intermittent use and when needed to supply a range of pressures and flow rates. The operating speed of axial compressors is considerably higher than for centrifugal and for this reason combined with the large number of vanes they tend to be noisier.

Discharge pressure and volume of air delivered are co-related. If at a given speed the back pressure is reduced, the volume delivered will increase and vice-versa. See Figures 29 and 30.

If the back pressure increases to a value above that generated by the compressor then a condition known as 'surge' occurs, and air flow is reversed. If the pressure then drops the pressure starts to deliver again and if this sequence continues serious damage could result. Practical operating limits are therefore terminated by a surging limit, see Figure 30. To avoid surging it is necessary to recycle a part of the compressed volume by means of a by-pass that opens automatically as soon as the surge point is reached. The anti-surge system should be independent of the other means of control and is controlled from the flow rate or the pressure difference. Provision should be made for isolating the compressor from the mains when operating conditions approach the surge point, and for restoring it when the surge pressure has fallen. With the latest forms of computer control it is possible to extend the stable range of both axial and centrifugal types.

Mixed flow units combine some of the advantages of axial and centrifugal units. For a given flow rate the mixed designs have a smaller diameter. There is less turbulence and therefore the efficiency is better. They are mainly used for single-stage compression.

COMPRESSOR PERFORMANCE

Practical compressors normally work on a polytropic process, *ie* intermediate between true isothermal and true isentropic – Figure 1. Isothermal remains the ideal, and much effort is made by compressor designers to get close to it. A few percentage points reduction in input power can make a significant difference to the economies of air compression.

Multi-stage compression with intercooling between stages is a widely used technique. Theoretically the most efficient design of a multi-stage compressor has the same pressure ratio for each stage. Multi-stage compression increases the volumetric efficiency and reduces the power consumption, the power input being a minimum if the total work is divided equally between the stages. When estimating the total power consumption and in evaluating the overall efficiency the power used in running the intercooling process has to be taken into account. The curves of Figure 2 demonstrate the argument. The first stage polytropic is close to the isentropic curve, but at the conclusion of this stage air temperature is reduced to its original value by intercooling and the volume is reduced. The second stage polytropic curve originates at the isothermal line and follows its own isentropic line. The shaded area represents the saving by intercooling. The desirability of multistage operation is greater with increasing final pressure and required capacity.

Oil or water flooding in a single-stage rotary compressor is another attempt to reduce the value of the polytropic exponent.

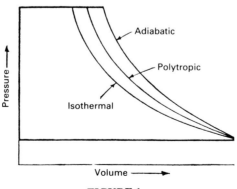

FIGURE 1

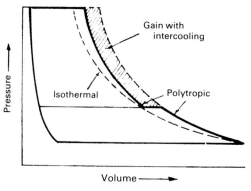

FIGURE 2

In order to compare the various compression cycles, it is useful to calculate the power required for the various cycles mentioned above and with a two stage compressor. For a pressure ratio of 8:1 the theoretical power per l/s of air compressed is 0.208 kW for isothermal, 0.284 kW for isentropic, 0.248 kW for polytropic (n = 1.2) and 0.242 kW for a 2-stage compressor. See Figure 3.

Multi-staging is valuable for other reasons than efficiency. It is usually necessary to limit the temperature to which the components of the compressor are subject and intercooling achieves that. A large pressure difference in a single stage can reduce the volumetric efficiency through excessive internal leakage. Mechanical complications impose a practical limit to the number of stages that are possible in a given application: mobile compressors, built for cheapness and to have a limited life are more likely to have a small number of stages than a fixed installation, built for economy in operation and a long life. Mobile oil-flooded screw compressors commonly have a pressure ratio of 8:1 and often as high as 13:1, whereas a comparable stationary unit would have two or three stages.

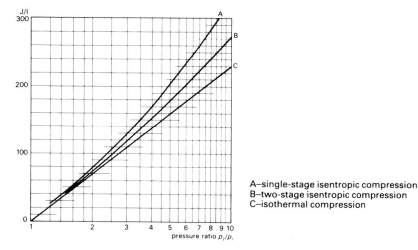

FIGURE 3 – Theoretical specific energy requirement curves for air.

In estimating the performance of compressors, it is essential to distinguish between the various types of efficiency, and the sources of power loss. From the designer's point of view the internal isentropic efficiency is important in estimating the thermodynamic efficiency of the design. A practical compressor is burdened with more than the thermodynamic losses. Further possible sources of losses are:

- Leakage past the piston or compression element.

- Heating of the inlet air, resulting in a reduced density and consequent reduction in capacity.

- Flow losses in the delivery passages due to restrictions in the delivery ports.

- Flow losses in the intake caused by the need for inlet valves, intake filters and restrictions in the inlet passages. Frequently manufacturers will quote performance values for the "bare air end", leaving out intake and delivery losses.

- Mechanical drive losses introduced by the gearing or by other drive mechanisms.

- Power absorbed by the cooling fan and by the lubrication pump.

TABLE 1 – Efficiency estimates for typical commercial compressors

	Internal isentropic	Overall isentropic
Reciprocating single-stage	0.90–0.95	0.83–0.94
Reciprocating multi-stage	0.76–0.83	0.72–0.78
Rotary screw compressors		0.80–0.85
Sliding vane		0.70–0.75
Single-stage centrifugal	0.80–0.86	0.77–0.82
Multi-stage centrifugal	0.68–0.77	
Axial compressors	0.85–0.87	
Turbo blowers		0.60–0.80
Roots blowers		0.60–0.65
Reciprocating compressors		

Reciprocating compressors

As explained in the chapter on Compressor Classification and Selection, reciprocating compressors can adopt a variety of different configurations and can be single- or multi-stage. The same method of performance estimation applies to them all.

The capacity can be readily calculated from the product of the total swept volume and the volumetric efficiency.

The next factor to be determined is the value of the index of compression, the polytropic exponent. In multistage compressors employing interstage cooling, the overall compression cycle approaches isothermal, but each individual stage is closer to isentropic, so it is convenient to analyse each stage separately. For modern high and medium speed compressors, the polytropic index can be taken as equal to the isentropic index (*ie* n = 1.4);

for low speed compressors a value of n = 1.3 is often used. The stage pressure ratio is determined by the pressure set by the delivery valves.

The stage temperature rise is

$$T_2 = T_1 \left(\frac{p_2}{p_1}\right)^{\frac{n-1}{n}}$$

The power absorbed by the stage is

$$W = 0.1p_1q_1 \; \frac{n}{n-1} \; \left[\left(\frac{p_2}{p_1}\right)^{\frac{n-1}{n}} -1\right]$$

See also Figure 3.

The inlet pressure and temperature of the inlet to the next stage depends on the degree of interstage cooling. In practice perfect cooling down to the inlet temperature is not possible. The cooling system necessarily absorbs power either in driving a fan or circulating the cooling water and this must be taken into account in determining the overall power required. Ten to 15% of the input power should be allowed for cooling purposes.

Volumetric efficiency

One characteristic of a reciprocating compressor is the clearance volume that remains in the cylinder at top dead centre after the delivery stage is complete. This is usually expressed as a proportion of the swept volume and typically has a value of 0.06 to 0.12. This volume is not discharged but expands until it occupies the cylinder volume. The effect of this is to modify the shape of the indicator diagram as in Figure 4. Theoretically this expansion represents no loss of energy if the index of expansion equals the index of compression (bearing in mind that the compression is isentropic and reversible, so the energy used is reclaimed).

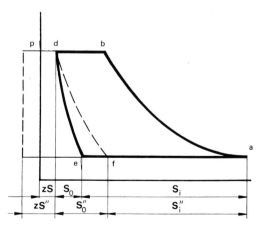

FIGURE 4 – Ideal and practical compression diagram. Theoretical diagram abdea. Practical diagram abdfa. Suction period: theoretical S_i, practical S_i''. Clearance volume: theoretical S_o, practical S_o''.

The volumetric efficiency of the compressor is a function of the first stage geometry, although for design purposes the separate stages can be separately analysed. It is given by

$$\eta_v = K \left[1 - V_c \left\{ \left(\frac{p_2}{p_1} \right)^{\frac{1}{m}} \frac{Z_1}{Z_2} - 1 \right\} \right]$$

p_1 and p_2 are the intake and delivery pressures in the first stage.

K is a factor depending on valve losses, intake air heating, leakage and pressure ratio. K can be taken as 0.96 for an initial estimate.

V_C is the relative clearance volume.

Z_1 is the compressibility factor at intake.

Z_2 is the compressibility factor at discharge.

m is the index of expansion, which as a first approximation is equal to γ, the isentropic index.

The expansion index varies with pressure according to the following:

For the first stage m = 1.20   For the fourth stage m = 1.35

For the second stage m = 1.25   For subsequent stages m = γ

For the third stage m = 1.30

A high volumetric efficiency is not necessarily better than a lower one; its significance lies in the calculation of the compressor capacity. If, however, the low efficiency is caused by leakage, then the design is at fault.

Other positive displacement types (screw and vane)

One general feature of these types is that they rely on the position of the delivery port to determine the built-in pressure ratio. (It would be more correct to refer to a built-in volume

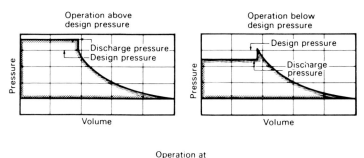

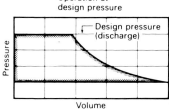

FIGURE 5 – Operation at pressures deviating from the built-in pressure ratio.

ratio, but the former is more common.) The position of the port is usually calculated on an assumed isentropic compression.

Operation at other than the design pressure, as for example when the pressure control valve in the delivery is set incorrectly, would result in the modified diagram of Figure 5 and wasted power.

Performance of screw compressors

The displacement of a screw compressor is calculated from

$$V = C D^3 (L/D) N$$

L and D are the length and overall diameter of the rotors. If the rotors are of different diameters, as is sometimes the case, by convention the male rotor is chosen as the reference.

C is the characteristic flute volume factor for the geometric shape chosen. A value of 0.4 to 0.5 is typical.

The volumetric efficiency can in practice only be determined from tests. It depends on:

- Geometry of the rotors
- Size of the blow hole
- Clearance gaps between rotors
- Volume occupied by the cooling oil
- Inlet restrictions
- Pre-heating of the inlet air

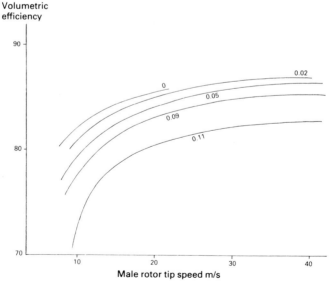

FIGURE 6 – Volumetric efficiency of 100 mm diameter oil-flooded compressor, for various values of inter-lobe clearance.

Note that there is no clearance volume as in reciprocating unit. Figure 6 shows a typical set of volumetric efficiency curves for a modern screw.

The theoretical power can be found, as in the case of a reciprocating compressor, from

$$W = 0.1 p_1 q_1 \frac{n}{n-1} \left[\left(\frac{p_2}{p_1} \right)^{\frac{n-1}{n}} - 1 \right]$$

It might be thought that the injection of massive quantities of oil into the compression chamber (typically, the ratio of the mass flow of the oil to that of the air is of the order 10:1) would result in an isentropic index close to unity. In fact, the flow velocity is so great that the actual heat transfer between air and oil takes place in the relatively static conditions in the oil reclaimer rather than in the compression chamber. The calculations on power can best be done by assuming isentropic compression and modifying the index from practical tests. The isentropic efficiency depends on the particular screw design. Typical curves are given in Figure 7.

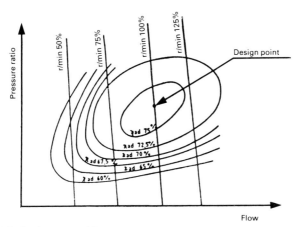

FIGURE 7 – Constant efficiency curves for the rotating screw compressor.

Performance of vane compressors

The swept volume of these units is readily calculated from the crescent area between the vanes.

The same remarks about isentropic efficiency apply as for screw compressors. When these units are oil flooded, oil churning represents one of the main power losses. There is little leakage across the outer edge of the vanes, because centrifugal force keeps them in intimate contact with the cylinder. There is, however, leakage across the ends of the vanes, which are not well sealed by the oil.

Performance of Roots and similar types

There is no built-in pressure ratio in these units and their efficiency is low, see Figure 8. Some improvement in efficiency is possible by adjusting the ratio of the volume in the

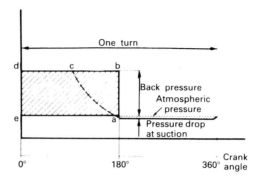

FIGURE 8 – The working diagram of a Roots blower. The compression work is represented by the area abdea. The area acdea represents the work required for compression in a piston type compressor.

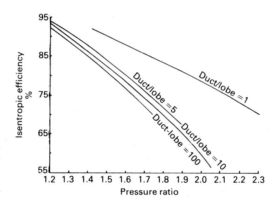

FIGURE 9 – Isentropic efficiency versus pressure ratio for various duct to lobe volume ratios.

delivery duct (*ie* the volume between the outlet port and the delivery valve) to the volume of the compressor lobes. See Figure 9.

Dynamic compressors

Any turbo machine operated at constant speed is capable of delivering a wide range of volumes with only a small variation in pressure. If the volume is decreased by throttling on the discharge side, pressure will also decrease. Pressure developed when the discharge is completely throttled will depend on the machine design and operating speed, but will be always lower than the maximum pressure. It is thus a characteristic of aerodynamic machines that dangerously high pressures cannot be built up under any operating conditions and so relief valves are not necessary in the discharge line. Regulation can be achieved either by throttling on the discharge side or variation in the impeller speed (or both).

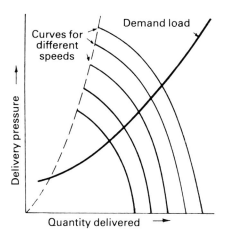

FIGURE 10 – Dynamic compressor operation.

The use of a valve offers the simplest form of control over pressure or capacity but throttling always represents an unrecoverable power loss. Speed variation is, therefore, a more economic form of control when the compressor characteristics can be expressed in the form of a series of curves which may be superimposed on the demand curve to establish the required operating point, as in Figure 10. In other words, by adjusting the speed, the compressor characteristics may be matched to any required values of pressure and capacity within the operating limits of the compressor and its driver. Efficiency will not necessarily be a maximum, unless this operating point corresponds to the design value for maximum efficiency.

The variable speed, capacity and adiabatic head can be related in terms of specific speed:

$$\text{Specific speed } (n_s) = \frac{Q \ \text{x} \ \text{r/min}}{H_a^{0.75}}$$

where Q = inlet capacity
and H_a = isentropic head per stage

This is a relative value only and has little significance other than as a design specific speed or value at which a given type of casing shows its maximum efficiency. It then serves as an index of classification for the form of the impeller in that type of casing. The compressor may be operated over a range of speeds, capacities and adiabatic heads, and thus a wide range of specific speeds. These actual or operating speeds are useful only as a basis for comparing performance relative to the performance obtained at the design specific speed.

The demand load itself will depend on the operating conditions and may involve working against a constant head or pressure, against frictional resistance, or a combination of both. In the former case the demand load will be constant and in the other two cases will vary with capacity – Figure 11. The resultant demand curve may also vary under actual working conditions due to a change in head or frictional resistance. The working point of

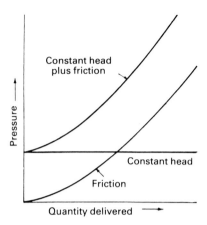

FIGURE 11

the compressor can be changed, *eg* by means of a butterfly valve installed preferably near the compressor inlet or near the discharge. Other methods of adjusting the compressor characteristics, and hence the pressure and capacity, include the use of adjustable diffuser vanes and adjustable inlet guide vanes.

Estimation of compressor performance with altitude

A number of factors combine to modify the performance of a compressor at altitude. Performance tests should always be used as the primary source of information. Where the results of such tests are not available, an estimate based on theoretical considerations can be made.

Refer to Table 6 in the chapter on Properties of Air and Gases, for details of the standard atmosphere. Note that the density, temperature and pressure all fall with increasing altitude. If the compressor is designed to produce a given amount of air at a given pressure at sea level its performance at altitude will be affected, not only because of the change in inlet conditions but also because it will no longer be able to operate at its optimum design efficiency.

There are two ways of looking at the change with altitude: when operating the same compressor at the same speed and same delivery pressure, the power to run it will be less at altitude; when it is desired to obtain the same useful output, the power required will be greater. The actual energy content of the compressed air is measured by the mass flow rather than the volume flow at inlet conditions, so a calculation of the extra power required at altitude can be made from an alternative form of the power equation:

$$W = 0.1 \; wRT \; \frac{n}{n-1} \left[\left(\frac{p_2}{p_1} \right)^{\frac{n-1}{n}} \right] - 1$$

As an example one can use this equation to compare the power at altitude with the power at sea level, when delivering air at a constant mass flow and a constant delivery pressure

TABLE 2

Compressor type	%Reduction for each 1000 m increase in altitude	
Piston type	Capacity	Power consumption
Medium size, air cooled	2.1	7.0
Large piston type, water cooled	1.5	6.2
Screw, oil injected	0.6	5.0
Large screw, water cooled	0.3	7.0

of 8 bar. It demonstrates that there is a power increase of 5% at 1000 m and 16% at 3000 m.

A further fall in performance will come from the change in volumetric efficiency, caused either by the clearance volume effect in a reciprocating compressor or from the change in the built-in pressure ratio in a rotary unit.

Table 2 gives some estimates of the observed reduction in the power and capacity in a variety of different compressor designs.

These considerations take no account of the reduction in power from the prime mover. Both electric motors and internal combustion engines give reduced output at altitude. Electric motors have a reduced performance of about 5% for every 1000 m increase in altitude through poorer cooling. For internal combustion engines, reliance has to be placed on the manufacturers' tests. Turbo-charged engines are better in this respect than naturally aspirated ones.

COMPRESSOR INSTALLATION

A compressor installation may be centralised where all the compressors needed are housed in a single compressor plant; or decentralised where individual compressors are located near the points of use. A centralised installation can show greater cost-effectiveness as well as making it easier to optimize intake conditions and silencing. There are a few situations where decentralised stations are to be preferred: if there is a requirement for different working pressures in some parts of the factory; where warm air is required, *eg* for pneumatic operation of forging tools; where part of the air is required at a considerable distance from the main installation and the pressure drop of a long line would be unacceptable.

Modern practice with centralised installations employs individual compressors with their associated ancillary equipment as completely separate units. It is then possible to shut down a single compressor for maintenance and leave the remainder operating. Most factories have a fluctuating demand for compressed air during the working day, so there is no need to keep all units running when only a proportion are needed to supply the factory. Even off-load, a compressor consumes a considerable amount of power so it makes sense to have the smallest number running at any one time. With modern control equipment compressors can be brought on stream rapidly when the demand for extra air is sensed. A decision about the number of compressors to be installed depends primarily on the importance of continuity of production. If it is acceptable to work with a reduced air supply in the event of an unscheduled shut down, then two compressors each capable of 50% of the total requirement may be satisfactory. If a partial shut down is not permissible then a reserve supply must be available, so that three compressors each capable of 50% must be installed. Generally the larger the compressor, the more efficient it will be, so the number of separate units should be kept small, consistent with these considerations.

In assessing the capacity of a compressor installation, attention must be paid to leakage which is an inescapable feature of practically all pipe runs; 10% should normally be allowed for unless extra care is taken in the initial design. Further possible increases in capacity should also be allowed for.

Ideally a compressor should be located where it can induct clean, dry and cool air; this requires that the air intake should be placed on the shaded side of its building, well away from its cooling tower; there must be no possibility that the air intake can draw in inflammable vapours for example from paint spraying or petrol fumes. Although an inlet filter must be incorporated, a site should be chosen where dust particles are not readily drawn in. There is no need to put a heavier burden on the filter than absolutely necessary. A filter will only extract a percentage of the dust in the air (although it may be as much as 98%); the remaining abrasive particles can cause serious damage to the lubrication system.

Another primary consideration is that a compressor is inherently a noisy machine and unless specifically designed or treated for noise control, should be located in a 'noisy' area of the factory or workshop. In the case of a centralised installation it is generally desirable (and virtually necessary in the case of larger machines) to have the compressor in a separate well ventilated building room or chamber, inducting the air from outside of the building. Each compressor should have a separate air intake, and the pressure drop over the intake system should be kept low.

Basic treatment of the intake system is the fitting of an air intake filter of a suitable type and size to ensure that air reaching the compressor is clean ; and a separate air intake silencer to reduce intake noise which is the predominant noise-generator. Further noise treatment may have to be applied both to the machine itself and the room (*eg* the provision of adequate ventilation can establish noise paths which may then require treatment). Further information can be found in the chapter on Compressor Noise Reduction.

In the case of reciprocating compressors, the intake flow is pulsatory. This can aggravate noise problems if the intake pipe is of such a length that there is resonance with the pulsation frequency of the compressor. Apart from the noise problem, the intake flow can also be adversely affected. As a general rule the chance of resonance occurring is greatest when the intake pipe length corresponds to one-quarter or three-quarters of the pressure pulse wavelength (λ) which for a single-acting compressor is given by:

$$\lambda = c/n$$

where λ =   wavelength in metres
        c   =   velocity of sound in air m/s
        n   =   compressor shaft speed r/s

For a double acting compressor

$$\lambda = c/2n$$

For dry air $c = 20 \sqrt{T}$ where T is the absolute temperature in K, which for an inlet temperature of 15°C equals 340 m/s. For a more accurate value, see the chapter on Properties of Air and Gases.

The intake lengths to be avoided are 0.17λ to 0.33λ and 0.67λ to 0.83λ.

High speed rotary compressors are comparatively pulsation-free and intake flow resonance is rarely a problem.

Compressor intake

The intake system must be designed with a low pressure drop in mind, that is with a limit on the air intake velocities. At the same time, intake flow should be smooth rather than

Every product we

MAKE

is designed
for the most cost-effective working

LIFE

and is backed by a sales
support organisation that couldn't be

BETTER

- Reciprocating compressors
- SSR rotary screw compressors
- SIERRA oil-free screw compressors
- CENTAC centrifugal compressors
- Dryers and filtration products
- Pipework installation systems
- Surveys and air consultancy services
- Compressor system design and build
- Fluid handling products and pumps
- Complete range of air tools
- Assembly line workstations
- Precision fastening systems
- Electric screwdrivers
- Air and electric angle wrenches
- Finishing and other specialist tools
- Tightness and torque auditing systems
- Air motors and air starters
- Hoists, winches and lifting products

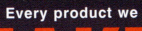

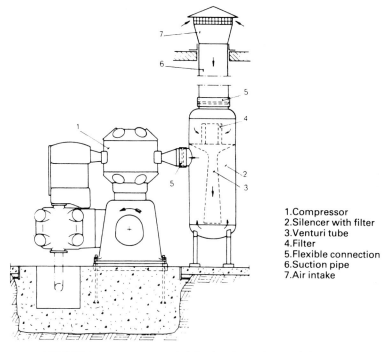

1.Compressor
2.Silencer with filter
3.Venturi tube
4.Filter
5.Flexible connection
6.Suction pipe
7.Air intake

FIGURE 1 – Arrangement of air intake pipe and venturi silencer.

pulsating, as this will appreciably reduce intake noise and resonance effects.

Low pressure drop is ensured by sizing the intake pipes such that the flow velocity is no more than:

5 to 6 m/s in the case of single-acting compressors

6 to 7 m/s in the case of double-acting compressors

Combined with a venturi-type intake silencer these intake velocities can be increased by up to 10%. A typical arrangement is shown in Figure 1.

Proper consideration must be given to the ventilation of the compressor house to dissipate the heat generated by the compressor and driver. This is particularly important in air-cooled units. With water-cooled compressors, or air-cooled compressors with water-cooled intercoolers, the amount of radiant heat to be dissipated is usually negligible since the majority of the heat is carried away by the cooling water.

Delivery line

Pressure drop in the discharge line does not materially affect the compressor efficiency, so the gas velocity can be 20 to 25 m/s. Pressure pulsations on the other hand can cause a considerable increase in compressor power as well as damage to the piping and valves. There are two ways of dealing with the problem of delivery line pulsations: tuning the pipe length between compressor delivery valve and the after-cooler; and inserting a pulsation

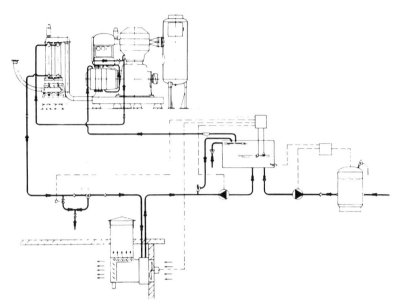

FIGURE 2 – A typical system for heat recovery from a water-cooled compressor plant. The cooling water is used partly for showers and washrooms, partly for heating adjacent premises.

damper in the form of a small air receiver.

Energy recovery

With water-cooled compressors it is possible to recover a high proportion of the intake energy in the form of hot water which can be used for space heating, feed water for boilers, etc in a large factory. A typical system for heat recovery is shown in Figure 2. With air-cooled compressors, the hot air leaving the intercooler and aftercooler can be directed to an adjacent room for space heating as in Figure 3.

Foundations

Installation is seldom a problem with compressors below a rating of 30 kW. Larger machines may need special foundations, the size and cost of such foundations increasing with compressor size, particularly in the case of reciprocating compressors with large out-of-balance forces.

Some layouts are less likely to generate out of balance forces. 'L', 'V' or 'W' configurations are better than vertical or horizontal types. Dynamic compressors and rotary screws seldom require anything extra by way of foundations over the need to support the compressor and driver.

Large compressors will require the use of lifting or hoisting equipment to erect them and to dismantle or remove sections for servicing.

In reciprocating compressors the gas pressures are balanced and therefore cause no

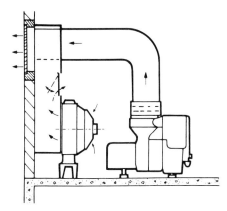

FIGURE 3 – Heat recovery from an air-cooled compressor installation.

TABLE 1 – Unbalanced inertia force for various compressor crank arrangements

| Crank arrangement | Forces | | | Couples | | |
| | Primary | | Second-ary | Primary | | Second-ary |
	Without counter-weights	With counter-weights		Without counter-weights	With counter-weights	
Single crank	F_1	$0.5 \cdot F_1$	F_2	0	0	0
Two cranks at 180°C						
Cylinders in line	0	0	$2 \cdot F_2$	$F_1 \cdot d$	$0.5 \cdot f_1 \cdot d$	0
Opposed cylinders	0	0	0	$f_1 \cdot d$	$0.5 \cdot F_1 \cdot d$	$F_2 \cdot d$
Two cylinders on one crank						
Cylinders at 90°	F_1	0	$1.41F_2$	$F \cdot d$	$F \cdot d$	$1.41 \cdot F_2 \cdot d$

In this Table:
F_1 = is the primary inertia force in N
F_2 = is the secondary inertia force in N
r   = is the crank radius in mm
n   = is the shaft-speed in r/min
M   = is the mass of the reciprocating parts in one cylinder in kg
l   = is the length of the connecting rod in mm
d   = is the cylinder centre distance in axial direction in mm

$$F_1 = \left(\frac{\pi n}{30}\right)^2 M r \, 10^{-3} \text{ and}$$

$$F_2 = F_1 r/l$$
   (the couples are expressed in N·mm)

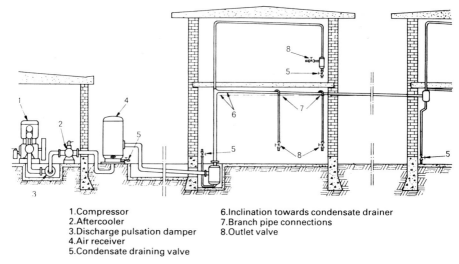

1.Compressor   6.Inclination towards condensate drainer
2.Aftercooler   7.Branch pipe connections
3.Discharge pulsation damper   8.Outlet valve
4.Air receiver
5.Condensate draining valve

FIGURE 4 – Principles for the construction of a compressed air pipe system.

external forces to be applied to the foundations. Such forces as are present are produced
by the reciprocating masses (pistons, connecting rods, crossheads). With careful design
these forces may be wholly or partially balanced. Analysis of the forces follows the same
method as for internal combustion engines, and specialist publications should be studied
for details of these techniques.

Small compressors (up to about 30 kW) are often single cylinder designs, which are
fundamentally incapable of internal balance; their small size and the magnitude of the
forces usually causes no problem however. For larger sizes, most compressor manufacturers provide foundation drawings, which in many cases require no more than a sturdy
concrete floor to which the compressor set is bolted.

The significant forces produced are sinusoidal at first order (compressor shaft speed)
or second order (twice shaft speed). Higher order forces are usually negligible for
foundation design. Both direct forces and couples are in general present. Table 1 gives the
out-of-balance forces for a variety of cylinder configurations.

The tendency today is to replace the traditional massive concrete foundations with
flexible mountings. The natural frequency of the compressor mass when supported on its
flexible mounting should be well below the forcing frequency at first or second order,
whichever is appropriate. One main advantage of flexible mounting is the degree of
isolation of vibration and noise that it provides.

Where a tuned delivery pipe is specified, its dimensions are supplied by the compressor
manufacturer and must be strictly adhered to. If this is not practicable, then a pulsation
damper must be installed with the shortest possible delivery pipe between the damper and
the aftercooler.

Ideally, delivery pipes for medium-sized compressors and large stationary compressors
should be welded and provided with flanged connections for fittings. With small

compressors, threaded pipe connections are more commonly used.

A complete compressor installation involves compressor and drive motor, aftercooler, moisture separator and receiver. The last needs to be located in a cool area and is frequently positioned outside the compressor house as is shown in Figures 4 and 5. However an overriding consideration is that the discharge pipe between compressor and receiver should be kept as short as possible. Where an isolating valve is fitted in the delivery pipework, the line on the compressor side should be protected by a safety valve of sufficient capacity to pass the whole of the compressor delivery without the pressure rising more than 10% above the set blow-off pressure.

Two further installations, typical of good practice are shown in Figures 6 and 7. See also

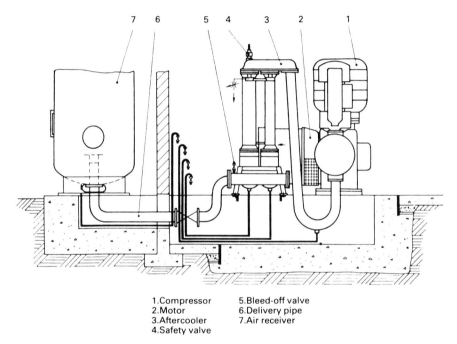

1. Compressor	5. Bleed-off valve
2. Motor	6. Delivery pipe
3. Aftercooler	7. Air receiver
4. Safety valve	

FIGURE 5 – Compressor with aftercooler and air receiver.

the chapters on Compressed Air Filtration and Compressor Controls.

There is an increasing trend for compressors to be made available as an integrated package, which incorporates the control system, heat reclamation, drying and filtration systems to meet the customer's requirements. They can be craned into a location, with only a power supply and pipe work required. These packaged units are necessarily limited in capacity although that too is increasing. For reciprocating and screw units, capacities up to 1000 l/s are available; dynamic compressors extend the range to about 1700 l/s. The packaged units are very efficient and are replacing many of the older central compressor stations. They are more flexible in application and their reliability means that there is little risk of an unscheduled breakdown disrupting the work of the factory. They can be robust

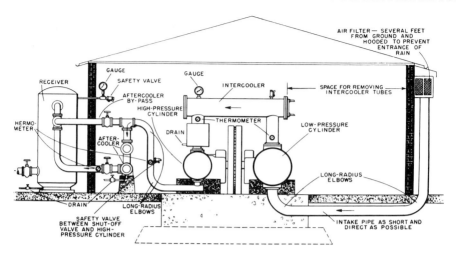

FIGURE 6 – Typical two-stage compressor installation.

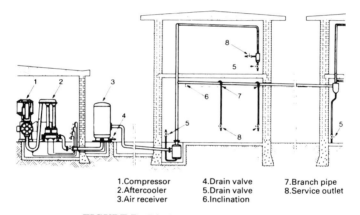

1.Compressor   4.Drain valve   7.Branch pipe
2.Aftercooler   5.Drain valve   8.Service outlet
3.Air receiver   6.Inclination

FIGURE 7 – Discharge side arrangement.

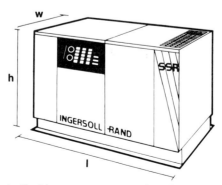

FIGURE 8 – Electrically driven rotary screw packaged compressor. *(Ingersoll Rand)*

enough and weather protected so can be located outside. One example of a packaged rotary screw is shown in Figure 8; l x w x h = 2.77m x 1.32m x 1.7m; the capacity is up to 450 l/s at 7.5 bar or 383 l/s at 10 bar; the weight is 2600 kg.

MOBILE COMPRESSORS

The expression "mobile compressor" refers to a compressor mounted on a chassis and supplied complete with an engine and all the necessary control systems, requiring only fuel or some other kind of primary power to generate compressed air. It can be mounted on wheels or skids. Usually it will also have a canopy to protect the working machinery from the weather and to reduce noise transmission. Figure 1 shows three examples of a two wheel rotary compressor capable of supplying air for a construction site. The pressure is up to 13 bar and the delivery up to 280 l/s.

The expression "portable compressor", although strictly a misnomer, is also commonly used; a compressor that can be easily lifted and moved from place to place is called "transportable". Such compressors can be carried on the back of a truck or mounted inside a van.

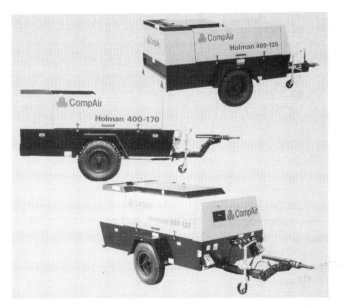

FIGURE 1 – Examples of portable compressor. *(CompAir Holman)*

Mobile compressors are frequently encountered by the roadside or on construction sites, where they are used to supply air for roadbreaking tools or drills and a wide range of contractor's tools. When used on the road, the chassis and wheels have to meet the appropriate legislation for road-going trailers. When used only on construction sites or in quarries, a simpler and cheaper chassis can be used. For smaller sizes (less than about 80 l/s capacity), a two-wheel chassis is standard; for intermediate sizes up to about 300 l/s both two- and four-wheel types are available. For large sizes and for rugged sites such as in quarries four wheel types are standard. A four-wheel chassis can either have turntable or automotive type steering on the front wheels. Independent torsion-bar suspension may be fitted for road use. A lifting eye or bale protruding through the canopy roof is fitted as standard.

Construction-use compressors usually have a storage facility to hold one or more road breakers and lengths of hose. Depending on the application, a towable compressor may incorporate the following features:

- A straight tow bar with levelling leg (U.K. use)
- A cranked adjustable towbar with jockey wheel (European models)
- Towing lights and indicators
- Towing eye or ball coupling
- Cable operated parking brakes
- Overrun and pneumatic brakes
- Security features such as a lockable control panel door and the lifting bale accessible only from within the canopy
- Integral lubricator
- Electric generator

Most canopies are made of steel, with gull-wing or side and front doors to give access to the controls for starting and for servicing. If silencing is incorporated, the canopy is lined with sound deadening material. Smaller road-going compressors may have canopies made of glass fibre reinforced plastic. Portable compressors are also available skid-mounted as an alternative to a wheeled undercarriage.

Mobile compressor types are frequently used in the dirty environment of a construction site or quarry, far removed from the comparatively clean atmosphere of a factory, so special attention has to be paid to intake filtration for both compressor and engine. Cooling arrangements depend on the type of engine used. Air-cooled engines rely on fan cooling, water-cooled engines have a conventional automotive type radiator. The oil cooler (in an oil-flooded compressor) will probably be a conventional shell-and-tube cooler.

There will usually be found safety features such as automatic shut-down when the air delivery temperature, the engine coolant temperature or engine oil pressure exceeds a safe maximum. Indicators will give a visual warning when filters need changing.

A fuel tank with a capacity equivalent to one shift continuous operation is standard. Most of the canopies of portable units are lockable for security reasons.

The exigencies of the equipment hire market, into which many mobile compressors are sold, leads to their use in very adverse conditions, often with only rudimentary servicing.

The price is highly competitive, so the extra cost of sophisticated control systems would be neither necessary nor desirable.

Pressure and capacity of portable compressors

The standard supply pressure of portable compressors is 7 bar, because practically all construction and hand-held tools are designed to operate at this pressure as a maximum. Most tools will work adequately (but with a reduced performance) at a pressure varying between 4 bar and 7 bar. This flexibility makes compressed air a very versatile medium for use by contractors.

Higher pressures, up to about 20 bar, are also available, intended for such specialist purposes as the supply of air for high performance quarrying drills.

The capacity of portable compressors varies from 1.5 m^3/min up to about 40 m^3/min. Note that the traditional designation for the capacity of portable compressors is ft^3/min, and even the products of European suppliers frequently have a reference designation which can be traced to the capacity in these units. (Note that 1 m^3/min = 16.7 l/s = 35.3 ft^3/min).

One frequently comes across the expression one-tool, two-tool up to five-tool compressor. The tool referred to is a concrete breaker or paving breaker with an air requirement of 1.5 m^3/min (25 l/s). Such compressors are capable of running that number of tools simultaneously.

Mobile compressor noise legislation

The mobile compressor is frequently blamed as one of the noisiest items of construction plant which affect the environment. In fact the public are probably unable to differentiate between noise from the compressor and noise from the tool connected to it. The tool is by far the largest producer of noise, and in a typical roadside site, noise from the tool swamps that from the compressor. Despite this, there has been much pressure on the legislators to introduce regulations to limit the permissible noise level of compressors. This has culminated, in the EC, in regulations forbidding the supply of compressors emitting more than a prescribed noise.

Similar regulations also prohibit the supply of noisy tools (see the Section dealing with contractor's tools), but the levels set for tools are much higher than for compressors.

TABLE 1 – Permissible sound levels for portable compressors

Nominal air flow q in m^3/min	Permissible sound power level in $dB(A)/1\ pW$
< 5	100
5 < q < 10	100
10 < q < 30	102
q > 30	104

The test method is described in EC Council Directive 84/533/EEC. It consists in placing the compressor on a concrete base in the centre of a hemispherical array of microphones,

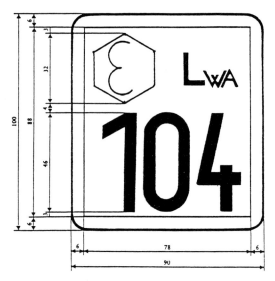

FIGURE 2 – Typical designation plate showing sound power level in dB(A).

then integrating the total sound power over the area of the hemisphere. There is a limited number of Government Approved test stations capable of carrying out a type test in accordance with the Directive. Arrangements differ in the various countries of the Community but a compressor approved in one can be freely imported and sold in any of the others. Figure 2 shows the designation plate that is attached to an approved compressor. The value 104 shown on that plate represents the sound power in dB(A) as measured by the testing station. It may be less than that required in the Directive; it may not, of course, be greater.

These regulations have no force outside the Community although other Western countries have similar, if not stricter, rules.

There may be some confusion about the designation of the sound emitted by a compressor (or any equipment for that matter). The approved method now is to specify the sound in terms of sound power against a base of 1 pW (10^{-12} W). This takes into account that a piece of machinery emits a different amount of noise in different directions, and therefore the only true assessment of its noisiness is to measure the total noise emitted over a surface enclosing it. However large the surface is, the total noise is the same. This is the principle behind the Directive. An alternative way of specifying the sound emitted by a compressor, which is frequently quoted in manufacturers' literature, is by the use of the Pneurop/CAGI Test Code. This is a simple method of noise measurement consisting of taking an average of four readings of sound pressure level at a distance of 7 m from the side of the compressor (not from its centre as in the EC Code). An approximate method of converting from sound power to sound pressure at 7 m using the Pneurop/CAGI Code is to subtract from the former 26 dB for a small compressor (up to 10 m³/min) and 27 dB for larger ones.

Out of the widespread use of the Pneurop/CAGI Code there arose the description of a compressor as standard, silenced or super-silenced. Standard means that no particular sound proofing has been applied, silenced means that the level of silencing is down to 75 dB(A) at 7 m, super-silenced means that a level of 70 dB(A) has been achieved.

If one wishes to determine how much noise an operator or a member of the public is exposed to, the local level of sound pressure has to be calculated from the sound power value and the distance from the centre of the compressor. The required relationship, ignoring such factors as the presence of reflecting objects and the directivity of the noise source, is:

$$L_p = L_w - 8.0 - 20 \log r$$

where L_p is the sound pressure, expressed in dB, at a distance r metres from the centre of the compressor

        L_w is the sound power level, expressed in dB

Methods of reducing the noise level of portable compressors

Most of the noise is generated by the engine, and since the engines used are standard industrial units with no particular attention paid to silencing for the compressor market, the burden is on the compressor manufacturer to apply sound reduction techniques to the canopy enclosure. The following techniques have been found useful:

- Line the canopy with sound-deadening material. The use of such materials as absorbent foam in the engine/compressor section where there could be contamination by fuel and oil leading to a fire hazard should be avoided. Double skinning could be used instead.

- Use flexible engine mounts between engine and chassis.

- Ensure the enclosure is complete. Use undertrays. Any opening doors should be well sealed. Instruments must be capable of being read from outside without opening the doors.

- The canopy should be soundly made, reinforced where necessary to prevent panel drumming and rattle.

- Extra engine exhaust silencing.

- Line the cooling air intake and exhaust passages with sound deadening material. This is an important point to observe – much noise can escape through the cooling air passages. Use silenced exhaust valves.

- Compressor intake from inside canopy.

- Mount canopy flexibly on chassis.

Some or all of these methods are capable of reducing noise level down to the required level. See chapter on Compressor Noise Suppression for further techniques.

In countries with high ambient temperatures, the incorporation of sound insulation material may have an adverse effect on the cooling of the compressor. Some of the insulation can be omitted but only for those countries where it is permitted.

Van mounted mobile compressors

Utility companies sometimes require the versatility of a compressor which, in addition to supplying compressed air to operate roadside equipment such as a road breaker, can also act as an electric generator for lighting purposes or as a supplier of hydraulic power. Such a unit can be mounted in a van, taking as its power source the van engine via a power take off from the main drive shaft. The compressor and generator can be under the floor of the van leaving the body free for use as a mobile workshop. There are a number of these available. Figure 3 shows diagrammatically how the various items are connected. A vehicle such as the one shown in the figure can supply 2.8 m³/min of air at 7 bar pressure and 6 kVA at a choice of voltages.

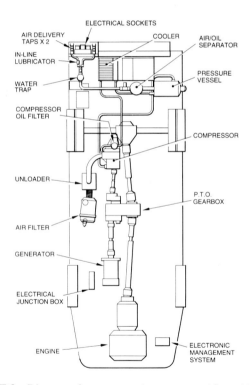

FIGURE 3 – Diagram of van mounted compressor. *(CompAir Holman)*

ACCEPTANCE TESTS

British Specifications covering the testing of compressors to verify guaranteed performance are to be found in BS1571 in a wholly metric form. Part 1 of that standard deals with type acceptance tests and Part 2 with simplified acceptance tests, calling only for measurements necessary to verify any guarantee given by the manufacturer to the purchase, in particular capacity, shaft input power or specific energy consumption and speed. Following is an outline of the procedure given in Part 2.

Preliminary tests should be taken for the purpose of:

- determining whether the compressor and associated systems are in a suitable condition for the acceptance tests to be conducted;
- checking of instruments;
- training of personnel.

After a preliminary test has been made, this test may, by agreement, be considered the acceptance test, provided that all the requirements for an acceptance test have been met.

The governing mechanism must be maintained in its normal working position.

During the test, the lubricant, the adjustment of the lubricating pumps, lubricators or other lubricating means must comply with the operating instructions. No adjustments other than those required to maintain the test conditions and those required for normal operation as given in the instruction manual shall be made.

Before readings begin, the compressor shall be run long enough to assure steady state conditions are reached so that no systemic changes occur in the instrument readings during the test.

The test report shall include the following information:

General information

Type of machine

If reciprocating whether

vertical or horizontal,

single- or double-acting,

cylinders are air- or water-cooled,

intercooled or aftercooled,

single- or multi-stage.

If rotary

the type of rotary elements employed,
if lubricated the type of lubricant used,
the type of filtration elements,
number of stages.

Manufacturer

Name and number.
Date of manufacture.
Type of prime mover.
Name of manufacturer of prime mover and its number if applicable.
Type of speed increasing or reducing gear if fitted.
Name of manufacturer of speed gear and its number.
Type of coupling used during the test.

Guaranteed performance on test and/or performance on type test

Specific intake pressure and temperature.
Capacity (FAD).
Delivery pressure.
Power consumption.
Speed.
Any other requirements of the standard, *eg* quantity of cooling water or temperature rise of cooling water.

Test observations

Only those measurements necessary to verify any guarantee given by the manufacturer to the purchaser.

Diameter of nozzle (mm).
Speed (r/min).
Barometric pressure (mm Hg).
Pressure at intake, absolute (bar).
Pressure drop across nozzle (mm H_2O).
Pressure on downstream side of nozzle (mm Hg).
Pressure at delivery, absolute (bar).
Temperature of air at intake (°C).
Temperature of air at nozzle (°C).
Power consumption.
Any other observations that are required to verify a guaranteed quantity of cooling water, inlet and outlet temperatures of cooling water, *etc.*

Results derived from observations

Nozzle constant.
Capacity (FAD).

TABLE 1 – Permissible deviations from guarantee conditions (BS1571)

Parameter	Deviation
Speed	±5%
Absolute intake pressure	±5%
Temperature of cooling water	±8%
Pressure ratio	±1%

TABLE 2 – Tolerance on capacity and specific energy consumption (BS1571)

Shaft input at normal load	100% Capacity		90% Capacity		
	Capacity	Specific energy consump-tion	Capacity	Specific energy consump-tion	No load power
	%	%	%	%	%
below 10 kW	6	7	—	—	20
10 kW to 100 kW	5	6	7	7	20
above 100 kW	4	4	5	6	20

Adjustments of results to guaranteed conditions

Corrected capacity (FAD)

$$= \frac{\text{test FAD x guaranteed speed}}{\text{test speed}}$$

If the speed for a specified FAD is not guaranteed, the correction does not apply provided that the machine gives the specified FAD at a speed which does not affect its service reliability (see also Table 2).

- Pressure at delivery. For positive displacement compressors no correction is necessary as it can safely be assumed that the specified delivery pressure can be obtained with specified intake conditions.
- Pressure at intake. For exhausters no correction is necessary within the permissible deviation, see Table 1.
- Corrected power consumption

$$= \text{power consumption} \ x \ \frac{\text{guaranteed absolute inlet pressure}}{\text{test absolute inlet pressure}}$$

$$x \ \frac{\text{guaranteed speed}}{\text{test speed}}$$

This is the only correction necessary provided that the deviations of test conditions from guarantee conditions do not exceed the values given in Table 1.

Acceptance tests to Pneurop/CAGI test codes

Three codes have been published under the joint auspices of Pneurop and the American Compressed Air and Gas Institute. They define acceptance tests for compressors which are constructed to specifications determined by the manufacturer and are sold against performance data published by the manufacturer. The codes are:

- PN2CPTC1 for Bare Displacement Air Compressors
- PN2CPTC2 for Electrically-driven Packaged Displacement Air Compressors
- PN2CPTC3 for I C Engine-driven Packaged Displacement Air Compressors

The publication of these codes enables the user to verify and compare the stated performance of compressors from different suppliers. In general, because their purpose is limited, they are less detailed than the national standards.

They specify the permissible variations in test conditions and the acceptable range of deviations from the manufacturers' claims in respect of flow rate, specific energy consumption and power consumption at zero flow. The permissible methods of determining the flow rate, shaft power, torque and electrical power are given.

They give the methods of calculating the correction factors that can be applied to the measured parameters when the test conditions do not agree with the specified conditions. These parameters include shaft speed, condensate formation, volume flow rates, power input and ambient pressure. These codes together form a useful guide to the measurement and comparison of compressor output.

COMPRESSOR NOISE REDUCTION

Noise as a hazard to employees

High levels of noise can produce permanent hearing damage and a conscientious employer will always bear this in mind when setting up a compressed air facility. Apart from his moral obligation, there is legislation to obey and codes of practice to which he has to pay attention. Some of these are:

- Health and Safety at Work Act 1974
- Control of Pollution Act 1974
- Noise at Work Regulations, 1989 SI 1790 (implementing EC Directive 86/188/EEC)
- Guidance to the Noise at Work Regulations. Noise Guides 1 to 8 published by HSE Books
- Personal Protective Equipment at Work Regulations 1992 SI 2966
- Guidance to the Personal Protective Equipment Regulations 1992 published by HSE Books
- EEC Article 100 Noise Directive 1986

Noise as an environmental problem

The manufacturer or the user of noisy equipment has to consider the effect it will have on persons other than the work force for whom he has primary responsibility. The European Community has taken the lead in ensuring that only equipment meeting specified maximum noise levels are acceptable in the countries of the Community. Some of the relevant legislation and Codes of Practice are:

- Planning Policy Guidance Note (PPG) 24 – Planning and Noise (supersedes in England only DoE Circular 10/73)
- British Standard 5228:1984 Code of practice for Noise Control on Construction and Demolition Sites
- EC Directives:
  84/533/EEC and 85/406/EEC Acoustic power levels admissible for motor compressors (implemented by Statutory Instrument 1985:1968)

Further information may be found in specialist publications, particularly the Handbook of Noise and Vibration Control (published by Elsevier Science Publishers Ltd) to which the reader is referred. This chapter is concerned mainly with the problems of noise reduction in compressed air installations.

The Health and Safety Commission take the view that an exposure equivalent of 90 dB(A) over an 8 hour day is the maximum level to which a worker should be exposed. Silencing techniques should be directed towards achieving rather better than that (to allow for deterioration over the life of the compressor). Bear in mind that the noise from a compressor will add to any other noise sources in the vicinity.

Techniques of noise reduction

The first step to take in noise reduction is to perform a survey of the sources of noise. A simple spectrum analysis of the noise in octave bands will indicate which frequencies predominate. This may be followed by a closer frequency analysis to determine the existence and magnitude of discrete frequencies. This survey should be done for all modes of operation, *eg* on-load, off-load and transition; the on-load condition is not necessarily the noisiest. The environment in which the compressor is placed can influence the noise levels. It is obvious that other sources of noise could confuse the readings, but what may be less obvious is that the reflectivity and the absorptive character of the walls or other structures may distort the spectrum from what would be produced by a compressor working in a free-field. It may not always be possible to perform the tests in an anechoic chamber but that would be the ideal.

Having performed the survey, it should be possible to identify the worst noise sources, which should be tackled in order of intensity.

Intake system

The intake system will include some or all of

- an intake filter
- an intake silencer
- piping and ducting
- control valves

Most compressors will incorporate an intake filter, which will by itself give some noise attentuation. This may be insufficient, in which case an intake silencer may be needed.

An intake silencer may be reactive or absorbent or both according to the frequency of the noise. For frequencies above 125 Hz, absorptive silencers are required and for frequencies below 125 Hz, reactive silencers are appropriate. If the 125 Hz band is straddled, then a combination silencer can be used.

A reactive silencer produces its effect by interference and reflection of the sound waves (an automobile silencer is reactive). A simple silencer would include an expansion chamber with a tail pipe, which may be sufficient by itself if one frequency dominates the spectrum. For multiple frequency suppression the number of chambers has to increase; design details of the method of analysis can be found in specialist publications. The lower

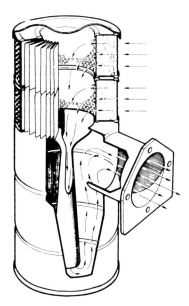

FIGURE 1 – Venturi silencer with two paper filter elements.

the frequency that is to be suppressed, the larger is the vessel required, so if low frequency noise is suspected at an early stage in the design, provision should be made for a reactive silencer and space can be set aside for its installation.

An absorptive silencer produces its effect by suppressing internal reflections with a porous ascoustic material lining the inside of the silencer cavity. When using this type of silencer, precautions must be taken to ensure that the material does not become contaminated by dust, water or oil which will not only reduce its effectiveness but may also be a place where fires start. Under severe vibration the material may disintegrate, discharging debris into the intake of the compressor.

Another type of reactive silencer is the venturi tube, which is particularly effective but expensive. The main advantage of a venturi is that the pressure drop through it is lower than with other kinds of restrictions, so it is particularly suitable for intake silencers where intake pressure losses could seriously affect the compressor efficiency. The flow velocity can be higher than with other silencers. An example of a venturi silencer combined with a filter is shown in Figure 1.

Inlet ducts

Guidance on the size and velocity of air flow in ducts is given in the chapter on Compressor Installation. A general rule to follow is: avoid coincidence of mechanical noise and acoustic frequencies, otherwise resonance of the pipe system can amplify rather than attenuate the noise.

It sometimes happens that a common intake system may supply more than one compressor. Such an arrangement should be avoided, as there is always the danger of amplified noise when pulsations of two compressors are in phase.

One should always attempt to design the pipework system so as to avoid undesirable frequencies. This is difficult, and it may turn out that when the installation is finished, it is unexpectedly noisy. One may then be faced with having to apply remedial treatment; this can be done by adding acoustic lagging to the outside of the pipes. Lagging should be made of a porous material such as glass fibre attached to the pipe, with an outer layer of dense cladding such as aluminium, steel or leaded plastics.

Intake valves

Compressor control systems that rely on throttling the intake may result in a high frequency note under part-load, partially throttled conditions. It may be necessary in this case to try a different valve.

Compressor casings

Casings can be very efficient radiators of noise, particularly if, in an attempt to reduce cost, the thickness has been reduced to a bare minimum. The thicker the sections and the more inherent damping there is in the material the better are its acoustic properties. Cast iron is a very good material in this respect. It is not generally very successful to try the application of acoustic cladding to a casing after manufacture. It is likely to be expensive and may also act as a thermal insulator, raising the temperature of the compressor. Water cooling jackets act as a natural noise barrier and it may be better in some circumstances to use water, rather than air, cooling.

An acoustic enclosure is likely to be the most effective form of noise barrier. Practically any required attenuation can be achieved by enclosure. Twenty five dB is perfectly possible whereas techniques of noise reduction at source will only give about 5 dB. In designing an enclosure, reference can be made to the Handbook of Noise and Vibration Control. By far the best enclosure is one made of solid brickwork or some other heavy material. This is not practical for a compressor which needs to be moved, so the next best

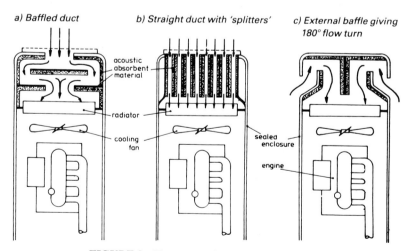

FIGURE 2 – Treatment of cooling air passages.

choice is a box made of heavy timber or steel partitions, lined with an absorbent material with no unsealed openings. It is the mass of the enclosure which is the main factor in suppressing the radiation of noise, but unless the interior is lined with an absorbent material, the level inside builds up with the result that the effect of the barrier is less. Cooling air should only be admitted through well-baffled ducts. The benefit of total enclosure can be easily vitiated by small openings for the passage of pipes.

Interstage silencers

These can be very expensive and should only rarely be considered. It is best to make sure that the interstage pipe is large enough to act as a pulsation damper; it should be a simple straight connection avoiding bends.

Delivery passages

After delivery, the air is at pressure so any enclosure is a pressure vessel with all the design limitations that that implies. Whether it be a simple damper to reduce the effect of compressor pulsation or a larger receiver to store a substantial amount of air to meet a variable demand, it will act as an acoustic barrier to noise in the delivery passages. To suppress vibration it may prove necessary to apply treatment to its exterior. Bear in mind that passages up to the cooler will be at high temperature and the treatment must be resistant to heat.

Flexible mountings

The compressor and its prime mover should be mounted on flexible supports, preferably on a massive base also. Certainly the enclosure should be isolated from the machine proper, otherwise its value as a barrier would be lost. There may be difficulties if the pipe work is rigidly attached to a compressor free to move on its supports, so flexible hoses or bellows should be interposed.

There are several types of isolator on the market. Simple composition pads have a small static deflection but present a high impedance to high frequency energy. Rubber mounts have a medium degree of static deflection, suitable for lower frequencies. Steel springs, and in severe cases air springs, can be used for very low frequency isolation.

Blow-off valve and relief valve noise

Blow-off valves should be equipped with their own silencers. Because they may be intermittent in operation, the noise that they make may not be noticed during a noise survey. An occasional noise may be more annoying to the neighbours than one continually present.

Safety valves only rarely discharge to the atmosphere. For safety reasons they must not be fitted with silencers or any form of piping. Indeed it may be desirable that they should make a noise as a warning. If a safety valve is continually blowing there is a fault in the control system.

Prime movers

Industrial installations normally have an electric motor to drive the compressor. Noise may be generated by the cooling fan or from the motor windings. The cooling air will come from inside the enclosure and probably share the same airflow as required for the compressor. Silencing an internal combustion engine is a study in itself, but the same type of enclosure as described for the compressor would apply here also. See also the chapter on Mobile Compressors.

Gear boxes

Step-up gears are needed for screw compressors and for turbines. Some screw compressors supplying dry air also have timing gears. Helical gears are quieter than spur gears and ground teeth than machine-cut teeth. Gear boxes can radiate noise if they are fabricated or of thin section. Cast iron is the preferred material of construction.

General

Most compressors are now available as packaged units. If the noise level produced by a standard package is satisfactory this must be the first choice. Only if particularly rigorous noise levels are required should one attempt the exercise of making a non-standard enclosure. As indicated above, the environment in which it is situated may completely alter the spectrum from that which has been measured on the same compressor in a free field.

Figure 3 shows an example of a well silenced piston compressor with a noise level of 61 dB(A). It delivers up to 370 l/min 10 bar pressure. The air receiver in this case is external. The noise reduction is achieved by using a high density sound proof material, an integral cooling fan and mounting the assembly on anti-vibration pads.

FIGURE 3 – Packaged compressor incorporating silencers. *(Ingersoll Rand)*

COMPRESSOR CONTROLS

Compressor flow rate must be regulated to match the system demand. Usually the discharge pressure is the controlled variable. The type of control depends upon the characteristics of the compressor, the prime mover and the system. The control may be manual or automatic; it may be discontinuous (on–off) or continuous through speed variation of the motor and/or inlet throttling. Unloaded start of the compressor must be provided for if the torque of the prime mover is not sufficient to turn the compressor under load.

Smaller stationary compressors may be run continuously with only a blow-off (pressure relief valve) on the delivery side which opens when a specified maximum pressure is reached to allow compressed air to be blown off to atmosphere instead of being delivered to the system (Figure 1). This is a wasteful way of operation, since the prime mover is always working at its full power even though there may be no demand from the system. If a receiver is included, a check valve may be used to prevent high pressure being blown back into the compressor (Figure 2). This is always necessary with valveless compressors

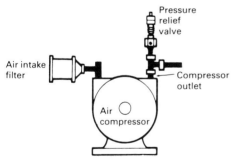

FIGURE 1 – Basic low pressure power supply system.

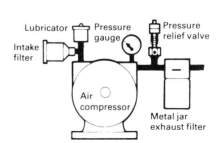

FIGURE 2 – Pressure relief valve adds basic control capability to simple system.

(*eg* vane or screw compressors) which might otherwise act as motors, driven backwards by the air in the system, but may not be necessary in piston or diaphragm types which have their own delivery valves. A check valve may still be used with the latter as a backflow protective device. A safety valve should also be incorporated in the system to protect against over-pressure. In this case, the safety valve would be set rather higher than the system operating pressure (about +10%).

For continuous duty it may be desirable to maintain a higher pressure in the receiver than is required in the supply line. In this case, the receiver is followed by a pressure regulator (Figure 3).

An alternative method of pressure regulation for constant speed motor driven compressors is by on–off cycling using an electrical pressure switch on the receiver to cut the motor supply at a predetermined 'high' pressure level, and back on again at a predetermined 'low' level (Figure 4). The receiver pressure is held between these pressures provided that the demand does not exceed the recharging capacity of the compressor.

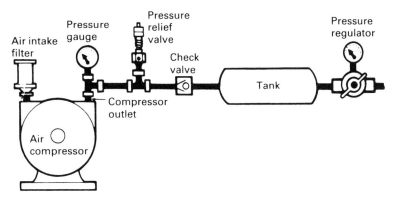

FIGURE 3 – Pressure regulator installed downstream of receiver compensated for built-up pressure higher than desired line pressure.

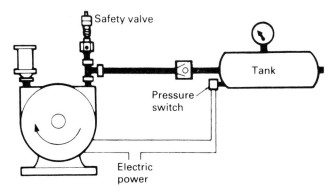

FIGURE 4 – A pressure switch controls receiver pressure independent of downstream pressure by on/off cycling of compressor drive motor.

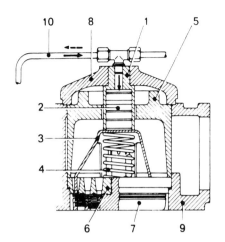

1. Unloading cylinder.
2. Unloading piston.
3. Unloading claw (depresses the valve discs and keeps the suction valve open during the unloading period).
4. Unloading spring (returns unloading claw at start of loading).
5. Valve retainer.
6. Seat of suction valve.
7. Suction valve guard.
8. Suction valve cover.
9. Cylinder head.
10. Compressed air to and from unloading device.

FIGURE 5 – Unloading device for piston compressors.

Unloading

Larger systems are commonly operated with the compressor running continuously and the capacity regulated by unloading. One mechanical method used with reciprocating compressors is shown in Figure 5.

Pressure from the receiver is transmitted to the underside of the piston of the device, which acts against a spring set at a predetermined pressure. When the pressure in the receiver exceeds this pressure, the spring compresses and allows air to flow to a piston on top of the suction valves which in turn depresses a claw which holds the suction valves off their seat, so that compression cannot take place in the cylinder. This is termed suction valve unloading.

When double-acting compressors are used, by having two air relays set at different pressures, it is possible to unload each side of the piston in turn, giving full load, half load and zero load. This is three step unloading.

A variation of this method is to allow the inlet valves to open normally to fill the cylinders but to keep them open for a timed interval of the compression stroke, after which compression can begin. This can provide stepless regulation, but is normally used only with process compressors.

In clearance pocket control, one or more pockets containing added clearance volume are connected to the first stage cylinder, thereby lowering the volumetric efficiency and the amount of air delivered. These pockets may be manually or automatically controlled. The pockets usually have a fixed volume, but may have continuously variable pockets. A normal method on double acting compressors is to have a combination of clearance pocket and suction valve unloaders to give 100%–75%–50%–0% unloading.

Reverse flow regulation

A method of stepless control suitable for reciprocating compressors is reverse flow regulation, by which the intake flow is caused to vary according to requirements, by

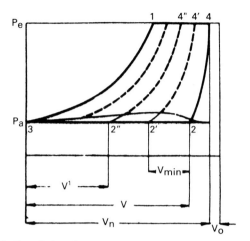

FIGURE 6 – Regulating diagram of a reverse flow regulation (theoretical).

delaying the closing of the suction valves. This delayed closing is achieved by means of special unloaders. These are actuated by the variable pressure of a piston built into the valve cover (usually a diaphragm piston in a servo-cylinder). A number of unloader springs are tensioned by means of an unloader plate, which in turn keeps the valve plate of the regulated suction valve open against the reverse flow of gas during the compression stroke of the cylinder. Part of this gas flows back into the suction duct and reduces the output capacity by an equal amount. During the compression stroke, the velocity of the gas flowing back rises steadily with increasing piston speeds as do the flow forces acting on the unloader and valve plate in a closing action. These forces vary with piston position and eventually exceed the hold-open force which has been adjusted to a certain value. This causes the valve to close (after a delay) and the reverse flow is terminated. The remaining gas in the piston is compressed and discharged in the usual manner. This permits adjustment to any desired capacity between 100% and 40% of full load (see Figure 6) and in certain cases stepless reductions down to 15% or less may be achieved.

A further method of controlling the output is by throttling the intake, but this requires more power at part load conditions than the methods described above.

Screw compressors

A different form of unloading is adopted in the case of screw compressors. The capacity is regulated by throttling the air intake and at the same time opening a blow-off valve. The pressure side is shut off with a non-return valve. Under load, the blow-off valve is closed and the throttle valve fully open. The unloading device is operated by a spring loaded automatic air relay of the same type as that used on reciprocating compressor controls. A complete system is shown in Figure 7.

Slide valve control for screw compressors

This type of control incorporates a third bore in the compressor casing, at the cusp of the rotor bores. In this bore is positioned a slide valve at the discharge side of the rotors.

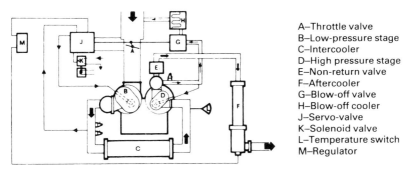

A–Throttle valve
B–Low-pressure stage
C–Intercooler
D–High pressure stage
E–Non-return valve
F–Aftercooler
G–Blow-off valve
H–Blow-off cooler
J–Servo-valve
K–Solenoid valve
L–Temperature switch
M–Regulator

FIGURE 7 – Capacity unloading of screw compressors. *(Atlas Copco)*

As the compressor is off-loaded, the valve moves progressively opening up ports which connect the compressed air back into the inlet. This form of control is only suitable when the compressor is driven by a constant speed device such as an electric motor. Speed control is usually preferable, as well as being simpler.

Multi-pressure capability for screw compressors

It is sometimes convenient to supply a compressor with different pressure capabilities to meet the varying demands of the plant hire industry. It is clearly not possible in an installation with a fixed prime mover to supply a range of pressure capabilities while delivering the same output. However it is possible in a limited way to exchange delivery volume and pressure. This can be achieved by what amounts to a shortening of the L/D ratio of the screw elements. A portion of the inlet length of the screws can be by-passed through one of a series of passages in the compressor casing, which vent some of the inlet air according to the desired pressure setting. The operating condition is set initially (which can be key locked to avoid improper use). Thus, in one particular compressor it is possible to have the following variations:

Delivery in l/s	Gauge pressure in bar
360	6.8
314	8.5
276	10

Unloading and speed control system in an engine driven screw compressor

Compressor air output is controlled by a combination of an unloader valve which regulates the amount of air entering the compressor and a speed controller which varies the engine speed. An engine is not, of course, capable of running down to zero speed; typically a diesel engine can operate down to 50% of its maximum rated speed and still give a useful output, say from 2000 r/min down to 1000 r/min. It could still idle at less than 50% but the torque obtainable would not be sufficient to turn a screw. Usually the engine speed is reduced first

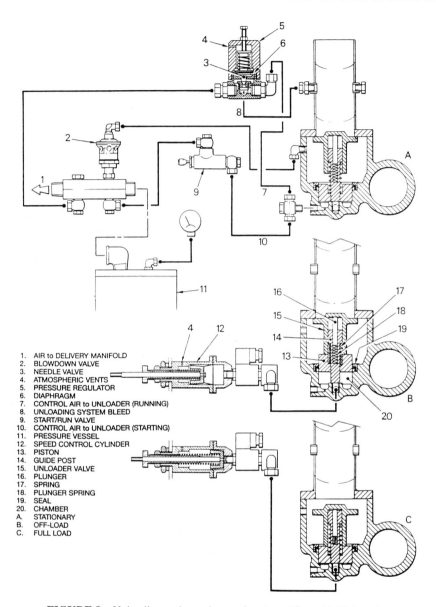

1. AIR to DELIVERY MANIFOLD
2. BLOWDOWN VALVE
3. NEEDLE VALVE
4. ATMOSPHERIC VENTS
5. PRESSURE REGULATOR
6. DIAPHRAGM
7. CONTROL AIR to UNLOADER (RUNNING)
8. UNLOADING SYSTEM BLEED
9. START/RUN VALVE
10. CONTROL AIR to UNLOADER (STARTING)
11. PRESSURE VESSEL
12. SPEED CONTROL CYLINDER
13. PISTON
14. GUIDE POST
15. UNLOADER VALVE
16. PLUNGER
17. SPRING
18. PLUNGER SPRING
19. SEAL
20. CHAMBER
A. STATIONARY
B. OFF-LOAD
C. FULL LOAD

FIGURE 8 – Unloading and speed control system. *(CompAir Holman)*

down to its minimum and intake throttling then takes over. Figure 8 shows a modern control system governing an engine driven compressor set at 7 bar pressure. This system includes a Start/Run valve for the warm up period, a pressure regulator valve set for the maximum compressor pressure and a blow down valve to release system pressure on shut down.

The unloader valve is situated in the inlet to the compressor. When the compressor is started the vacuum created in the unloader body causes the valve 15 to move against the spring 18. Air is then admitted to the compressor which builds up pressure in the pressure vessel 11. This pressure is piped to the regulator 5 and the Start/Run valve 9. On start up, pressure passes through this valve and acts on the unloader piston 13 and speed control cylinder 12. When the pressure reaches 2.75 bar it moves the unloader valve to close off the air intake and moves the speed control cylinder to reduce engine speed to idling while the engine warms up. When the engine has reached operating temperature, the Start/Run valve is pressed which cuts off the air passing through it. The unloader valve piston then admits air to the compressor and the engine speeds up, to deliver air at a maximum of 7 bar. When that pressure is reached, the unloader valve and speed control cylinder move until air demand is exactly matched.

When the compressor is shut down correctly, the delivery cocks are first closed and the machine is then in the off-load condition. When the rotors stop, pressure in the vessel is prevented from getting to the inlet by the closed unloader valve. The remaining pressure then operates the blow-down valve to release all pressure in the system, which is then ready for the next start up. The blow-down valve ensures that the compressor cannot start against system pressure.

Control in multi-compressor installations

There are some advantages in industrial installations in having a number of separate compressors coupled together and feeding into the same supply line. Apart from the flexibility of being able to take an individual compressor out of service for maintenance and repair, there are possibilities inherent in this arrangement for designing a very efficient control for part-load operation. By adopting such a control system it has become possible to design rotary sets which approach the off-load efficiency of reciprocating units. In recent years the availability of cheap electronic control and sensing techniques has brought this type of approach into every day use.

There are several possibilities for this type of control, which in increasing order of complexity are:

- Control using system pressure sensed by a pressure switch to switch on or off individual compressors in a multi-set installation. Suitable for those installations with three or more compressors; this can be expensive because each compressor needs its own control card. The compressors could be switched in cascade or in ring sequence (to balance compressor use).
- Simple stop/start suitable for small units (up to about 20 kW each) whereby a system pressure switch activates a timer which stops a compressor if the signal persists after say one minute.
- Auto-idling where the receiver pressure is vented in response to a signal from the system pressure switch after a predetermined time. After a further preset time the set is switched off completely.
- Pulse-width modulation. The principle behind this development ensures that the

individual compressor can operate at either full-load or idling. The system pressure is sensed as in the previous methods and the compression chamber pressure is switched between system pressure and atmospheric in a series of pulses with varying time widths. The receiver pressure is maintained at system pressure and isolated from the compression chamber by an unloader valve. A further modification of this system is possible if the cycling time turns out to be too rapid. In this case the PWM is tempered by operating at intermediate steps of say 66% and 33% of full load.

Variable speed electric motor controls

Most electrically driven compressor sets use a constant speed motor, relying on intake throttling for capacity control. There is now coming on to the market an alternative form of capacity control using a variable speed electric motor drive (VSD). This works practically in much the same way as the engine speed control described above. The motive power comes from a variable speed, variable frequency induction motor. This has two major advantages: it is more efficient in part load operation, when compared with a compressor where the capacity control relies on modulation or on a basic load/no load regulation system; and it eliminates the need for a gear train. This form of control includes a frequency convertor in the compressor package and has other features which are possible using advanced electronic monitoring and control. The acceleration and deceleration levels can be controlled, which reduces the mechanical stress on the components of the

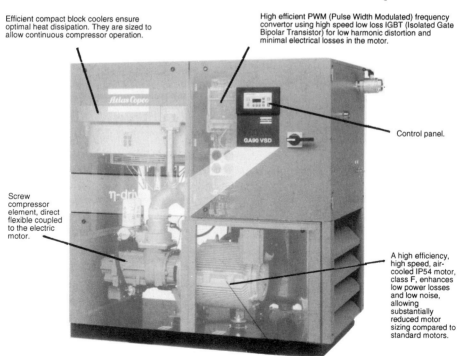

Efficient compact block coolers ensure optimal heat dissipation. They are sized to allow continuous compressor operation.

High efficient PWM (Pulse Width Modulated) frequency convertor using high speed low loss IGBT (Isolated Gate Bipolar Transistor) for low harmonic distortion and minimal electrical losses in the motor.

Control panel.

Screw compressor element, direct flexible coupled to the electric motor.

A high efficiency, high speed, air-cooled IP54 motor, class F, enhances low power losses and low noise, allowing substantially reduced motor sizing compared to standard motors.

FIGURE 9 – Variable speed compressor. *(Atlas Copco)*

drive. The concept accurately measures system pressure and monitors the frequency convertor and speed so as to keep the delivery pressure constant within a narrow band.

While the system is more economical in running costs in an application where a substantial part of the time is at part load conditions, this must be offset against the extra capital costs. Figure 9 shows a completed packaged unit.

Compressor protection

Critical parameters which need to be maintained to provide basic protection for a compressor are coolant water flow or temperature, discharge air temperature and lubricating oil pressure or level (in the case of lubricated compressors). Typical minimum parameters, *ie* danger levels, might be:

- coolant water flow 2/3 of normal flow rate
- discharge air temperature 160°C
- lubricating oil pressure 2 bar

Temperature sensors and pressure sensors can provide cover for these. If temperature and pressure operated switches are used, they can provide automatic protection either as alarm devices or control devices, shutting down the system when a critical parameter level is reached.

Where compressors employ closed-circuit water cooling (either thermo-syphon or

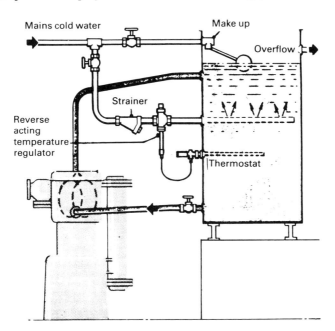

FIGURE 10 – Temperature regulator controlling introduction of water to holding tank. *(Spirax Sarco)*

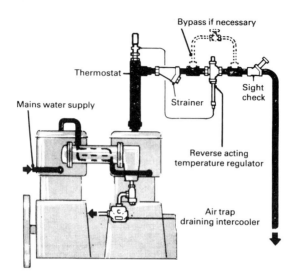

FIGURE 11 – Temperature regulator controlling mains water cooling of two-stage compressor. *(Spirax Sarco)*

forced circulation), additional cooling water can be provided from the main or other cold water supply via a temperature regulator. This allows cold make-up water to be supplied to the cooling tank at a controlled rate, hot water being displaced through an overflow. A typical installation of this type is shown in Figure 10.

Automatic temperature control can be applied to mains water cooling systems, the purpose in this case being to eliminate wastage of water and to prevent excessive cooling which could cause condensation within the cylinder. A typical installation applied to a two-stage compressor is shown in Figure 11. It is important that the temperature regulator cannot shut off the supply of cooling water completely. This can be guarded against by incorporating an adjustable bleed in the valve itself (ensuring a small flow when the valve is shut) or the use of a bypass, the flow through which is controlled by a needle valve.

In the case of larger compressors operating in a closed cycle, cooling is accomplished by passing the water through a mechanical cooler or cooling tower. Again temperature control can prove advantageous, particularly to avoid too low a temperature being developed in the second stage cylinder.

An important consideration in the case of cooling towers is the possibility of freezing, with the consequent cessation of water circulation calling for the compressor to be shut down. This possibility can be avoided by using a temperature regulator valve commanding a supply of steam to a heating coil i n the sump. The regulator can be set to open when the water temperature falls below, say 2°C, so that no steam is used until near freezing conditions are approached (Figure 12). To guard against the possibility of the cooling tower freezing solid, the line from the diversion control valve can be arranged to return water from the compressor direct to the sump instead of to the top of the tower under low temperature conditions.

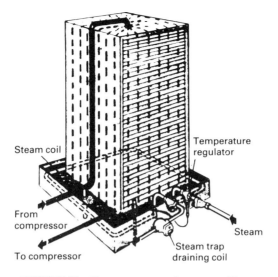

FIGURE 12 – Temperature regulator controlling temperature of cooling tower sump. *(Spirax Sarco)*

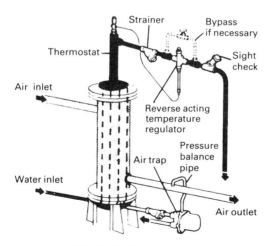

FIGURE 13 – Temperature regulator controlling aftercooler. *(Spirax Sarco)*

Temperature regulators can also be used to advantage on water cooled after coolers. The use of a temperature regulator enables the water consumption to be set to the point of maximum efficiency and water economy. An example of the use of a temperature regulator controlling main water supply to an aftercooler is shown in Figure 13.

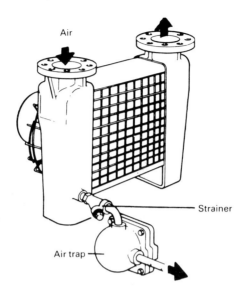

FIGURE 14 – Air blast aftercooler. *(Spirax Sarco)*

Water cooled aftercoolers are usually the most efficient type; the lower the air temperature they produce, the better. Air cooling is also widely used, particularly where cooling water is expensive or not readily available; portable units almost exclusively use air cooling. An air blast aftercooler is usually the first choice. A motor driven fan directs a stream of air over a bank of finned tubes, through which the compressed air passes, see Figure 14.

Fluid Power Technology

is our speciality
applying
COMPRESSED AIR & PNEUMATICS
to meet any requirement in any industry

Components ▶

Equipment ▶

Plant ▶

Machinery ▶

Systems ▶

Supplies ▶

Manufacture ▶

Survey ▶

Installation ▶

Servicing ▶

PNEUMATIC ALARMS
... the **'AIR-AUDIO'** range
Panel Mounted Air Horns
Controls & Interfaces
Complete Annunciator Systems

FLUIDIC DRIVES
... the **'TRANSPORTOR'** range
No moving parts transfer &
processing of flowable media.

▶ ▶ ▶ ▶ ▶ ▶ ▶ ▶

TRAINING RIGS
... the **'CUBIC'** range
3-D Working Units for
Manufacturing Automation &
Process Control Simulation

Other products in the
'Pneumethods' *brand range :-*

Air Boosters - Counting Systems
Sequential Timers - Pneumatic Sequencers
Pressure Intensifiers - Interlocked Push-Buttons
Differential Pressure Switches - Level Controls

Speciality Systems : **Fluidics, Fluid Logic, Pneumatic Circuitry**

WE PLACE THE EMPHASIS ON SPECIALTY, from the design and development of new
devices to the manufacture of complete fluid power systems tailor-made to your needs.

Tel: 01946 810179
(8 lines)
Fax: 01946 813660
01946 812739

FRIZINGTON RD. INDUSTRIAL ESTATE,
FRIZINGTON, CUMBRIA CA26 3QY

AIR RECEIVERS

A compressor plant normally includes one or more pressure vessels known as air receivers which provide the following functions:

- Equalisation of pressure variations in the supply lines.
- Storage of compressed air to meet heavy demands in excess of the compressor capacity.
- A source of additional cooling and a collection point for residual condensate and oil droplets.
- Prevention of rapid loading and unloading of the compressor in short cycle duties.

The usual form of receiver is illustrated in Figure 1. The detailed design of such a vessel is covered by a number of standards (see below).

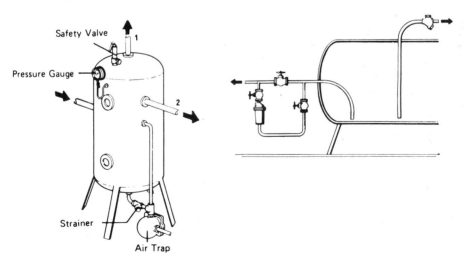

Typical vertical receiver with air trap.   Small horizontal receiver with alternative trapping arrangements.

FIGURE 1

TABLE 1 – Useful capacity of air receivers (cubic metres of free air)

Receiver volume m³	Permissible pressure drop — bar							
	0.25	0.5	0.75	1	1.25	1.5	2	2.5
1	0.25	0.5	0.75	1	1.25	1.5	2	2.5
2	0.5	1	1.5	2	2.5	3	4	5
3	0.75	1.5	2.25	3	3.75	4.5	6	7.5
4	1	2	3	4	5	6	8	10
5	1.25	2.5	3.75	5	6.25	7.5	10	12.5
6	1.5	3	4.5	6	7.5	9	12	15
7	1.75	3.5	5.25	7	8.75	10.5	14	17.5
8	2	4	6	8	10	12	16	20
9	2.25	4.5	6.75	9	11.25	13.5	18	22.5
10	2.5	5	7.5	10	12.5	15	20	25

Storage capacity

The useful capacity of a receiver is the volume of free air which can be drawn from it at its design pressure; this is, of course, different from its actual internal volume. The size of the air receiver depends on the duty it is required to fulfil. The simplest duty required is one in which the demand is constant and reasonably close to that of the compressor output. A rule-of-thumb guide for receiver size in this situation, where the pressure is less than 9 bar, is six times the compressor flow rate per second or 0.1 times the flow rate per minute. This rule applies only where the control system involves automatic valve unloading, and where the difference between the loading and reloading of the system is no less than about 0.4 bar. If the pressure is higher than 9 bar, a rather smaller receiver will be acceptable.

If the demand is variable, the calculation of receiver size depends on the acceptable pressure drop and the number of cycles of loading and unloading that are permissible per unit time. For a very variable demand, the receiver volume should be about three times that calculated for uniform demand.

In a situation where the air demand is intermittent, requiring a fairly large volume flow for a short period of time with no demand at all for the remainder of the time, it is worthwhile performing a calculation to determine the most economical combination of compressor and receiver. Take as an example an application in which 60 m³/min at 7 bar is required to be delivered for 10 minutes every hour. Clearly one solution would be to have a compressor capable of 60 m³/min working for only those ten minutes. Since the mean flow is only 60 ÷ 6 = 10 m³/min, another solution would be to have a smaller compressor working continuously and storing its excess in a large receiver. The smaller compressor would have a higher pressure rating and the excess flow for 50 minutes would be stored in the receiver at that pressure.

The fundamental equation relating pressure drop to receiver volume is:

$$\frac{V}{V_R} = \frac{\Delta P}{P_0} \qquad [1]$$

where V_R is the receiver capacity in m^3
         V is the air requirement for a given operation in m^3
         ΔP is the pressure drop (bar) experienced in the receiver during the operation
         P_O is atmospheric pressure in bar

If the operation is identified by a flow rate q in m^3/s over a time t, this equation can be rewritten as:

$$\frac{qt}{V_R} = \frac{\Delta P}{P_o} \tag{2}$$

These equations assume that no air is being added to the receiver during the operation. If the compressor is simultaneously supplying air at a rate q_c m^3/s, the latter equation becomes:

$$\frac{(q-q_c)t}{V_R} = \frac{\Delta P}{P_o} \tag{3}$$

In the above example of intermittent use, the acceptable pressure drop has to be chosen. Taking two possible choices for the maximum pressure, case (a) 20 bar and case (b) 35 bar, both from compressors delivering 10 m^3/min.
    Applying equation [3] to case (a)

$$\frac{(60-10) \times 10}{V_R} = \frac{20-7}{1}$$

so V_R = 500 ÷ 13 = 38.5 m^3 and the actual volume of the receiver will be 38.5/21 = 1.83 m^3.
    Similarly for case (b) V_R = 18 m^3 and the actual volume of the receiver will be 18 ÷ 36 = 0.5 m^3.
    In deciding which choice to make the most economical solution is when the total cost of the compressor, receiver and fuel is a minimum.

Receiver volume calculations in automatic start/stop control

If the control system involves stopping and starting of the motor, then a receiver much larger than that for automatic unloading is required so as to avoid too rapid a cycling of the controls. The primary requirement is that the receiver must be sized so that no more than about ten starts per hour, evenly spread, take place; this is so that damage to the starter system and the motor windings is avoided. The spread of pressure should be as large as can be tolerated to further reduce the number of cycles. The compressor manufacturer will normally calculate the correct receiver size based on the control system used.
    Equation [3] can be rewritten as

$$V_R = \frac{q_c(1-\psi)\,\psi t P_o}{\Delta P} \tag{4}$$

where $\psi = q/q_c$
and, in this instance, t is the time between successive starts of the compressor
       ψ is the actual air demand as a fraction of the amount delivered by the compressor. It

turns out that the worst condition for frequency of stops and starts is where $\psi = 0.5$, *ie* when the demand is half the supply. If the demand is greater or less than a half, the number of stops is less frequent so this condition is usually taken as the design point. Replacing ψ by 0.5, equation [4] becomes:

$$V_R = \frac{q_c t P_o}{4 \Delta P}$$

For t = 6 minutes and $\Delta P = 1$, the relationship becomes;

$$V_R = 1.5 \, q_c.$$

Compare this with the formula for the receiver size quoted above for unloading systems of control:

$$V_R = 0.1 \, q_c$$

This comparison between the two values for the volume indicates the importance of assessing the correct receiver size to meet the requirements of the system.

Air receiver legislation in the U.K.

There is a large legislative framework governing the whole area of receiver design and certification which needs much study before it can be applied with confidence.

The Health and Safety at Work Act 1974 places a general duty on a number of persons involved with matters of safety; the duties are shared by designers, manufacturers, suppliers, importers employers, employees *etc*. Regulations in specific branches of industry have been made under the Act. The primary document on pressurised systems (of which air receivers are usually the most critical element) is The Pressure Systems and Transportable Gas Containers Regulations 1989 (SI 1989 No 2169), all of which are now in effect. These Regulations implement EC Directive 76/767/EEC. The Health and Safety Executive have prepared a Code of Practice – Safety of Pressure Systems — which gives practical guidance on these Regulations . The British Compressed Air Society have gone further and have produced a comprehensive set of Guidance and Interpretation Notes which should be studied alongside the two official documents.

It should be noted that these Regulations repeal much of the Factories Act 1961 insofar as it related to pressure vessels, however vessels already in use and bearing markings made in accordance with the Act will not have to be changed unless they are modified subsequently.

The Regulations do not give detailed design methods (reference must be made to the appropriate Standards for these) but are concerned with ensuring that pressure systems falling within their scope are properly made, inspected, maintained and marked so as to safeguard those persons affected. The intention is to prevent the risk of injury from the release of the stored energy in a compressed air system. They are in fact wider in scope than compressed air and encompass steam, gases and mixed liquids and gases. The complete system includes the pressure vessels, pipework and any protective devices, where the pressure is in excess of 0.5 bar. If the complete system does not incorporate an air receiver, it may not be covered by the Regulations but it would be unwise to assume that to be the case without expert advice.

In the Regulations a system can be classified as Minor, Intermediate or Major, depending on the product of the gauge pressure of the system in bar and the internal volume in litres (*ie* the product PS.V).

A minor system has a maximum operating pressure below 20 bar (gauge), the largest pressure vessel in the system has a PS.V less than 2 x 10^5 bar litres and the operating temperature between –20°C and 250°C.

An intermediate system is one which either

(a) has a system pressure of 20 bar or above where PS.V is less than 1 x 10^6 bar litres
     or

(b) has a system pressure less than 20 bar and a PS.V greater than 2 x 10^5 but less than 10^6 bar litres.

A major system is one in which PS.V is greater than 1 x 10^6 bar litres.

If PS.V is less than 250, certain of the Regulations do not apply (particularly in respect of marking), although such a vessel may come within the scope of the Simple Pressure Vessels Regulations, see below.

Most compressed air systems will fall into the Minor category. The Regulations place a different burden on the organisation responsible for carrying out the inspection and operation of the system according to the category in which it falls and defines the levels, qualification and experience of the "competent person" who draws up an examination scheme.

The Simple Pressure Vessels (Safety) Regulations 1991 (SI 2749) apply to smaller pressure vessels. These Regulations are the national implementation of EC Directive 87/404/EEC (amended by 90/488/EEC). A simple pressure vessel is defined as one which is welded and made of non-alloy steel or non-alloy aluminium or non-age hardening aluminium alloy; the simple construction implies that it is to be cylindrical with convex ends or hemispherical; it must have a working pressure not greater than 30 bar and the product of the working pressure and the volume is not to exceed 10 000 bar litres. A pressure vessel coming under these regulations can be classified as A1, A2, A3 or B, depending on the product of the gauge working pressure and the internal volume (PS.V):

Classification	PS.V (bar litres)	
A1	> 3000	≤ 10 000
A2	> 200	≤ 3000
A3	> 50	≤ 200
B	≤ 50	

Different Regulations apply to the different categories, the main differences being the type of certification procedure which must be applied.

Meeting the requirements of SI 2749 may be less onerous than SI 2169, so it is important to determine into which category and sub-category a particular pressure vessel falls and establish the inspection and certification procedure accordingly.

Most compressed air systems containing a receiver will fall within the scope of the Simple Pressure Vessels Directive.

Countries other than the U.K.

For members of the European Community other than the U.K., the same EC Directives apply, which makes for simplicity in design and certification and ensures that a pressure vessel acceptable in one European country will be equally acceptable to the others. For other countries general reference should be made to Boilers and 'Pressure Vessels – An International Survey of Design and Approval Requirements' published by BSI. This is a comprehensive reference to the national regulations, design codes and approval organisations of a large number of countries. Fortunately European and American Codes are widely acceptable. In a subject which is constantly changing, the latest edition of this publication should be studied.

Design of pressure vessels

There are three British Standards to which reference should be made for information about design, manufacture and inspection – BS 5169:1992, BS 5169:1975, BS 5500:1994 and BS EN 286.

- BS 5169 gives information specific to fusion welded steel air receivers (excluding vessels covered by SI 2749).
- BS EN 286 replaces BS 5169:1975 and is for vessels covered by SI 2749. It is for simple unfired pressure vessels designed for air or nitrogen.

These two are mutually exclusive.

- BS 5500 is for unfired fusion welded presure vessels.

BS 5500 is a comprehensive work with a wealth of material on all aspects of welded pressure vessels and should be studied by anyone embarking on the design. This chapter can only give an outline of the straightforward principles of design of receivers. For more complex matters such as the design of flanges, reinforcement of access openings, numbers and location of bolts and welding sizes, the original standards must be referred to. Computer programmes are available which remove some of the burden of calculations.

Materials for manufacture

A limited range of materials is permissible to meet the requirements of the standards. Reference should be made therein. Broadly certain steel grades contained in BS 10207 and BS 10208 are acceptable for plates, strip and bar; material for tubes and forgings are specified in ISO 2604. For aluminium and aluminium alloys refer to the standard.

Design in accordance with BS 5169

BS 5169 specifies three classes of air receivers:
Class I – No limitation is placed on the design but all welded seams require non-destructive testing.

Class II – The design pressure shall not exceed 35 bar and the product of design pressure and internal diameter shall not exceed 37 000 bar mm.

Class III – The design pressure shall not exceed 17.5 and the product of design pressure and internal diameter shall not exceed 8800 bar mm.

Wall thicknesses

Simple theory for determining the thickness of a cylinder subject to internal pressure produces the relationship:

$$t = \frac{pd}{2f - p} \qquad [5]$$

where: t is the wall thickness
         f is the nominal design stress
         p is the internal pressure
         d is the internal diameter

This equation is true for any system of consistent units.

The equivalent equation given in BS 5169 states that the thickness of the shell plate shall not be less than

$$t = \frac{pd}{20fJ - p} + 0.75 \qquad [6]$$

where: t is the thickness of the wall in mm
         p is the pressure in bar
         d is the inside diameter of the shell in mm
         f is the design stress in N/mm^2
         J is the joint factor, which may vary from 0.4 to 1.0
         0.75 is the allowance in mm for corrosion,

providing that in no case shall the thickness be less than 4 mm, or, in the case of Class III receivers where the internal diameter does not exceed 300, the thickness shall be not less than 2.5 mm or 0.01 D, whichever is the greater.

   Where the ultimate tensile stress (uts) is known, the design stress is

$$f = uts / 3.5$$

Design of ends

Ends shall be hemispherical or semi-ellipsoidal and may have manholes or hand holes. In the case of Class II and Class III receivers, and Class I receivers where the internal diameter does not exceed 600 mm, flat ends are permitted. If semi-ellipsoidal ends are chosen, the ratio of the major axis to that of the minor axis shall not be greater than 2.25:1.

   For hemispherical ends the thickness of the end plate shall not be less than

$$t_e = \frac{pR}{20fJ - p} + 0.75 \qquad [7]$$

where: R is the radius of the hemisphere
         t_e is the thickness of the end

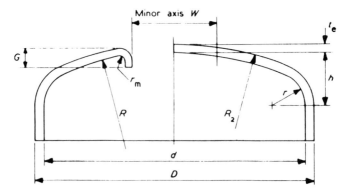

FIGURE 2 – Pressure vessel end showing a closed end to the right
and a typical opening on the left.

For other concave shapes

$$t_e = \frac{pdK}{20fJ} + 0.75 \qquad [8]$$

In both cases J = 1 for endplates made of one piece, and for plates made of more than one piece J is the appropriate factor obtained from the tables in the Standard. K depends on the ratio (h + t)/D, where h is the height of the height of the dished end and D is the outside diameter. K is read from Figure 3. In no case shall the thickness be less than that of the cylinder.

In some cases it is permissible to have flat ends rather than concave ends for an internal diameter not greater than 600 mm.

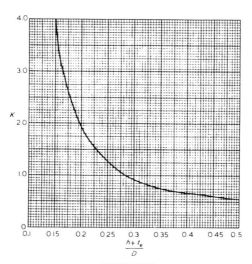

FIGURE 3

For flat plates welded to the cylinder

$$t_e = \sqrt{\frac{pd^2}{CS}} + 0.75 \qquad [9]$$

where: S is the minimum tensile strength of the plates
C = 14.3 for unflanged end plates or = 17.4 for flanged end plates.

Openings in shells and plates

Openings have to be provided in receivers for the purpose of inspection, see Table 2. All openings have to be ring reinforced to strengthen the vessel in the region of the material removed. A good design guide is to make the ring from as much material as was removed. Refer to the standard for details.

TABLE 2 – Examination holes and manholes for air receivers

Vessel internal diameter mm	Length of cylindrical section mm	Number of openings (minimum)				
		Small sight hole	Large sight hole	Hand hole	Head hole	Man-hole
0 to 300	0 to 1500	2	0	0	0	0
> 300 to 450	0 to 1500	0	2 see note 1	0	0	0
> 450 to 800	0 to 1500	0	2 or 1 see note 2	2 or 1 see note 3	0	0
	>1500 to 2000	0	0	2 or 1 see note 2	2 or 1 see note 3	0
	Over 2000	0	0	see note 4		0
>800 to 1500	0 to 2000	0	0	2 or 1 see note 2	see note 3	0
	Over 2000	0	0	see note 4		0
Over 1500	Unlimited	0	0	0	0	1

Note 1: Where the length is more than 1500 mm additional sight holes may be provided.

Note 2: In the case of a cylindrical body the holes should each be sited near the ends (within sight of the longitudinal joint and the base) or else in the centre of the ends.

Note 3: A single head hole or hand hole which should be situated in the central third of vessel length.

Note 4: The number of inspection holes shall be increased accordingly. For a length of less than 3000 mm it is, however, sufficient to site a head hole in the centre of the cylindrical body. On the cylindrical body the greatest distance between the head holes shall not exceed 3000 mm, and between hand holes 2000 mm. The latter shall each be located either near to or in the ends.

Design in accordance with BS EN 2186

The alternative procedure of this document is confusing to the designer used to the earlier standards because the notation and the way in which the equations are expressed are different. Air receivers made in accordance with this standard are of simple construction, primarily welded and serially made. If possible this standard is preferred provided that the conditions it covers are met.

The normal design stress is the lower value of 0.6 R_{eT} or 0.3 R_m ,where R_{eT} is the guaranteed minimum tensile yield strength and R_m is the guaranteed minimum tensile strength.

The normal thickness (e) takes account of corrosion and tolerance in the specification of the material thickness.

$$e \geq e_c + s + c$$

where e_c is the calculated thickness
s is the corrosion allowance
c is the absolute value of the negative tolerance on the material thickness.

There may also be an extra allowance to cater for thinning in production.

$$e_c = \frac{PD_o}{20f + P} K_c \qquad [10]$$

In this relationship,

D_o is the *outside* diameter in mm

P is the design pressure in bar (< PS, the maximum working pressure)

f is the nominal design stress in N/mm²

K_c is a calculation coefficient which depends on the welding process and the inspection method (varies from 1 to 1.15).

For hemispherical ends

$$e_c = \frac{PS\, D_o}{40f + PS} \qquad [11]$$

For flat ends without openings

$$e = C\ D\sqrt{\frac{PS}{10f}} \qquad [12]$$

where: C is a coefficient determined from a consideration of the end design
D = design diameter of the flat ends. For a simple welded end this is the internal diameter.

The above treatment is only an outline of the general approach to the design of air receivers. Any serious designer should study the full standards; they contain a great deal of useful material. A competent person, as defined in the Regulations must demonstrate a full understanding of the appropriate standards.

COMPRESSOR LUBRICATION

Lubrication is an extremely important parameter in compressor performance, its function being to dissipate frictional heat, reduce wear on sliding surfaces, reduce internal leakage and protect parts from corrosion. The lubricant can also act to flush away wear products, contaminants in the air and moisture. There is a tendency with modern compressors to dispense with oil in the compression chamber so as to avoid having to separate out the oil carried over with the air into the supply passage. This serves to remove the risk of contamination at source, which is so important for air needed for processing (food preparation, for example). Lubrication of the moving parts away from the chamber is usually necessary.

TABLE 1 – Factors affecting lubrication in the cylinders of reciprocating compressors

Cause	Effect
Deposits on valve seats	Increase in discharge temperature
Deposits in discharge passages	Increased discharge temperatures and pressures
High discharge temperatures	Increased oil oxidation and deposits
Contamination by hard particles	Abraded cylinder surfaces Interference with valve and piston seating
Contamination by solid particles on oil-wetted surfaces	Deposit build up on discharge valves and passages
Excess oil feed	Deposit formation
Moisture and condensation (cylinder overcooling)	Displaced oil films on metallic surfaces
Rust	Increased wear and oil oxidation
Water droplets entrained in suction gas between stages	Wash off lubricant high in cylinder Excessive wear. Rust
Oil droplets entrained in suction gas between stages	Oxidised oil repeatedly exposed to oxidising conditions causing rapid build up of deposits
High cylinder pressures on piston rings	Increased rubbing pressure Ruptured oil films

TABLE 2 – General guide to compressor lubricants in accordance with BS 6413

Type of compressor	Type of oil	Remarks
Reciprocating compressors	ISO grades within the range ISO 68 – ISO 150 ISO-L-DAA light duty ISO-L-DAB medium duty ISO-L-DAL heavy duty SAE 20	Moderate temperatures and pressures. Dry air. Higher or lower viscosities depend on the ambient temperature
Rotary, straight lobe	Premium hydraulic or turbine oil	Low temperature ISO 100 Normal temperature ISO 150 High temperature ISO 220
Rotary dry screws (timing gears and bearings)	Premium oil ISO 32 – ISO 46 with anti-wear and good demulsibility	Select for gear lubrication
Rotary flooded	ISO grades within the range ISO 32 – ISO 46 ISO-L-DAG light duty ISO-L-DAH medium duty ISO-L-DAJ heavy duty	
Rotary sliding vane (dry)	Premium oil anti-wear and good demulsibility ISO 100 – ISO 150 SAE 30 – 40	Condensation conditions
Centrifugal and axial flow (bearings)	Premium circulation ISO 32 or ISO 36	Select for bearings
Centrifugal and axial flow (gears)	Premium circulation oil ISO 68	Select for gear lubrication

Notes: Oils should be rust and oxidation inhibited.

ISO-L grades will be adopted as the future standards; detailed specifications will shortly be published.

Separate ISO classifications exist for vacuum pumps and gas compressors.

good condition. Siphon wick oilers may be used when it is important that there should be no undue variation in oil level in the container. Grease lubrication may be provided at certain points (*eg* to lubricate rolling bearings). In this case the lubricating points are either fitted with grease nipples or screw-down oil cups.

In the case of cylinders, the ideal oil feed is that which will provide efficient and effective lubrication with a minimum amount of oil and at the same time maintain a good piston seal – an excess of oil is undesirable because it then has to be removed. A practical check is to service the discharge valves periodically and examine their appearance. They should not be dry nor show signs of rust (insufficient oil) or be excessively wet (excess of oil). They should have a wet appearance and be oily to the touch. Excess of oil can also be noted by the appearance of pools of oil lying in the cylinders or surplus oil on the connecting rods.

Figure 2 shows the various lubrication flows for a typical reciprocating compressor. This is for a compressor supplying industrial air in which a small amount of oil is acceptable, so the cylinders are lubricated. If oil-free air were required, the cylinder lubrication would be omitted, but crankcase lubrication would remain. It is customary to incorporate an externally mounted gear pump with a pressure relief valve which controls the pressure to the moving parts. Oil is drawn into the pump through a submerged oil strainer and from the pump passes through a pressure filter with a disposable element. Oil is fed to the main bearings, through the connecting rods to the small-end bearings and crosshead and drains back into the crankcase.

In this design, the cylinders are lubricated by a crankshaft-driven force feed lubricator drawing oil from a separate reservoir. The valves seldom require separate lubrication.

Vane compressors

Two kinds of vane compressors are available, oil-free and oil-flooded. In the former kind, adequate lubrication is necessary to minimize friction and wear on the vanes and/or wear ring, and to assist in sealing. Quite high bearing pressures can be generated by the force of the vanes on the slots, calling for a lubricant forming a tenacious oil film.

Bearing lubrication is normally independent by means of a force-feed lubricator. Oil for lubricating the vanes may be fed to the casing via the air intake, direct to the casing or through the shaft. The first method is usually preferred, loading the air with oil from an atomizer; air at discharge pressure may be used to inject oil into the intake. The other two methods involve force feeding , either direct to the ends of the rotor or through a hole in the centre of the shaft. In the latter case, interconnecting radial holes drilled along the length of the shaft throw out oil under centrifugal force. The vanes in these compressors are usually made from cast iron which has good self-lubricating properties.

In the second kind of vane compressor, the compression chamber is flooded with oil, for the purpose of removing the heat of compression. Much more oil is supplied than is needed for lubrication or for sealing. Vanes in this type of compressor are usually made of a reinforced resin material, and are much lighter than metallic vanes, reducing the centrifugal bearing forces.

Rotary compressors of the dry type do not require internal lubrication because the rotors operate with a positive clearance. Only shaft bearings and the timing gears require lubrication; plain bearings may be lubricated by ring oilers and timing gears by splash lubrication. When rolling bearings are used, bearings at the timing end may be lubricated by oil and those at the other end by grease.

Oil-flooded rotary compressors (vane or screw)

Rotary compressors require lubrication of the bearings, the step-up gears and the timing gears where fitted (although in most rotary screw compressors, the screws drive each other and no timing gears are necessary). The oil-flooded types use the same oil for both lubrication and cooling injected into the compression chamber. Usually the oil is circulated by an externally mounted gear pump, drawing oil from the oil separator. The oil in the separator is at supply pressure, which is high enough in itself to circulate the oil without a pump. Some manufacturers dispense with a pump and rely solely on pressure circulation. This is bound to be a compromise for two reasons: at start-up, when the pressure is low the circulation flow is lower than desirable; and at varying speeds the oil flow is constant rather than proportional to speed. For mobile units this seems to work well, but industrial units usually include a pump. Where the pump is driven off the main drive, circulation stops when the rotors stop. But where the circulation is done by pressure alone, oil flow would persist as long as air pressure was present in the reclaimer, with the result that oil would be discharged back into the compression space. This is clearly undesirable so this kind of pumpless circulation has to have a non-return oil-stop valve to prevent it; such a valve is pilot operated from a pressure signal in the delivery casing.

Figure 3 shows a typical oil circulation system.

The heat of compression, absorbed by the oil in the compression chamber has to be removed by an oil cooler, usually of an air-blast type. The same cooling fan draws air

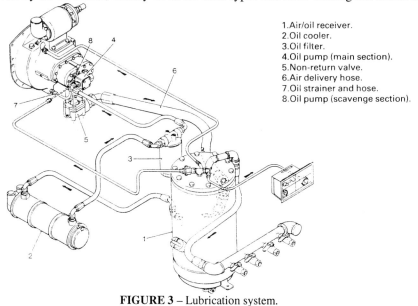

1. Air/oil receiver.
2. Oil cooler.
3. Oil filter.
4. Oil pump (main section).
5. Non-return valve.
6. Air delivery hose.
7. Oil strainer and hose.
8. Oil pump (scavenge section).

FIGURE 3 – Lubrication system.

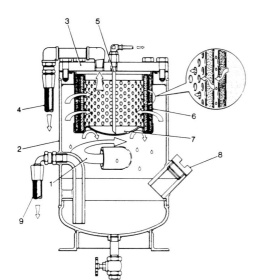

1.Air/oil mixture (from compressor).
2.Receiver casing.
3.Cover.
4.Air delivery pipe.
5.Scavenge tube.
6.Separator element.
7.Baffle.
8.Oil filler cap.
9.Oil feed to cooler.

FIGURE 4 – Oil separation and recovery.

through the oil cooler and the air aftercooler (where necessary).

The oil separator is one of the critical items in an oil flooded type of compressor. Its operation is shown in Figure 4. The method of separation is partly through mechanical impingement and partly through agglomeration in the filter pack.

Turbocompressors

In a turbocompressor only the bearings and gears require lubrication. The duty is less severe than for a positive displacement rotary compressor. The oil system may include an immersion heater as well as a cooler to maintain a stable oil temperature.

Compressor fires

One problem which sometimes occurs both in reciprocating and in oil-flooded units is the occurrence of internal fires and explosions. This phenomenon is not fully understood, but appears to be associated either with the use of an incorrect oil, running with a low oil level or not keeping strictly to the correct maintenance procedure. The presence of rust, aluminium or magnesium in the system can contribute to the danger. Combustion can start in the reclaimer vessel of a flooded unit where the oil is present in a finely divided form. The build-up of carbon deposits appears to serve as a starting point for fires. Regular cleaning and strict attention to manufacturer's recommendations will usually prevent this problem. BS 6244 should be studied for guidance on safety and includes information on avoiding the potential for fires.

Synthetic lubricants

There has been much interest in recent years in the use of synthetic fire resistant lubricants based on phosphate ester, di-ester, glycol or silicone, which are used on their own or blended with mineral oils. They reduce the combustible mists and minimize carbonaceous residues on valves and piping; they are claimed to have a longer life and superior lubricating properties. One should be wary of changing to the use of such lubricants without reference to the compressor manufacturer, and certainly one should never attempt to mix them with each other or with conventional mineral oils, despite any claim that the two are compatible.

For high output reciprocating compressors, dibasic ester fluids are recommended; for refrigeration compressors, alkyl benzenes are used, and for rotary compressors poly-alpha-olefines (PAO) are preferred.

Some of the problems that may arise are
• Incompatibility with metallic components.
• Incompatibility with the elastomers used in the diaphragms or seals.
• Plastics used in sight glasses or separator bowls may be affected.
• Paint used to coat the inside of pressure vessels or delivery passages may be softened and flake away, causing blockage of the filters.
• The cost will probably be higher (up to five times) than that of mineral oils.
• The lubrication properties may not be as good in the presence of water.
• The change periods may be shorter (although some suppliers claim an extended life).
• They tend to be more toxic, so precautions should be taken where the air is discharged into working areas.

However they appear to be a good answer to the problem of fire hazards, and should be considered where this is a problem.

Safety aspects of compressor lubricants

Observation of the following guidelines should guarantee safe use of compressor lubricants:
• Always use the oils recommended by the manufacturer.
• Never attempt to compensate for wear by the use of higher viscosity oil.
• Change all filters at recommended times. If solid particles are present in the filter element, locate their source. Rust or finely divided metal particles can lead to fires.
• If there is suspicion of inadequate lubrication, the tissue paper test can be used. Wipe the surface of the internal parts with a thin piece of tissue paper; it should be evenly stained, not soaked or dry (suitable for non-oil-flooded machines only).
• Keep records of amount of lubricant used and try to identify exceptional consumption.
• Ensure regular cleaning of oil and sludge from separators, reservoirs, aftercoolers and intercoolers.
• Never clean the inside of compressors with paraffin (kerosene) or carbon tetrachloride.
• An open flame should never be used to inspect the interior of a compressor or pressure vessel.
• If overheating is suspected, allow the unit to cool well below the spontaneous ignition temperature of the oil vapour (about 315°C).

HEAT EXCHANGERS AND COOLERS

All practical compressors work under adiabatic conditions with some heat loss. The amount of heat produced by compression is considerable which may not be an immediate disadvantage at low compression ratios, except that the air will be saturated with water vapour, some of which can be removed by aftercooling. Unless the machine specifically requires aftercooling, a low compression ratio machine may work satisfactorily uncooled (*eg* a single-stage centrifugal blower). If the machine does require cooling (*eg* in a reciprocating compressor), the necessary cooling may be achieved by finned cylinders with or without forced circulation. Unless this form of cooling is effective, the incoming charge of air will be preheated by transfer from the cylinder walls which may seriously reduce the efficiency of the compression. In general, cooling may be advantageous or necessary to achieve the following:

- Maintain working clearances on machine components.
- Limit the amount of heating applied to the inducted charge, thus improving the efficiency of the compression and increasing the volumetric capacity.
- Cool the compressed air between stages to improve compression efficiency of the second and subsequent stages.
- Cool the delivered air to remove water (aftercooling).
- Reduce the temperature of the air for use when operating hand tools.

Interstage cooling

Where more than one stage of compression is involved, it is desirable from the point of view of maximum efficiency that each stage does an equal amount of work. This can only be realized if the temperature of the air into the second stage is similar to that of the air inducted into the first stage. Cooling by means of a water jacket or finned cylinders will only remove a proportion of the heat generated by compression, because not all the air passing through the compressor will come into contact with the cooled surfaces of the cylinder walls. With an efficient water cooling system applied to a reciprocating compressor cylinder, only about 10 to 15% of the heat generated will be removed. Extra cooling between stages will be needed to bring the temperature down to an acceptable value. This will involve a separate heat exchanger.

Aftercoolers

An after cooler is a heat exchanger used to cool compressed air after leaving the compressor. Cooling reduces the air temperature and precipitates out water droplets which can be drained off. Aftercoolers incorporate a moisture separator for this purpose. Aftercooling reduces the volume of the air, but this presents no power loss (other than that required to run the heat exchanger).

As a general rule, air-cooled aftercoolers are used with air-cooled compressors, and water-cooled aftercoolers with water-cooled compressors. Temperature of the discharged air is normally about 10 to 16°C above the ambient when air cooling is used, and 10°C above the temperature of the water when water cooling is used. It is, of course, possible to reduce the temperature of the air to any desired level by refrigeration techniques.

Air cooling

In the case of small single stage compressors, natural air cooling may be sufficient using finned cylinders. The performance of such a compressor can be expected to vary with changes in the ambient temperature, as the effectiveness of the cooling varies with the difference in temperature between the cylinder and the ambient. Forced draft cooling is to be preferred, generated by an engine driven fan or from a power take-off on the compressor shaft, as this can provide more effective heat dissipation and more consistent heat transfer. If the compressor is located in a warm atmosphere it may be advantageous to draw the cooling air from a lower temperature source, *eg* through ducting led to the outside of the building. The amount of cooling required by a compressor is higher than that considered optimum for an internal combustion engine. The aim is to lower the compressor temperature as far as possible rather than to maintain it at an efficient running temperature as in the case of an engine.

Forced draft air cooling may be adequate for both small and large compressors, particularly if the latter are low capacity or multi-stage machines. For the latter, the increase in temperature per stage is lower than that of a single-stage machine of the same pressure ratio. Air cooling has its limitations when the compressor has a high capacity and is run continuously.

Water cooling

Most larger single-stage compressors and many multi-stage compressors are water cooled. In reciprocating compressors, the principle adopted is basically similar to that of internal combustion engine cooling, using water jackets for the cylinder barrels and cylinder heads. They can be either cored into the cylinder casings or added as fabricated jackets.

The temperature of the cooling water should be as low as possible with a high flow rate so as to keep the compressor temperature low. In a closed system, this will require a further heat exchanger to cool the circulating water, which can follow automobile practice, *ie* a radiator exposed to forced draft cooling from a fan. In larger installations other types of cooling can be adopted such as cooling towers. Typical water cooling requirements are shown in Table 1.

Centrifugal compressors are cooled by three possible methods:

- Diaphragm cooling, which implies that the air passages are water jacketed with a supply of coolant through the jackets. The diaphragm is a stationary element between the separate stages of a compressor and is kept cool by heat transfer from

TABLE 1 – Typical cooling water requirements (litres per cubic metre free air)

Component	Pressure bar (gauge)	Water temperature			
		16°C	21°C	27°C	32°C
Compressors:					
Cylinder jacket:					
Single-stage	2.8	0.8 to 1.0	1.15	1.3	1.45
	4.2	1.2 to 1.3	1.6	1.8	1.95
	5.6	1.45 to 1.95	2.1	2.25	2.4
	7.0	2.0 to 2.4	2.6	2.75	2.9
Two-stage	5.5 to 8.5	1.6	1.8	1.95	2.25
Intercoolers:					
In series with cylinder jacket	5.5 to 8.5	4.5	5.4	6.5	7.3
Separate	5.5 to 8.5	4.0	4.85	5.7	6.5
Aftercoolers:					
Single-stage	5.5 to 8.5	6.5	7.3	8.5	9.7
Two-stage	5.5 to 8.5	4.0	4.85	5.7	6.5

the water in the jacket. This is not a particularly efficient form of cooling and can be supplemented by an external intercooler

- Injection cooling involves spraying a suitable liquid into the return channel of the compressor stage. In the case of a refrigeration compressor, the refrigerant itself can be used. For air compressors, water can be used, providing that the increase in water vapour can be tolerated.
- Intercooling, which may be placed between any two of the stages (*eg* between the third and fourth stage of a six stage machine).

Shell-and-tube coolers

Shell-and-tube heat exchangers can adopt a variety of different forms, as in Figure 1. The

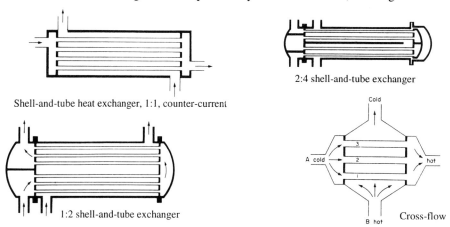

Shell-and-tube heat exchanger, 1:1, counter-current

2:4 shell-and-tube exchanger

1:2 shell-and-tube exchanger

Cross-flow

FIGURE 1 – Configurations of shell-and-tube heat exchangers.

flow through them is designated as either counter-current, in which the liquid to be cooled travels counter to that of the coolant; co-current in which it travels in the same direction; or cross-current in which it travels at right angles. Generally counter-current is more efficient because the temperature difference between the two fluids is greatest, but practical heat exchangers often combine all three so as to produce maximum cooling efficiency in a minimum of space.

If a heat exchanger has one pass of the shell-side fluid and one pass of the tube-side fluid it is a 1:1 exchanger; if there is one shell-side pass and two tube-side passes it is 1:2 and so on.

In studying the theory of heat exchangers the fundamental equation is:

$$Q/t = K \, A \, T$$

where: Q/t is the rate of heat transfer (W)
K is coefficient of heat transmission $W/(m^2 \, K)$
A is the effective cooling area (m^2)
T is the logarithmic mean temperature difference (K)

The logarithmic mean temperature difference (LMTD) needs some explanation. It will be apparent that in a heat exchanger the temperature difference between the fluid to be cooled and the coolant varies throughout the device, even under steady flow conditions. The LMTD is a way of determining an average temperature for use in the above equation. Figure 2 gives the values for LMTD when the temperature differences between the two streams at entry and exit are known. Strictly, different curves should be drawn for different configurations of tube and shell, but Figure 2 drawn for a simple counter-current form is usually good enough, particularly when the other parameters are not known with any great accuracy.

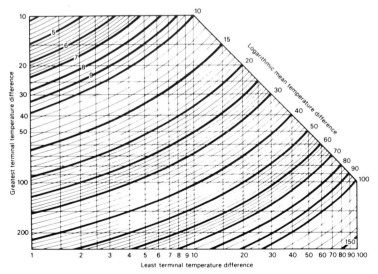

FIGURE 2

The value of K is hard to determine theoretically, and it is best when performing calculations to use test results. A brief explanation of the method of analysis may be helpful where tests are not available. Reference to a standard work on heat exchanger theory is recommended if further information is needed.

K is a combination of three factors – the convective coefficients of each fluid and the conductance of the partition between the two. The equation relating K to the constituent coefficients is

$$\frac{1}{K} = \frac{1}{h_1} + \frac{1}{h_2} + \frac{d}{k}$$

where: h_1 and h_2 are the convective coefficients of the two fluids in contact with the heat exchanger partition in W/(m² K)

d is the thickness of the partition wall in m

k is the thermal conductivity of the material of the partition in W/(m K)

If h_1 and h_2 are very different in magnitude, the larger one has little effect on the value of K, so it is good practice in efficient coolers to make the two coefficients close in value. One way of doing this is to make the two flow velocities approximately equal. It will be apparent that the velocity of the shell-side stream is likely to be smaller than that of the tube-side. It can be effectively increased by making the fluid follow a tortuous path of smaller cross-section by the insertion of baffles in the shell, Figure 3.

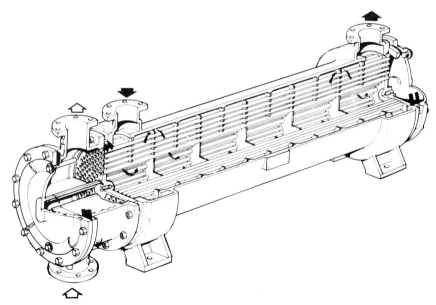

FIGURE 3 – Cut away view of commercial shell-and-tube
heat exchanger. *(Serck)*

Often the contribution of the term d/k can be neglected, so we then have:

$$K = \frac{h_1 h_2}{h_1 + h_2}$$

The computation of the values of h_1 and h_2 presents a complex problem in heat transfer theory. It can be shown from dimensional analysis that the convective coefficient of the fluid is primarily dependent on the Nusselt number, N_u. This is a non-dimensional quantity which plays a role in heat transfer theory analogous to that played by Reynolds number in determining the friction flow in pipes. It is given by:

$$N_u = \frac{\overline{Q}D}{kT} = \frac{hD}{k}$$

whre $\overline{Q}$ is the rate of heat loss per unit area (= $Q/(tm^2)$ and D is the significant linear dimension of the solid.

The Nusselt number has been determined experimentally for a number of cases of practical interest, and is quoted in a variety of forms according to the geometry of the heat exchanger. Some relationships, which are commonly quoted for shell-and-tube exchangers experiencing turbulent flow (the usual situation), are

$$N_u = 0.023\ R_e^{0.8}\ P_r^{0.4}\ \text{for flow parallel to the tubes}$$
$$N_u = 0.26\ R_e^{0.6}\ P_r^{0.3}\ \text{for flow at right angles to the tubes}$$

where R_e is the Reynolds number of the flow regime, and P_r is the Prandtl number for the fluid. The Prandtl number is easy to determine since it merely depends on the properties of the fluid.

$$P_r = \frac{c_p \mu}{k} = \frac{c_p \rho \upsilon}{k}$$

TABLE 2 – Prandtl numbers for heat exchangers.

Substance	Dynamic viscosity Pa s	Thermal conductivity W/(m K)	Specific heat J/(kg K)	Prandtl number
Air				
0 C	17.3 x 10⁻⁶	0.0241	1009	0.72
50 C	19.6 x 10⁻⁶	0.0279	1009	0.71
100 C	22.0 x 10⁻⁶	0.0317	1009	0.70
Water				
0 C	1.79 x 10⁻³	0.561	4217	13.4
50 C	0.55 x 10⁻³	0.624	4180	3.67
100 C	0.28 x 10⁻³	0.686	4215	1.76
Oil Cornea H68				
0 C	0.876	0.128	1821	12 462
50 C	0.0394	0.123	2009	643
100 C	0.0078	0.119	2198	144

Some values of the Prandtl number for fluids of interest are given in Table 2.
Reynolds number, R_e, is discussed in detail in the chapter on Pipe Flow. It is defined as:

$$R_e = \frac{vD}{\upsilon}$$

v is the velocity of flow in the pipe
D is the diameter of the pipe
υ is the kinematic viscosity

All the physical constants, with the exception of the velocity, in the Reynolds and Prandtl
numbers are properties either of the construction of the heat exchanger or of the fluids, so
for any given design it is possible to express the value of the Nusselt number as

$$N_u = const \ x \ v^m$$

The last step in determining the individual k values is the relationship between h and N_u.

$$h = \frac{N_u k}{D}$$

Note that D for the shell is not its overall diameter but rather a dimension obtained by
approximating the space between the tubes to an equivalent diameter. It is calculated as
four times the ratio of the area of flow to the wetted perimeter.

The appropriate value of h is found for both shell- and tube-side flows. These two are
combined with the wall coefficient as in the equation quoted above to give K.

It should be emphasised that the above theoretical treatment can do no more than give
an indication of the principles involved in design and most designs rely on practical
experience. No account has been taken in the theoretical treatment of corrosion or the
presence of layers of oil or scale on the surfaces. Regular cleaning and inspection will
ensure that the heat exchanger is kept at its maximum efficiency.

Fail-safe heat exchangers

The fail-safe heat exchanger is a variation of the tube assembly heat exchanger, Figure 4.

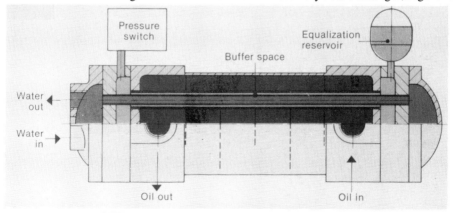

FIGURE 4 – Fail-safe heat exchanger. *(HPC Engineering)*

One tube is fitted inside another such that a space exists between them, which is filled with a harmless heat transfer medium and kept at a pressure of at least 0.5 bar by the expansion tank. In the case of leakage through corrosion or breakage, either the water or the oil mixes with the heat transfer medium but they cannot mix with each other; contamination is impossible. This type of heat exchanger can be used in the food or pharmaceutical industry, or for providing hot drinking water.

Air blast coolers

Portable and other compressors frequently use automobile type radiators. One form of construction uses corrugated metal sheets, laid together and dip-brazed; another type uses an aluminium cooling matrix. The designs are based on the principle of providing a large exposed area in a high conductive material.

Air is forced through by a motor driven fan. Analysis of this type of cooler is best done by the use of characteristics obtained from tests. Suppliers of coolers have comprehensive design charts for selecting the appropriate type, and much practical experience in choosing the ideal design parameters.

SECTION 3

Energy and Efficiency

ENERGY

PIPE FLOW CALCULATIONS

CLOSED LOOP (TWO LEVEL) SYSTEMS

HEAT RECOVERY

ENERGY GENERATION AND DISPERSAL

The power needed to compress air was discussed in the chapter on Compressor Performance. It is instructive to compare the equivalent power for some alternative energy forms, in particular for hydraulic and electrical energy.

Energy balance in compressed air systems

Consider as a theoretical concept a system in which it were possible to use the compressed air at the temperature at which it came out of the delivery port of an isentropic compressor, with all the pipework thermally well insulated and with all the energy possessed by the air regainable at the point of use. One would then have a perfectly efficient cycle, which would need the same power to drive it as would an efficient hydraulic or electric system. This can be seen to be so by remembering that isentropic compression is, by definition, reversible, which means that all the energy put into it can be regained. Unfortunately, a practical pneumatic system falls far short of this ideal and is much worse overall than its hydraulic equivalent.

In a simple pneumatic system consisting of a compressor directly connected to a power tool, much of the energy supplied to drive the compressor is rejected as heat into the cooling system and eventually into the surroundings. Only if one can use this heat for space or water heating or for processing will it not be wasted.

For most purposes compressed air cannot be used if its temperature is much higher than ambient. Portable compressors which have a minimum amount of cooling sufficient to supply usable air will have a discharge temperature about 60°C above ambient (where ambient has a maximum of 40°C). Stationary compressors will be equipped with the appropriate size aftercooler to meet the temperature and humidity requirements of its use in the factory.

At the other end of the power chain, ie at the point of use, there will be a rotary or percussive power tool, a pneumatic cylinder, a rotary motor, a paint spray gun or one of the many other ways of using air. If the output device produces a measurable form of mechanical power, it becomes possible to calculate an overall energy balance. This is an instructive exercise for demonstrating the overall efficiency of compressed air systems.

Consider an isentropic compressor, working under theoretically ideal conditions, producing air. The energy required has been previously calculated in the chapter on Compressor Performance and can be obtained from Figure 3 of that chapter. At a pressure ratio of 7:1, *ie* at a delivery pressure of 6 bar gauge, the power required is 0.26 kW per l/s FAD, or 0.26 x 8 (= 2.08) kW per l/s of compressed air. If an hydraulic pump were able to produce the same compressed volume flow at the same pressure, the power required would be 0.6 kW.

This calculation shows that the ratio of hydraulic power to pneumatic power is 0.6/2.08 = 0.288 or 29%.

Output energy available

It could be rightly argued that a calculation of this kind takes no account of the pressure energy content of the compressed air, but only of its kinetic energy or energy of transport. The former can be reclaimed only if the output machine is designed to work expansively. Unfortunately, from the point of view of the energy conservationist, few devices do work this way or are able to use much of the energy of compression; most tools and other types of machine work almost as if they were hydraulic machines. The air is discharged at practically the same pressure as it entered. Some modern tools are designed to allow a degree of internal expansion, but in most such cases the reason is not so much economy as the need to reduce the noise caused by a sudden expansion of the air.

There are two reasons why pneumatic machines are designed on an hydraulic basis. The first and main reason is that they would have to be much larger for the same power output; the greater the need to maximize the power/weight ratio, the less is the design expansion ratio. Another reason is that if the air were to expand fully its temperature would fall and the moisture content in it would freeze, with the likelihood of the flow passages becoming blocked; this is already a problem for many tools in the region of the exhaust port.

TABLE 1 – Power output of pneumatic equipment

Type of equipment	Power produced in kW per l/s
Piston motors	0.05 – 0.055
Small vane motors	0.04
Large vane motors	0.06
Gear motors	0.055
High performance percussive drills	0.025
Small hand held percussive drills	0.04 – 0.045
Road breakers	0.035 – 0.045
Pneumatic cylinders, small	0.05
Pneumatic cylinders, > 100 mm dia	0.04
Theoretical "hydraulic" power	0.06

Notes: Performance is at 6 bar gauge pressure.

Energy output for percussive tools is the product of the blow energy and the frequency of operation. Note that the output efficiency of high performance drills is low because they are designed to produce maximum output in a minimum size drill.

To demonstrate the energy needed by some typical machines, Table 1 has been prepared from published values of power output.

It is instructive to calculate the overall energy usage of a typical compressed air system. Take as an example a rotary motor, capable of producing 0.05 kW per l/s of air FAD at 6 bar pressure. As determined above, to produce 1 l/s requires 0.26 kW at the compressor. The overall efficiency is therefore 0.05/0.26 = 0.19 = 19%. This calculation demonstrates the basic feature of compressed air – its comparative inefficiency. It shows that 81% of the power of the prime mover is converted into an unusable form of energy, *ie* heat, and to a lesser extent into friction and noise; it also indicates the importance of ensuring that the air is not wasted.

Waste heat usage

If one is designing a compressed air system, it is worthwhile investigating the possibility of using some part of the 81% for other purposes such as space heating, for hot washing water or, if the factory can use it, for processing. Paint drying, paper drying and baking ovens are possible applications that could use some of the waste heat. It is quite possible to produce hot water at a temperature of 90°C. This subject is discussed in detail later. It has been calculated that 10% of all electrical power used in industry produces compressed air; the opportunity of using 80% of that for heating presents a considerable challenge to the designer.

The subject of Compressor Control has been dealt with previously. Maximum efficiency in the control system will ensure that power is not wasted. A compressor that runs partially or totally unloaded for any appreciable time may consume up to 75% of its full load power because of the minimum power needs of the compressor, the lower power factor and reduced motor efficiency. It is worthwhile incorporating an automatic control system that stops the compressor when it no longer needs to provide air. For installations where there are wide fluctuations in demand, the use of several smaller compressors connected to a multi-set control enables the shutdown of one or more units when demand for air is low, significantly reducing the power consumption.

Comparison with other energy media

It might be asked why, if compressed air is so inefficient, it is such a widespread form of energy. There are several reasons for this. One of the main uses of compressed air is the operation of pneumatic tools and power cylinders for factory automation. Percussive tools are particularly suitable for being driven by air; and despite the many attempts to develop equivalent tools working on hydraulic power or electricity, in the main they have not been successful.

By their nature, percussive tools have a rapidly varying demand for air. They operate at up to 3000 cycles per minute; during each cycle the demand varies from zero to maximum, which the compressibility of the medium can accommodate. An hydraulic tool, working with an incompressible medium has artificially to create compressibility in the supply line by including gas accumulators in the circuit close to the tool.

Pneumatic tools are cheaper, lighter and easier to handle than electric or hydraulic ones, and although they may not make such good use of the input power, ergonomic factors outweigh considerations of cost. Provided that reasonable precautions are taken, compressed air is a clean and safe medium. Even when considering cost, the simplicity of construction of pneumatic tools, their tolerance to misuse and lack of maintenance often means that a pneumatic system may cost no more overall than its competitors.

Another feature of compressed air as a power medium is that it does not matter if the output device stalls. A pneumatic motor has a stall torque as large as or larger than its running torque, and can sustain that with no chance of damage. An electric motor would soon burn out under these conditions, and an hydraulic motor would blow off at its pressure relief valve.

Air does not make a mess on the floor when it escapes. A leaking hydraulic system is soon noticed and remedied but a leak of air will continue to be ignored. Anyone who has visited a factory after the shift has finished cannot fail to have noticed the hiss of a leaking air main. The waste that this represents can be easily and cheaply remedied.

Overall advantages of compressed air

The use of compressed air in low cost automation using simple pneumatic cylinders and valves is widespread. It is possible to introduce quite sophisticated logic using air as both the control medium and the power medium. The connections are simply made with nylon tube, so it is easy to experiment on different circuits with a minimum of fuss, in a way that would not be possible with another medium. Pneumatic logic circuits today make widespread use of manifolds and sub-bases, which make for compact connections and reliability. This is discussed elsewhere in the text.

Compressed air easily lends itself to central generation; the transmission pipes are simple to install, being made of steel, copper, aluminium or ABS.

There are other uses for air, for which no other medium would be acceptable – for pneumatic conveying, for aeration, for paint spraying, for blast-cleaning and many other commercial processes.

As well as forming an indispensable part of a factory layout, it is also widely used on contractors sites and in mines and quarries. For operating a range of equipment from small tools like road breakers, rock and concrete drills and sump pumps up to large track mounted quarry drills it has unparalleled advantages. Although for powering high performance rock drills it has to some extent been superseded by hydraulic power, it will still form a major energy source for the future.

There is a natural tendency, when first making a choice of a compressor to run an installation, to go to some trouble to obtain the most efficient one on the market. Compressor manufacturers spend much of their development effort in seeking a few extra percent improvement in efficiency, which is an obviously desirable aim. But very often the extra efficiency in the generation of the air is not matched by a similar effort put in to economising its transmission and use. It will be found that good housekeeping of the transmission system and proper use and maintenance of equipment will be amply repaid in fuel savings.

The remaining chapters of this section will deal in some detail with the matters touched on above.

PIPE FLOW CALCULATIONS

Compressible gas flow

All gases are, by definition, compressible and their flow pattern is governed by very complex mathematical relationships. The accurate determination of the flow characteristics of such a medium can be extremely difficult. The aerodynamicist is concerned with the nature of air flow in a completely free medium where the solid boundaries are so far away as to be irrelevant to the solution of the flow problem; his main interest is in finding information about the local pressures and the flow velocities. The pneumatic engineer on the other hand is concerned primarily with the flow behaviour in a pipe. He wants to determine the average flow velocities and pressure drops in pipes of different sizes and materials, and the effect of putting in valves and fittings. He is less concerned with the details of the flow patterns across a pipe diameter or around the complex geometry of a valve. It is quite sufficient for him to know how much pressure is available to him at the end of a pipeline to operate a piece of equipment. To secure this limited objective, major simplifications to the theory are permissible.

This chapter is intended to be of help to the designer in estimating how much air (or other gas) can be expected to flow through a pipe. The aim is to supply practical answers; so, where assumptions can be made which produce results accurate enough for engineering purposes, those assumptions are indicated. The subject is not one which can be relied upon to give precise answers, so in this, as in many topics in pneumatics, there is no substitute for carrying out tests on an actual installation. In planning a layout for a factory or other form of fixed installation, the designer is well advised to work with generous factors of safety, not only to allow for uncertainties in the analysis but also to cater for future expansion of the system.

A determination of the pressure drop requires a knowledge of the relationship between pressure and volume. This is not easily determined in a particular case, but it is usual to consider two extremes, which are the same two as were considered for compression of air, namely isothermal and isentropic. Isentropic flow (or more accurately, in this instance, adiabatic since we are dealing with an irreversible process) applies to short, well insulated pipes. Isothermal is the more usual assumption, firstly because it is easier to analyse, but also because it more closely represents the actual situation in pipelines, where the air quickly reaches the ambient temperature and remains there, even when the pipes are well lagged. The descriptions isothermal and isentropic would have no relevance unless the fluid were compressible, that is unless it were possible to define a relationship between its pressure and its volume. It is usual to study first the formula for pressure drop that would apply to an incompressible fluid because it turns out that this applies equally well to many important cases of compressible flow. This equation is known as the Darcy formula or the Darcy–Weisbach formula:

$$\Delta P = \frac{\rho f L v^2}{2D} \qquad [1]$$

where the variables have their usual definitions.

Another form of the equation, which is usually found when dealing with the flow of liquids, is:

$$h = \frac{f L v^2}{2g}$$

In this formula h is the loss of head in metres. The concept of a head loss is readily understood in liquids, where pressure can be equated to the height of a liquid column. In gases it is not so relevant. The preferred relationship, therefore, is equation [1].

The variable f is known as the Darcy friction factor, which has to be obtained experimentally. The formula applies whether the flow is laminar or turbulent, but always with a value of f appropriate to the flow regime. Fortunately, a large number of tests have been done on pipes of various materials, sizes and internal roughnesses, so it is possible to look up the value of f on a chart when certain other parameters are known. The main factor governing the value of f is the Reynolds number for the flow. It does not matter whether the fluid is a gas or a liquid; if the Reynolds numbers are the same, the value of f will be the same.

The complete isothermal formula

The equation for compressible flow, which takes into account that the pressure at the end of a long pipe may be significantly smaller than that at the beginning, is:

$$\frac{\Delta P}{L} = \frac{\rho f v^2}{2D} + \rho v \frac{dv}{dL}$$

This is the same as the Darcy equation, with an added expression to allow for the drop in pressure necessary to decelerate the flow. The extra expression is significant for pipes with a large pressure drop.

The solution of this equation is usually written as:

$$w^2 = \frac{\rho A^2}{f L/D + 2 \log_e(P_1/P_2)} \frac{P_1^2 - P_2^2}{P_1} \qquad [2]$$

w is the mass flow.

This is correct for any consistent set of units. In particular, in conformity with the standard nomenclature, it applies where L, D and A are expressed in metres, P in N/m^2 and w in kg/s.

This solution is developed on the basis of the following assumptions:

- Isothermal flow
- No mechanical work done on the system
- Steady flow
- The velocity is the average across the section
- The gas obeys perfect gas laws, *ie* the compressibility factor is unity
- Friction factor is constant
- The pipe is straight and level.

For gas flow it is permissible to add one further assumption:

- Forces due to acceleration of the flow can be neglected.

Equation [2] now becomes:

$$w^2 = \frac{\rho DA^2}{fL} \frac{P_1^2 - P_2^2}{P_1} \qquad [3]$$

It can be seen that this is equivalent to the incompressible Darcy equation if the pressure difference between the ends is small. It can also be seen that if the pressure drop is less than 10 % of the inlet pressure and is based on the conditions at either inlet or outlet, the formula is accurate to about 5 %, which is satisfactory for most purposes. If the pressure drop is more than 10% and less than 40%, the formula can still be used provided the average value of ρ is used. Note that this procedure may involve iteration if the first guess of pressure drop proves to be inaccurate. For long pipelines, where the pressure drop is greater than 40%, two techniques can be used; one is to divide the pipe into shorter sections and then by a process of iteration solve for the complete pipe; the other is to use equation [2]. The latter procedure cannot take into account differences in the value of ρ along the length. If a significant pressure drop of this magnitude is found to occur in a normal compressed air layout, there is quite obviously a fault in the design of the system which should be remedied.

Equation [3] expressed in the usual units becomes

$$w = 316.23 \sqrt{\left[\frac{\rho A^2}{fL/D} \frac{p_1^2 - p_2^2}{p_1} \right]} \qquad [4]$$

where p_1 and p_2 are in bar, and A, L and D in metres.

Determination of Reynolds number

The friction factor f is primarily dependent on the value of Reynold's number R_e. This is a non-dimensional quantity which governs the flow regime in pipes and heat exchangers as well as in aerodynamic theory. It is defined as

$$R_e = \frac{vD\rho}{\mu} = \frac{vD}{\upsilon}$$

for any consistent set of units. In particular in SI units, v the local velocity of the flow is in m/s, D the inside diameter of the pipe is in metres, υ the kinematic viscosity of the gas is in m^2/s, μ the dynamic viscosity is in Pa s and ρ the density is in kg/m^3. Note that the value of μ for air is 17 x 10^{-6} Pa s at 0 °C. It varies with absolute temperature in such a way that at 200 °C it has a value of 26 x 10^{-6} Pa s. Linear interpolation between the two values is adequate for most purposes; for more accurate values and for other gases, refer to a standard textbook of physical constants such as Kaye and Laby "Tables of Physical and Chemical Constants" (published by Longman).

The first important result that comes from the determination of Reynold's number is whether the flow is laminar or turbulent; indeed Osborne Reynold's (after whom the number is named) is noted for his experiments on finding the transition point between the two kinds of flow. If the number is less than 2000, the flow is laminar; if more than 4000 the flow is turbulent. Between the two is a critical region where it may be one or the other, depending on the roughness of the pipe, changes in section and other destabilising factors. The uncertainty in this statement rarely leads to practical difficulties in the field of pneumatics, except when studying the flow in low pressure fluid logic elements. The flow in pipes is usually well into the turbulent regime.

Determination of the friction factor

For laminar flow the friction factor f is given by

$$f = 64/R_c$$

If this value is substituted into equation [1], one obtains the relationship

$$\Delta P = \frac{32Lv\mu}{D^2} \qquad [5]$$

This is known as Poiseuille's law (or Hagen–Poiseuille's law) for laminar flow.

For turbulent flow the position is more complex. Different formulae are used for different industries, particularly for the analysis of pressure drop in long natural gas pipelines; two frequently quoted ones are the Weymouth formula and the Panhandle formula. The flow of liquids through pipes is considered in detail in Valves, Piping and Pipeline Handbook (published by Elsevier) which also contains a discussion of various other standard formulae.

For consistency when studying the flow of air and other gases in the short runs typical of compressed air mains, it is recommended that the values of the friction factor calculated by Moody are used ("Friction Factors for Pipe Flow", Trans. Amer. Soc. Mech. Engrs., Vol 66, Nov. 1944). Figure 1 reproduces the Moody results. The friction factor f can be obtained when both Reynold's number and the pipe roughness are known. The absolute roughness is the height of an average surface defect on the inside of the pipe, but the significant parameter is the relative roughness, *ie* the ratio of absolute roughness to pipe diameter.

Moody obtained results for a variety of pipes, some of which are rarely found now, such as wood and rivetted steel, and for others such as concrete which are of little interest to the

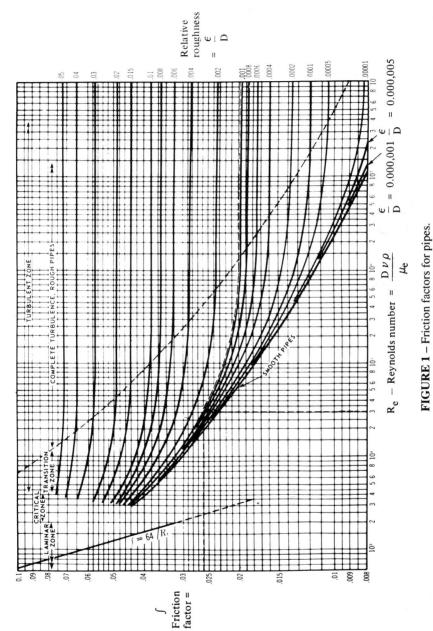

FIGURE 1 – Friction factors for pipes.

Example: Friction factor for pipe with relative roughness 0.001 at flow Reynolds number of 30.000 = 0.026

TABLE 1 – Absolute roughness for various pipes *(Hydraulic Research Station, U.K.)*

| | Absolute roughness (mm) | | |
Pipe material	Good	Normal	Poor
Cast iron, uncoated	0.15	0.3	0.6
Cast iron, coated	0.06	0.15	0.3
Steel, uncoated	0.015	0.03	0.06
Steel, coated	0.03	0.06	0.15
Wrought iron	0.03	0.06	0.15
Drawn pipes, steel, copper, etc		0.003	
Alkathene, etc		0.003	
UPVC, cemented joints		0.03	
UPVC, O-ring joints		0.06	
Glass fibre		0.06	
Concrete	0.06	0.15	0.6

pneumatic designer. The only pipes that the designer is likely to use today are commercial steel, drawn tubing and rubber and plastic hose. Typical values of the absolute roughness are given in Table 1. It is important to emphasise that the values presented in that table are for clean, new pipes. Rust and other deposits will form over a period of time; these cause an increase in the absolute roughness and in the worst cases will narrow the inside diameter. Due allowance should be made for these factors at the design stage. About 25% to 50% should be added to the theoretical friction factor to allow for deterioration.

At very large Reynolds numbers, the friction factor is primarily dependent on the value of the relative roughness, as can be seen from Figure 1. A condition of complete turbulence is reached when the friction factor is dependent solely on the roughness and no longer on the Reynolds number.

Pressure drop through bends and fittings

Pressure losses in a piping system are caused by a number of factors:

- Pipe friction as explained above
- Changes in direction through bends
- Obstructions such as valves
- Cross-section changes, which may be gradual or sudden

There are three common ways of dealing with the last three factors. The first way is to consider the pressure loss due to inserting a fitting as being equivalent to adding an extra length of pipe of the same diameter as the fitting. Many suppliers of fittings will provide pressure drop data in this form. One of the difficulties in this method is that the equivalent pipe length is related to a particular kind of pipe and unless one uses that pipe, the whole system has to be analysed on the basis of replacing the actual pipe lengths with a set of completely artificial pipe lengths, which can cause confusion.

The two other methods start from the Darcy equation, [1] above.

$$\Delta P = \frac{\rho f L v^2}{2D} = \frac{\rho K v^2}{2}$$

TABLE 2 – Equivalent pipe lengths of valves and bends *(Atlas Copco)*

Item	Equivalent pipe length in m										
	Inner pipe diameter in mm										
	25	40	50	80	100	125	150	200	250	300	400
gate valve, fully open	0.3	0.5	0.6	1.0	1.3	1.6	1.9	2.6	3.2	3.9	5.2
half closed	5	8	10	16	20	25	30	40	50	60	80
diaphragm valve fully open	1.5	2.5	3.0	4.5	6	8	10	-	-	-	-
angle valve, full open	4	6	7	12	15	18	22	30	36	-	-
globe valve, fully open	7.5	12	15	24	30	38	45	60	-	-	-
swing check valve fully open	2.0	3.2	4.0	6.4	8.0	10	12	16	20	24	32
bend R = 2d	0.3	0.5	0.6	1.0	1.2	1.5	1.8	2.4	3.0	3.6	4.8
bend R = d	0.4	0.6	0.8	1.3	1.6	2.0	2.4	3.2	4.0	4.8	6.4
mitre bend, 90°	1.5	2.4	3.0	4.8	6.0	7.5	9	12	15	18	24
run of tee	0.5	0.8	1.0	1.6	2.0	2.5	3	4	5	6	8
side outlet tee	1.5	2.4	3.0	4.8	6.0	7.5	9	12	15	18	24
reducer	0.5	0.7	1.0	2.0	2.5	3.1	3.6	4.8	6.0	7.2	9.6

factor appropriate to pulsating flow is greater than that calculated on the basis of the average flow requirement. The general recommendation is to increase the factor by at least 50% to allow for this.

Table 6 can be used to obtain the pressure drop in flexible hoses. This table takes into account the end fittings and the presence of bends.

Orifice flow

So far, the consideration of pressure drop has been restricted to pipes carrying air at moderate flow velocities. The analysis adopted would fail at velocities approaching that of sound. At that velocity the flow becomes choked and any further increase in upstream pressure would not increase the flow. The velocity of sound is given by:

$$v_s = \sqrt{(\gamma RT)}$$

The relationship between the actual flow velocity and the speed of sound is the Mach number, M, where

$$M = v/v_s$$

R for air has a value of 320 J/(kg K). At a temperature of 273 K (0°C), v_s equals 296 m/s, which is much greater than any likely flow velocity in a pipe. Sonic conditions are important, however, when considering discharge through an orifice or other small restriction situated either at the outlet of a chamber, Figure 2 (where the upstream velocity is zero) or in a pipe (where the upstream velocity has some positive value). The change in velocity at the restriction is so sudden that there is no time for any heat exchange with the surroundings, so the behaviour is adiabatic.

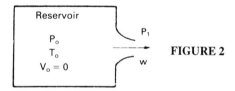

FIGURE 2

Nozzle equations

The exact formula for the mass flow of air through a nozzle is:

$$w = AP_o \left\{ \frac{2\gamma}{RT(\gamma-1)} \left[\left(\frac{P_1}{P_o}\right)^{\frac{2}{\gamma}} - \left(\frac{P_1}{P_o}\right)^{\frac{\gamma+1}{\gamma}} \right] \right\}^{0.5} \quad [6]$$

where: w is the mass flow (kg/s).

This relationship is found written in a variety of ways.

This formula applies only up to the critical pressure ratio, *ie* when M = 1 and when the flow velocity is a maximum. The critical pressure ratio is given by:

$$\frac{P_1}{P_o} = \left(\frac{2}{\gamma+1}\right)^{\frac{\gamma}{\gamma-1}}$$

Above that ratio, the formula for mass flow is:

$$w = AP_o \left\{ \frac{2\gamma}{RT(\gamma+1)} \left(\frac{2}{\gamma+1}\right)^{\frac{2}{\gamma-1}} \right\}^{0.5} \qquad [7]$$

Equations [6] and [7] are unwieldy for practical calculations, so it is usual to adopt simpler forms, which give very close approximations to the true values

$$w = 0.0404 \frac{C_p A p_0}{\sqrt{T}} \left\{ 1 - \left(\frac{p_1/p_2 - 0.528}{1-0.528}\right)^2 \right\}^{0.5} \qquad [8]$$

for subcritical flow, and

$$w = 0.0404 \frac{C_p A p_0}{\sqrt{T}} \qquad [9]$$

for supercritical flow. The boundary between the two flows is when $p_1/p_0 = 0.528$.

A is in m^2 and p_1 and p_o are in Pascal. The discharge coefficient, C_D, has been included to allow for flow through other than a perfect, well rounded orifice. The formula applies in this form for air with a value of γ equal to 1.4.

To obtain the volumetric flow in litre/s of free air, with p_o in bar and an air temperature of 15 °C, the equation for supercritical flow is:

$$V = 0.197 \times 10^6 \times C_D A P_0$$

Table 7 is based on equations [8] and [9] and gives values of discharge through an orifice, where the mass flow is converted into FAD in litres/s. This table can only be used

TABLE 7 – Discharge of air through an orifice

Gauge pressure before orifice in bar	Diameter of orifice (mm) Discharge of air in litres per second at 1 bar abs, and 15°C										
	0.5	1	2	3	4	5	7	10	15	20	25
0.5	0.054	0.217	0.869	1.956	3.477	5.433	10.65	21.73	48.90	86.83	135.8
1.0	0.076	0.303	0.213	2.729	4.851	7.579	14.86	30.32	68.21	121.3	189.5
2.0	0.114	0.455	1.819	4.093	7.276	11.37	22.28	45.48	102.3	181.9	284.2
3.0	0.152	0.606	2.425	5.457	9.702	15.16	29.71	60.64	136.4	242.5	379.0
4.0	0.189	0.758	3.032	6.821	12.13	18.95	37.14	75.79	170.5	303.2	475.7
5.0	0.227	0.910	3.638	8.186	14.55	22.73	44.57	90.59	204.6	363.8	568.5
6.0	0.265	1.061	4.244	9.550	19.98	26.53	51.99	106.1	238.8	424.4	663.2
7.0	0.303	1.213	4.851	10.91	19.40	30.32	59.42	121.3	272.9	485.1	757.9
8.0	0.341	1.364	5.457	12.28	21.83	34.11	66.85	136.4	307.0	545.7	852.7
9.0	0.379	1.516	6.063	13.64	24.25	37.90	74.28	151.6	341.1	606.3	947.4
10.0	0.417	1.667	6.670	15.01	26.68	41.69	81.71	166.7	375.2	667.0	1042.0
12.0	0.493	1.971	7.882	17.74	31.53	49.27	96.56	197.1	443.4	788.2	1232.0
15.0	0.606	2.426	9.702	21.83	38.81	60.64	118.8	242.6	545.7	970.2	1516.0
20.0	0.796	3.183	12.73	28.65	50.93	79.58	156.0	318.3	716.3	1273.0	1990.0

Table 7 is based on a 100% coefficient of flow.
For well rounded entrance, multiply values by 0.97;
for sharp edged orifice multiply by 0.65, for approximate results.

where the discharge is to atmosphere (1 bar). For other downstream pressures, the formula must be used.

These equations can be further generalised to:

$$w = p_0 C \left\{ 1 - \left(\frac{p_1/p_0 - b}{1 - b} \right)^2 \right\}^{0.5}$$

for subcritical flow, and

$$w = p\,C$$

for supercritical flow.

Expressing the flow equations in this form enables C and b to be defined for any other kind of valve or restriction. They can be experimentally determined (according to the method described in BS 5793) and used in subsequent analysis. See also ISO 6358.

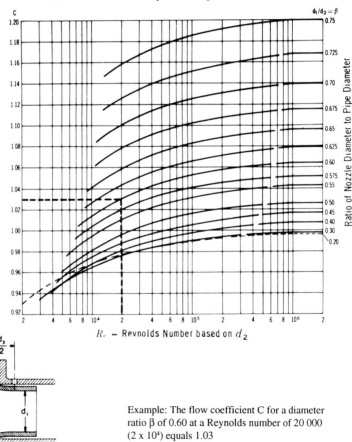

Example: The flow coefficient C for a diameter ratio β of 0.60 at a Reynolds number of 20 000 (2 x 10⁴) equals 1.03

$$C = \frac{C_d}{\sqrt{1 - \beta^4}}$$

FIGURE 3 – Flow coefficient C for nozzles.

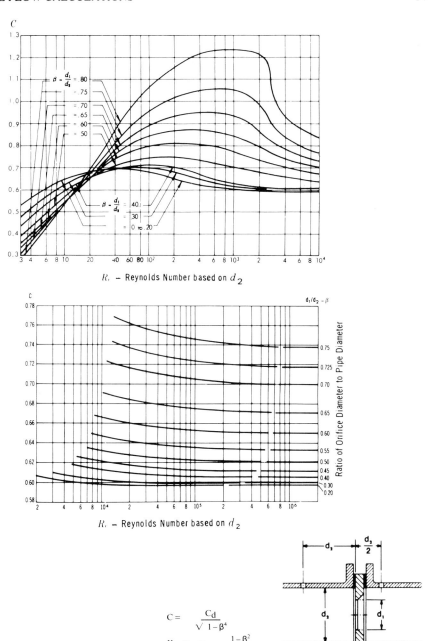

FIGURE 4 – Flow coefficient C for square-edge orifices.

The above theory is based on the assumption that the flow starts in a reservoir which is so large that the effect of the upstream velocity can be neglected. This is generally not the case; most orifices or other restrictions are situated in a pipe carrying gas where the velocity is appreciable. This effect can be taken care of by modifying the value of the flow coefficient C_D. Figures 3 and 4 can be used, based on the value of the upstream Reynold's number. Figure 2 gives the C value for rounded orifices or nozzles and Figure 3 is for sharp edged orifices. Usually sharp edged orifices are typical for valves, even though actual section change may have a small radius. Note that the nozzle coefficient, C, can be directly related to the resistance coefficient discussed earlier in this chapter.

General considerations in determining flow losses

The data presented in this chapter can be used to assess pressure drop in most cases of interest to the designer of pneumatic systems. As previously emphasised, the subject is one in which a number of assumptions have to be made to obtain useful results, and the calculations made at the design stage should be verified by tests on the finished layout. The following is a check list of items in a layout which can contribute to the overall pressure drop. The list is, as far as one can generalise, in order of importance:

- Sharp edged orifices
- Changes in section, especially at entry and exit of a pipe or hose run
- Valve insertion losses
- Bends and tees
- Lengths of small bore pipe or hose supplying tools
- Main pipe runs.

For further information on pipe flow and the choice of materials for pipe runs refer to the chapter on Compressed Air Distribution.

CLOSED LOOP (TWO LEVEL) SYSTEMS

One of the more interesting suggestions made in recent years for improving the efficiency of compressed air generation is the closed loop or two level system.

The conventional compressed air system generates air at about 7 bar gauge pressure and when its energy has been extracted exhausts it to the atmosphere. As has previously been pointed out, much of the energy content of the compressed air is wasted through rejection of heat to the atmosphere and through the inability of most output devices to use the air expansively. The new development generates air at a much higher pressure than the conventional 7 bar and releases it, not to the atmosphere, but to an intermediate pressure, so that the pressure differential remains at 7 bar. A typical two level system might be one in which the main operating pressure was 12 bar and the discharge pressure was 5 bar; the 7 bar differential is thereby retained. The 5 bar pressure would then be carried back to the compressor inlet by a separate line.

It is not immediately obvious why this is better than discharge to atmosphere. The reason lies in the thermodynamics of the compressor. The ratio of compression in this example is $(12 + 1)/(5 + 1) = 2.2$, instead of 8, which makes for a more efficient use of energy in the compressor. The power needed to drive a compressor is primarily dependent on the pressure ratio, but the energy available at the point of use is dependent on the pressure difference between inlet and exhaust. As indicated in the chapter on Energy Generation, most tools and other output devices work as if they were "hydraulic", *ie* they are unable to use the compressive energy of the air. There is an advantage in salvaging some of that energy and putting it back into the compressor, a principle that is adopted in the two level system.

The theory of two level systems has been developed by the Fluid Power Centre (FPC) at Bath University under a contract from the British Compressed Air Society and the Department of the Environment. So far as is known, no company has yet ventured to install a practical embodiment of such a system, but because of its theoretical importance for power savings it is worth discussing it in detail in this volume. The BCAS are anxious to cooperate with a potential user. Unfortunately, until a company is found willing to invest in its practical development, it will remain an interesting academic exercise. Much of the

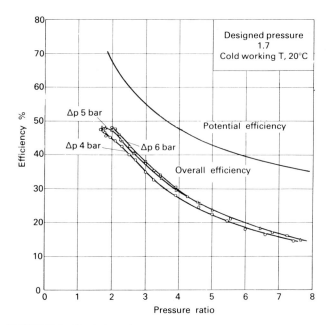

FIGURE 1 – Variation of compressor efficiencies with pressure ratio
for two level mode.

A plot of the potential efficiency is given in Figure 1.

On the same figure is a plot of the overall efficiency measured from experiments
performed at FPC on a modified vane compressor. The overall efficiency, taking into
account mechanical losses in the compressor drive, is rather less than the potential
efficiency.

When comparing theoretical efficiency with the overall efficiency, it appears that the
mechanical efficiency alone is of the order of 67%. On the face of it this seems rather lower
than one would expect, but the values are obtained at an early stage in the development
of the concept, and it is likely that they would improve with experience.

Experiments with two level operation

Some of the experiences with various kinds of compressors and output machines are worth
listing as a measure of the achievements so far.

1. There will always be a certain amount of leakage from the system even if all
    unnecessary leaks are eliminated. No tool can be perfectly sealed and it is expected
    that there will always be some usage of air from the low pressure line. This loss is
    taken care of by having a make-up compressor in the low pressure line.

2. Most of the conventional types of displacement compressors have been examined.
    With some modifications, it appears that they can all be used. With rotary
    displacement machines the built-in pressure ratio would need to be correct for the

pressure difference between high and low. Throttling of the inlet would be an inappropriate method of control, so an onload/offload system seems the obvious choice.

3. Cylinder actuators present the least problem, provided that the pressure rating is adequate. In conventional operation, choking the exhaust is used to give speed stability; it appears that two level systems are equally suitable in this respect.

4. Percussive tools. Redesign is necessary, but tests so far performed on a chipping hammer indicate that this would present no major problem. Ordinary percussive tools are very leaky. The air escapes to atmosphere anyway, so there is no motivation to reduce the leakage rate. In a two level system, leakage would be much more serious, so FPC have modified their chipping hammer to ensure that leakage to atmosphere is practically eliminated.

5. Rotary motors show a loss in power through inlet losses, which could be remedied by increasing port sizes.

6. Circulation of the lubricating oil is inevitable in a closed system. An oil-flooded compressor would seem to be an appropriate choice, since it could easily cope with the extra oil. Oil-free compressors would probably need to have an oil separator at their inlet. FPC claim to have had no problem in this respect.

General conclusions on two level systems

There are problems to be overcome, but none appears to be insuperable. The rewards in terms of fuel economy are considerable. FPC claim that a reasonable figure for energy savings is of the order of 40%. There will be capital savings in the compressor, aftercooler and dryer, only partly offset by the extra cost of running a second high pressure mains line.

The concept is a challenge to an enterprising factory manager who is considering a new installation. BCAS are willing to give technical assistance to any company who is interested in pursuing the idea.

HEAT RECOVERY

Heat exchangers have been discussed in an earlier chapter. This chapter discusses how some of the heat generated during compression can be utilised for other purposes so as to improve the overall economy of the installation.

Many, if not most, of the industrial compressed air systems in use in the U.K. have been designed with little regard to economy of operation. Initial first cost is often the prime consideration, with little regard given to the possibilities of integration within the overall energy needs of the factory. Frequently as the needs of the factory expand, organic growth of the air system results in a further deterioration of what may have been ill considered at the outset.

Types of compressor cooling

There are three basic types of compressor cooling:
- external supply;
- air cooled radiators;
- water cooled heat exchangers;
- cooling towers.

External supply

Water from the sea, rivers, wells or the mains water authority are possible external sources of cooling. It will only rarely be possible to take advantage of these sources and rarer still will the heat extracted be usable or recoverable, although it is possible to imagine circumstances where some use can be made of it such as in fish farming or horticulture. Heating of the incoming mains water can be used to provide a supply of wash water, although getting the correct balance between supply and demand may be difficult.

Air cooled radiators

Water is recirculated without contamination or loss. If the compressor is oil-flooded there may be a heat exchanger between the oil and the cooling water circulating through the radiator. This is a very common form of cooling, the main disadvantage being that extra power is needed to drive the fan. It is hardly practical to cool the water to below 10°C above ambient, without extravagant use of fan power and a large radiator.

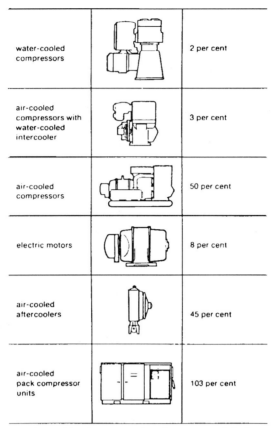

water-cooled compressors		2 per cent
air-cooled compressors with water-cooled intercooler		3 per cent
air-cooled compressors		50 per cent
electric motors		8 per cent
air-cooled aftercoolers		45 per cent
air-cooled pack compressor units		103 per cent

FIGURE 1 – Energy given off to the room as heat in per cent of shaft power. *(Atlas Copco)*

If a compressor set is situated on its own room, the heat generated will pass into the air in the room and has to be removed by a fan. Figure 1 shows the heat given off by a selection of different types of compressor. In a compressor that is cooled by an external water cooler very little of the heat passes into the room (2% in Figure 1). Note the rather surprising figure of 103% for a freestanding compressor. This figure takes account of the extra fan power to drive the aftercooler. The ventilation of the room must be sufficient to ensure that the heat exchange works correctly and that no damage is done to the other equipment in the room. Insufficient cooling shortens the life of the electric motor.

To estimate the ventilation flow needed the following relation may be used.

$$q = \frac{Q}{\rho c_p \Delta T}$$

where q   is the flow of the cooling air (m³/s)
      Q   is the heat released (W)
      ΔT   is the temperature rise above that of the cooling air
      ρ   is the density of the cooling air (ambient)

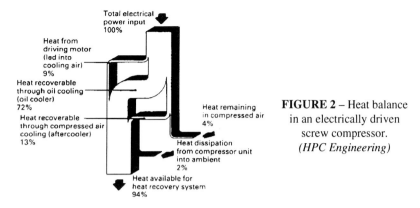

FIGURE 2 – Heat balance
in an electrically driven
screw compressor.
(HPC Engineering)

The ventilation flow should be such as to keep the temperature less than 10°C above ambient. When c_p = 1000 J/(kg K), ρ = 1.2 kg/m^3 and ΔT = 10°C, the relationship becomes:

$$q = Q/\ 12 \qquad m^3/s$$

In this expression, Q is the heat in kW.

A typical heat flow diagram is presented in Figure 2; the actual heat flow will depend on the particular circumstances.

Almost all the heat can be passed into heating the air in the adjacent rooms (or even further away through insulated heating ducts). This heat is only valuable for winter heating

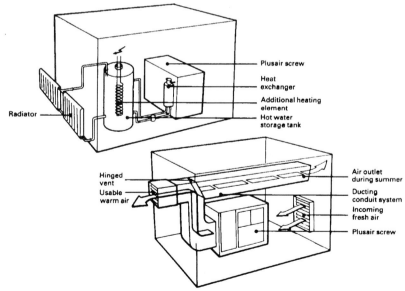

FIGURE 3 – Upper diagram: The system provides a simple and economical method of producing hot water, using a water cooled oil heat exchanger in series with the oil cooler of a screw compressor. Lower diagram: A ducted warm air system permits the redirection of warm air from the compressor to the work area in cold weather, or to the atmosphere in summer. *(HPC Engineering)*

and has to be rejected to outside in summer, as in Figure 3. A heat recovery unit of this kind would incorporate thermostatically operated shutters.

Water cooled heat exchangers

A more versatile use of the heat is a combination of water heating and hot air, the proportions of the two varying according to the factory installation. In Figure 2, which applies to an oil-flooded screw compressor, a maximum of 72% is recoverable for the production of hot water and a minimum of 22% for the production of hot air. The hot water can be produced at 70°C, which can be further boosted by an extra heating coil. It is unlikely that there will, at all times, be a balance between heat produced and heat used, so the control system must incorporate a means for diverting the water flow and ensure that adequate cooling of the compressor is always available. Some of the heat may have to be rejected to ambient or passed into the hot air.

Central heating is needed during the summer months only so the yearly savings have to be calculated on 6 to 8 month basis if that is the main use to which the heat is put. As well as central heating, hot water can be sent to wash rooms or used for product cleaning and process uses such as galvanising. The heat exchanger can be as simple as the industrial equivalent of a domestic hot water tank in which case hot water is produced at a maximum of 55°C. For a rather more efficient package a plate heat exchanger can produce water at 70°C. the temperature quoted should not be taken as other than indicative of the possibilities. Actual temperatures need to be calculated when the cooling requirements of the compressor are known.

If any danger of water pollution must be avoided, the use of a safety heat exchanger is

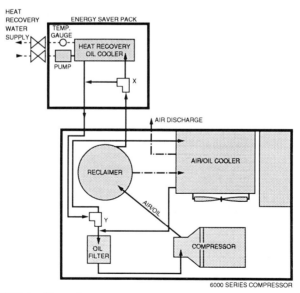

FIGURE 4 – These diagrams show the operation of the Energy Saver Pack as a free standing unit. *(Broom Wade)*

recommended. This is a type of exchanger, either tube or plate, incorporating a buffer space between the two flow elements and draining grooves which allow pollution to flow to the outside. See chapter on Heat Exchangers and Coolers.

An example of a system of heat reclamation is shown in Figure 4, which illustrates an Energy Saver Pack (ESP). Using this pack both space heating and water heating are possible. The Energy Saver Pack is supplied as a stand-alone unit or incorporated into the compressor housing.

The following calculation is typical of the savings that are possible which the ESP:

Heat recovery

Compressor power	55 kW
Heat rejected to water	45 kW
Water flow	29 l/s
Water temperature leaving ESP, compressor at full load, ambient temperature 20°C	60°C
Annual heat rejection (3000 hrs operation)	135 MW h

Energy savings per year (£ Sterling)

Electricity	135 MW h at 4.2p/kW h	£5670
Heating oil	2640 gals at 63p/gal	£1663
Gas	4590 therms at 31p/therm	£1423

Payback period

Based on the above savings and assuming a typical price for the ESP 1 unit of £2300, the payback periods are:

Electricity	$\dfrac{£2300 \times 12}{£5670}$	5 months
Heating oil	$\dfrac{£2300 \times 12}{£1663}$	17 months
Gas	$\dfrac{£2300 \times 12}{£1423}$	19 months

There are other unquantified savings such as the space heating and the unused capital cost of the water heating equipment.

Cooling towers

This is another form of cooling, which can also be used in compressor installations. Cooling is achieved by spraying hot water over a wet surface while air also passes over the surface. Hot water enters the top of the cooling tower and comes into contact with moving air as it passes over the wet surface and falls into the collecting sump.

The major cooling effect is due to evaporation, so the performance depends upon the cooling load imposed and the wet-bulb temperature of the incoming air. Cooling towers can cool to within 3°C of the wet bulb temperature. Water is lost through evaporation, drift and blow-down, as in Table 1.

TABLE 1 – Water losses in per cent of water flow

Δt_{water}	10K	20K	30K
evaporation	1.2	2.7	4.0
drift	0.2	0.5	0.8
blow down	1.0	1.0	1.0
total water loss	**2.4**	**4.2**	**5.8**

Two types of cooling towers are available, open and closed. They are illustrated in Figures 5 and 6.

The open type is prone to contamination as the air and fresh water enters the tower. The closed type can accept either oil or water in the circuit with no risk of contamination and is a little more complicated.

It is not possible to reclaim the waste heat from a cooling tower. There are, however, circumstances in which a cooling tower would form part of a heat reclamation circuit. No heat recovery unit is so well balanced that the total heat generated in the compressor is available for use and in this instance the extra heat may be extracted in a tower.

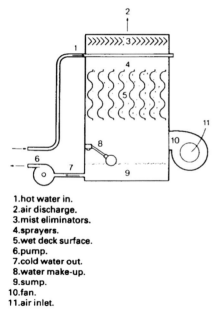

1. hot water in.
2. air discharge.
3. mist eliminators.
4. sprayers.
5. wet deck surface.
6. pump.
7. cold water out.
8. water make-up.
9. sump.
10. fan.
11. air inlet.

FIGURE 5 – Schematic operation of open system.
(*Atlas Copco*)

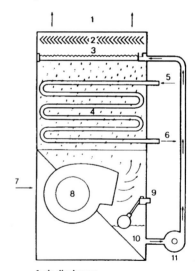

1. air discharge.
2. mist eliminators.
3. water distribution system.
4. cooling coil.
5. hot fluid in.
6. cold fluid out.
7. air inlet.
8. fan.
9. water make-up.
10. sump.
11. pump.

FIGURE 6 – Schematic operation of closed system. (*Atlas Copco*)

Moisture separators

All heat exchangers need to incorporate a means of draining the water which is condensed out from the cooled air. Most water separators contain an automatic feature which discharges the condensed liquid. These are discussed in detail in the section on Air Treatment.

SECTION 4

Compressed Air Transmission and Treatment

COMPRESSED AIR FILTRATION

BREATHING AIR FILTRATION

STERILE AIR AND GAS FILTERS

AIR DRYERS

COMPRESSED AIR DISTRIBUTION

CONDENSATE TREATMENT AND DRAIN VALVES

PIPE MATERIALS AND FITTINGS

REGULATING VALVES

PRESSURE GAUGES AND INDICATORS

TOOL LUBRICATION

COMMISSIONING AND SAFETY

SYSTEM MAINTENANCE

COMPRESSED AIR FILTRATION

The properties of compressed air make it a versatile, secure and economic medium that is economic to produce and handle. Compressed air is used either as a carrier medium for the transport of energy to the point of use where its potential and kinetic energies can be converted into a driving force for pneumatic equipment , or as a process medium itself (*eg* breathing air) or for processing purposes (agitating, mixing, packaging, conveying and pressurising). These applications can be found discussed in other chapters in the book. For most applications of compressed air, a requirement of air quality exists, expressible as a permissible level of contaminants.

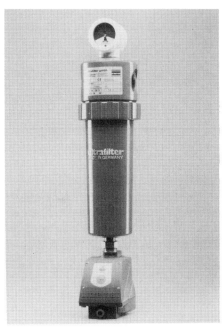

FIGURE 1 – Compressed air filter with differential pressure gauge and condensate chain. *(ultrafilter)*

The purity demands of high precision complex and fully automated pneumatic systems and of chemical, biochemical, biomedical, electronic, pharmaceutical and food processing applications increase year by year, requiring a continuously improving level of air treatment. The degree of treatment of air for tool use is clearly not so critical, so the air treatment equipment must be regulated to the end use.

Research indicates that contamination to the extent of 190 million dirt particles per cubic metre (even more in heavily industrialised areas) is commonplace. As much as 80% of these particles are smaller than 2 microns and pass straight through a compressor intake filter. This contamination is supplemented by water vapour and unburnt hydrocarbons from aviation, industrial and domestic heating and vehicle fuels. When this poor quality of air is compressed, the particle content increases to an excess of 1 billion particles per cubic metre.

The contaminants likely to be found in compressed air fall into the following categories:

- atmospheric dust, smoke and fumes inducted by the compressor
- atmospheric bacteria and viruses
- water vapour inducted and passed through the compressor
- gas generated in the compressor
- oil carried over from the compressor
- solid contaminants generated within the system

Compressed air quality classes for solids, water and oils are given in Table 1, in accordance with ISO 8573

TABLE 1

Quality class	Dirt particle size (μm)	Water pressure dewpoint °C at 7 bar	Oil mg/m³
1	0.1	-70	0.01
2	1	-40	0.1
3	5	-20	1
4	40	3	5
5	–	7	25
6	–	10	–

To specify the quality class for a particular application, quote the three classes in turn, *eg* Quality class 2.2.2 is dirt 1µm, water -40°C, oil 0.1 mg/m³.

Refer to Table 2 for quality recommendations for typical applications.

Particles smaller than one-third of the smallest clearance they will have to pass through can generally, from purely mechanical considerations, be accepted. For average use in air lines supplying mechanical equipment, a filtration rating of 40 micron is usually adequate.

The degree of treatment required for removing contaminants from a compressed air system depends on the application. For general industrial applications, *eg* compressed air mains for general factory use, partial water removal by aftercooling followed by filtering

Econometer:
Service life twice as
long as other filters

Unique ultrafilter now with ultramat 'intelligent' technology within the price

Housing:
Three part construction
for easy handling

Condensate Drain:
Zero Loss Technology

For more information. Telephone: (+44) 01827 58234 Fax: (+44) 01827 310166

ultrafilter
international
ultrafilter and more

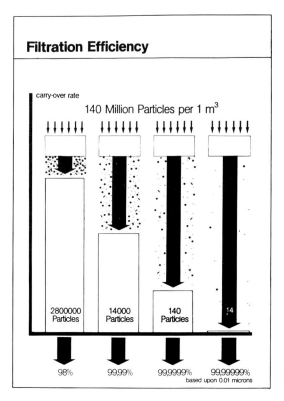

Filtration Efficiency

FIGURE 2 – Relationship between filtration efficiency and particles/million. *(ultrafilter)*

Inducted contaminants

Industrial stationary compressors are normally installed in separate rooms away from factory contaminants, drawing in air from the outside atmosphere as in Figure 3. The level of dust concentration is likely to be of the order of 10–50 mg/m³. It is standard practice to fit the compressor with an intake filter, having an efficiency of the order of 99.9%, based on the dust concentration present in the ambient air (see Tables 3 and 4). The intake filter can be expected to pass all particles smaller than 5 micron as well as a proportion of the larger sizes. It will also pass atmospheric water vapour and all gases, bacteria and viruses.

TABLE 3 – Dust Concentration

Dust size (μm)	% Total weight
5	12
5 to 10	12
10 to 20	14
20 to 40	23
40 to 80	30
80 to 200	9

TABLE 4 – Contaminants present in air

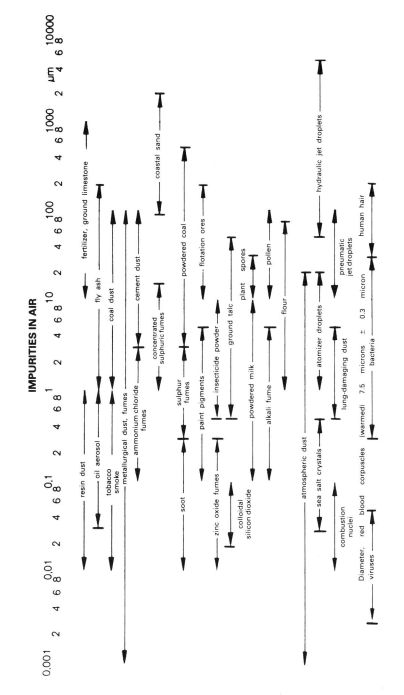

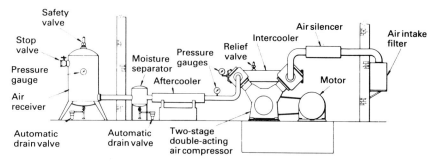

FIGURE 3 – A typical compressed air installation.

Types of intake filters

Intake filters (Figure 4) commonly used are:

- Paper filters – with renewable elements used on all types of compressors. Filtering efficiency is high (over 99%) with a typical pressure drop of 2.5 to 3.5 mbar if correctly sized. They are not recommended for use with reciprocating compressors

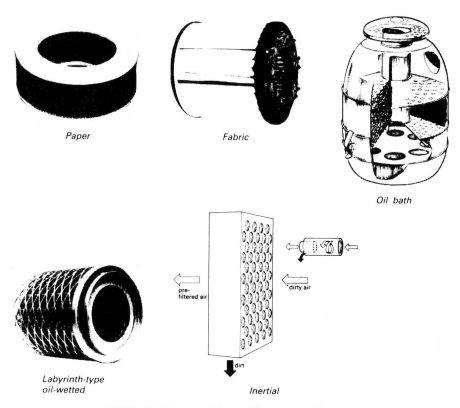

FIGURE 4 – Types of intake filters. *(Atlas Copco)*

unless a pulsation chamber is incorporated between the filter and the compressor intake. They are not suitable for handling air in excess of 80°C.

- Fabric filters – stronger than paper and generally cleanable by back flow of compressed air and sometimes by washing.

- Oilbath filters – which have a higher solid retention capability than the above, with a capacity for collecting impurities equal to the mass of the oil. They are not suitable for use on piston compressors unloaded by closing the air inlet valve, unless installed with a bypass or check valve to prevent oil being blown back out of the filter with reversed air flow.

- Labyrinth type oil wetted filters – mostly used on small compressors inducting relatively clean air. They require periodic cleaning.

- Inertial filters – reverse flow or cyclonic, suitable for coarse filtering of large volumes of air. They are normally used as pre-filters in conjunction with one of the other types mentioned above.

The intake filter is normally specified by the compressor manufacturer based on the operating conditions.

Water vapour

Ambient air always contains water vapour. After it leaves the compressor, it is fully saturated with water vapour, the actual amount present being directly proportional to the temperature and inversely proportional to the pressure, see Table 5. The liquid water is best removed when the temperature of the air is lowest and the pressure is highest,

TABLE 5 – Moisture content of saturated compressed air

Air temp °C	Water vapour content of fully saturated compressed air. Grams of water vapour per cubic metre (g/m³) of air at standard atmospheric pressure of 1013 m/bar (0 bar gauge) and saturation conditions and when compressed to the pressure and temperature shown.												
	Air pressure in bars												
	0	0.4	0.63	1.0	1.6	2.5	4.0	6.3	8.0	10.0	12.5	16.0	20.0
0	4.82	3.45	2.97	2.42	1.87	1.39	0.97	0.67	0.54	0.44	0.36	0.29	0.23
5	6.88	4.93	4.24	3.46	2.68	1.99	1.39	0.95	0.77	0.63	0.52	0.41	0.33
10	9.41	6.74	5.80	4.73	3.66	2.72	1.90	1.30	1.06	0.87	0.70	0.56	0.45
15	12.7	9.08	7.83	6.39	4.94	3.67	2.56	1.76	1.43	1.17	0.95	0.76	0.61
20	17.4	12.5	10.7	8.75	6.77	5.02	3.51	2.41	1.95	1.60	1.30	1.04	0.84
25	23.6	16.9	14.6	11.9	9.18	6.82	4.77	3.27	2.65	2.17	1.77	1.40	1.14
30	30.5	21.8	18.8	15.3	11.9	8.81	6.16	4.22	3.43	2.81	2.29	1.81	1.47
35	39.0	27.9	24.0	19.6	15.2	11.3	7.87	5.40	4.38	3.59	2.92	2.32	1.88
40	49.6	35.5	30.6	24.9	19.3	14.3	10.0	6.87	5.57	4.55	3.72	2.95	2.39
45	63.5	45.5	39.2	31.9	24.7	18.3	12.8	8.79	7.13	5.84	4.76	3.77	3.06
50	81.0	58.0	49.9	40.7	31.5	23.4	16.4	11.2	9.10	7.45	6.07	4.82	3.90

Example: 1 m³ of air at atmospheric conditions fully saturated at 20°C contains 17.4 g of water vapour. When compressed to 6.3 bar and 50°C it can only retain 11.2 g as vapour, therefore 6.2 g (17.4 – 11.2) is released as liquid water. If cooled down to 25°C a further quantity of 7.93 g (11.2 – 3.27) of water will condense out.

ie immediately after the compressor and its aftercooler. Standard practice is to follow the compressor element with an after cooler of sufficient capacity to reduce the temperature to within 10°C of the temperature of the cooling water or air. In the former case, approximately 20 litres of water will be required for every 2.5 m³ of free air being cooled. Further cooling is possible if a reasonably large receiver is fitted.

Refer to the chapter on Air Receivers to assess the proper volume. It is often quoted that the size of the receiver should be approximately equal to 30 times the rated free air delivery of the compressor, for 7 bar applications. A receiver should incorporate a condensate drain, preferably of an automatic type.

Further cooling is likely to occur in the distribution main, so these should be laid out with a pitch in the direction of flow so that gravity and air flow will carry the water to drain legs, located at appropriate intervals. These also should be fitted with automatic drain valves to prevent them being flooded. Down loops in the distribution main should be avoided where possible, but if this cannot be avoided, they must incorporate drain legs at the bottom of the down loop. All take-off points from the distribution main should be located at the top of the main to prevent water getting into the take-off line.

Humidity control

Removal of water from the compressed air to a level where no new condensation can be formed in the compressed air system (*ie* the dew point is made lower than the ambient temperature) can have advantages in any industry and in some industries is essential.

In pneumatic tools, where the energy is extracted from the air expansively, the drop in temperature can result in the formation of ice in the expansion passages, which can be so serious as to choke the exhaust and stop the tool working. Further advantages of the use of dry compressed air are:

- Air tools can be lubricated more efficiently.
- Lubrication of all pneumatic components is improved when dry compressed air is used and servicing intervals are increased.
- The use of dry air in spray painting eliminates any risk of damage to paint finish from water droplets.
- In blast cleaning the reliability of equipment is improved and risks of icing under outdoor conditions are reduced.
- In a dry compressed air system, there is no corrosion which could lead to leaks and loss of pressure.
- There is a no need for draining of condensed water.

Air drying is also desirable in conditions of high ambient temperature and local humidity, where it may be difficult to secure sufficiently low compressed air temperatures to provide ideal conditions for air line filters. Different dew points are required depending upon the final use, the ambient temperature and humidity.

Pre-filters

Downstream from the compressor, aftercooler and receiver, a pre-filter can be installed. Their use is recommended where heavy contamination of oil, water or dirt is expected and is the first stage in air purification. Porous filter elements such as sintered bronze, sintered stainless steel, polythene or polypropylene with pore sizes of 5 to 25 micron will remove the heaviest contamination and protect the heat exchangers in the dryer. The air flow is from inside to out, with the coarse particles of dirt and scale retained on the inside of the filter. Oil, water and the remaining fine particulate matter then pass through the dryer to a sub-micron filter. The latter has become an essential part of achieving oil-free compressed air.

Theory of coalescing filters

Oil is the most difficult contaminant to remove from a compressed air system. Modern techniques of oil separation use the principle of coalescing, which is a term used to describe the action of forcing aerosols to combine into droplets which become large enough and heavy enough to be drained away by gravity. There are three different processes occurring in a coalescing filter, depending on the particle size. These are illustrated in Figure 5.

1. Diffusion

Particles, both solid and aerosols from 0.001 to 0.2 micron follow random motion, moving erratically within the airstream. When they collide with the filter media fibres, solid particles adhere strongly by intermolecular attraction, but liquid aerosols which do not adhere so strongly can migrate with the airstream and are then forced to agglomerate into larger particles. Diffusion is affected adversely when the particle velocity exceeds the ideal, which will vary with pressure, temperature and nature of the aerosol.

2. Interception

For particles in the range 0.2 to 2 micron, interception is the dominant mechanism. They follow the direction of the flow and are intercepted when they pass close to the media fibre where the inertia forces can be overcome and interception can occur. Provided that a smooth laminar flow is maintained, interception is not greatly affected by the variables

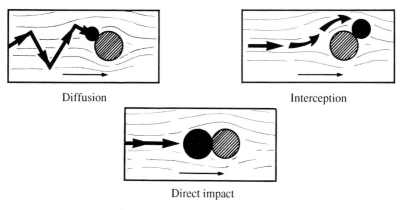

Diffusion   Interception

Direct impact

FIGURE 5 – Coalescing filter principles.

affecting diffusion. Filtration equipment should be designed to operate at air velocities well below those which could give rise to turbulence.

3. Direct impact

Particles larger than 2 micron have sufficient mass and inertia to leave the airstream as it deviates around the fibres and can impact directly on the fibres.

Filtration efficiency is measured by the percentage of contaminants of a particular micron size which are captured by the filter. Efficiency is important because it determines not only contaminant removal but also filter life. Selection of the appropriate grade of filter is important, because contaminant removal higher than required for the application results in higher filter costs. Generally the higher the efficiency the shorter the life. Filtration efficiency ranges from 90% to better than 99.99999%.

Oil removal

The problem of oil removal is complicated by the fact that oil can exist in three forms: liquid, oil/water emulsions and oil vapour. Special filters are required to remove oil vapour and oil aerosol.

Modern oil removal filters are of the coalescing type and commonly use glass fibre elements. Oil particles impinge on and adhere to the fibres, resulting in a gradual build-up of coalesced droplets. These droplets are driven to the outside of the material by the airstream, where the oil is stopped by a porous sock covering the element. It then flows by force of gravity to the bottom of the sock where it drops to the filter bowl. The oil is automatically drained from the filter. These filters are capable of removing the oil content in compressed air down to a level of 0.1 mg/m^3 or less.

Glass microfibres are considered to be an ideal filter medium for coalescing of liquid aerosols. This material is neither absorbent or adsorbent and consequently retains its original properties. It is superior to natural wool which has been used in the past. Glass microfibres are hydrophobic (water repellent) and so water tends to form on such fibres in droplet form rather than as a film, a condition which is favourable to filtration efficiency. Unfortunately, neither glass nor any other material is oleophobic (oil repellent) and oil will form as a film on glass microfibres, increasing their effective diameter. Allowance for this increase can be made and the oil film will not appreciably detract from the filtration efficiency, once the filter medium has been wetted. Glass microfibres in the range 0.5 to 0.75 micron diameter yield the best results as filter coalescing media. The depth of the fibre bed and the ratio of the void space to fibre space are of critical importance in the proper operation of this type of filter. The application of coalescing filtration techniques to the removal of oil aerosols, although relatively new, is overcoming a long standing difficulty in the use of compressed air as a process and control medium. The coalescing filter has shown that it is a solution to the growing need for oil-free air.

The mechanical sandwich construction of the two stage filter element, embedded between stainless steel support sleeves, assures high filtration efficiency even under changing pressures and flow directions. Due to the coalescing effect of the filter medium, the elements are self-regenerative in the removal of liquids (oil or water). It is advisable to fit pre-filters capable of removing dirt particles down to 5 micron or less, otherwise the

TABLE 6 – Media grading system

Grade	Colour code	Coalescing efficiency (%)	Pressure drop (bar)		Largest particle passed (micron)	
			Dry	*Wet*	*Aerosol*	*Solid*
2	Green	99.999+	0.1	0.34	0.4	0.1
4	Yellow	99.995	0.085	0.24	0.6	0.2
6	White	99.97	0.068	0.17	0.75	0.3
8	Blue	98.5	0.034	0.09	1	0.4
10	Orange	95.0	0.034	0.05	2	0.7
12	Black	90.0	0.017	0.03	3	1

Application of grades

Grade 2   Extremely fine particulate and coalescer; filtration of electronic grade gases.
Grade 4   Very high efficiency coalescer for protection of instrument quality systems.
Grade 6   High efficiency coalescing grade for removal of liquid aerosols and suspended fines; protection of conveying and breathing systems.
Grade 8   Good coalescing combined with high flow rate; protection of valves, cylinders, etc.
Grade 10   Pre-coalescer or pre-filter for grade 6 to remove gross contamination.
Grade 12   Pre-filter for heavy contamination; removing large amounts of liquid; protection of main-line filters.

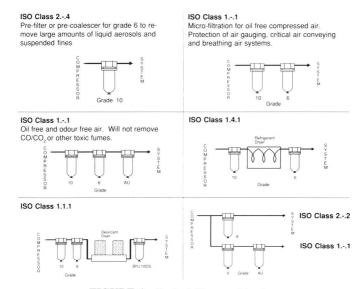

FIGURE 6 – Typical filter combinations.

coalescing filter may become choked with dirt. High efficiency filters should be installed downstream of a dryer where used. Note that for some applications it may be necessary to combine coalescing filters with activated carbon filters to remove the residual oil vapour.

Table 6 gives some typical filter grades. Manufacturers have their own methods of designation, and it is wise to consult with them on specific applications. Figure 6 illustrates

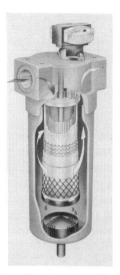

FIGURE 7 – Coalescing filter showing differential pressure gauge and
automatic drain.

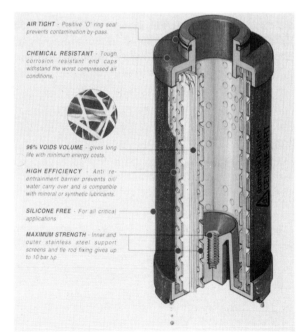

AIR TIGHT - Positive 'O' ring seal
prevents contamination by-pass.

CHEMICAL RESISTANT - Tough
corrosion resistant end caps
withstand the worst compressed air
conditions.

96% VOIDS VOLUME - gives long
life with minimum energy costs.

HIGH EFFICIENCY - Anti re-
entrainment barrier prevents oil/
water carry over and is compatible
with mineral or synthetic lubricants.

SILICONE FREE - For all critical
applications

MAXIMUM STRENGTH - Inner and
outer stainless steel support
screens and tie rod fixing gives up
to 10 bar Δp

FIGURE 8 – Construction of oil coalescing filters. *(ultrafilter)*

some combinations. Two typical forms of construction are shown in Figures 7 and 8. Most
filter bodies can incorporate a differential pressure gauge and an external automatic drain
device.

Oil vapour

Oil-free air will still contain hydrocarbon vapours and odours which in the food, dairy, pharmaceutical and brewing industries must be removed. Activated carbon filters can be used for this purpose. These are incorporated downstream of the oil removal filters.

Activated carbon adsorption

Filters based on activated carbon have been used historically to deodorize compressed air and are particularly effective for the removal of oil vapour contaminants. Activated carbon is not effective for oxides of carbon, methane, ethylene, ammonia and sulphur compounds for which chemical or catalytic methods can be used.

Activated carbon is excellent for adsorbing oil vapour and is suitable for purification of compressed air. The activated carbon must be in a finely granulated form to present as much surface area as possible for adsorption. This is done in a granulated tube, which has a large surface area and a deep bed to increase the contact time.

Activated carbon has a selective preference for oil vapour over water vapour, which is of vital importance for even instrument quality compressed air has more water vapour than oil vapour. The efficiency of activated carbon filters should be such that there is no trace of hydrocarbons or odour in the delivered air.

Solid contaminants

Standard air line filters for distribution lines generally remove particles down to about 50 micron. Finer filters can be used for better protection, where required, but it is better to use these for second stage filtering at individual take-off points. Filters can be placed in four categories:

- Rough filters for distribution mains, capable of removing particles down to 50 micron.
- Medium efficiency filters in the range 5 to 50 micron.
- Fine filters in the range 1 to 5 micron.
- Ultra-fine in the range down to 0.1 micron.

Fine and ultra-fine should be considered for second and third stage filtering and should be protected by a coarser filter to protect the finer filter from gross contamination.

Air line filters

A well designed air line filter of the correct size will effectively remove water but cannot affect the vapour content of the air. If the air is subject to further cooling down stream of the filter, more water may condense out. If a complete absence of water is required, the air must be treated so that its dew point is lowered below any temperature reached by the air in use.

The operation of an airline filter is simple. Air entering the filter passes through louvres which directs the air in a swirling pattern. Centrifugal force throws the liquid and particulate matter outward to the inside of the filter bowl, where it falls down to the bottom of the bowl to be drained away. A baffle prevents the turbulent air from picking up the

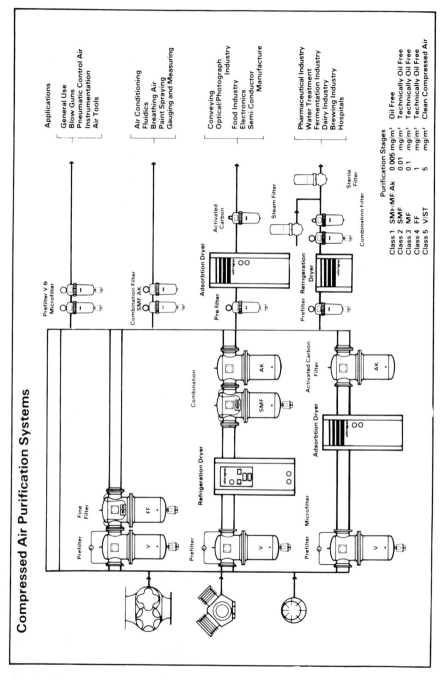

FIGURE 9 – Examples of complete filtration systems for specific application.

Filter Tips

Microfilter

Make the most of the experience of a leading worldwide specialist in compressed air purification.

Millions of Microfilters in use is evidence of their guaranteed reliability.

Get to know the ZANDER Microfilter range. With their unique 4 phase filtration, the filters ensure considerable savings in energy and costs.

ZANDER UK Ltd.
P. O. Box 2585
Tamworth
Staffordshire B 77 4 QZ
Telephone (0 18 27) 26 00 56
Fax   (0 18 27) 26 11 96

ZANDER®

FILTRATION
ADSORPTION
CONDENSATION
WASTE WATER
TECHNOLOGY

liquid and returning it to the system. As the air leaves the bowl it passes through a filter element which removes more of the remaining impurities.

Airline filters can be equipped with an automatic drain to save manual operation of the drain valve. All filters should be mounted vertically, with a 5 mm drain line. A filter of this kind is often followed by a line lubricator to add the correct lubricant to the air line when used for tools or circuit elements.

When selecting the appropriate filter for an application, a performance and cost should be properly balanced. The initial cost may be low, but if the element needs to be replaced frequently, replacement costs may exceed the initial saving. A filter with a high efficiency rating may cause an unacceptable pressure drop, which may further increase the operating costs. Excessive pressure drop may result from incorrect pipe sizes or excessive flow through the filter. Never select a filter on the basis of the pipe size alone; they should be chosen on the basis of maximum flow and system pressure.

In order to prevent failures in air operated devices, monitoring of the filter for element replacement is necessary. Some filters rely on pressure drop through the filter as an indication of element saturation. Others rely on a visual check of the element, or observation of a colour change.

Element life

In coalescing filters, element life is determined by pressure drop, in absorption filters, element life is governed by saturation. It is worthwhile changing the filter elements at prescribed periods whether or not the pressure drop indicates it to be due. Elements should be changed at least once a year.

Element life depends on the type of compressor supplying the system. An oil-free rotary compressor has a minimum of oil carry-over; a reciprocating compressor is next best.

In a rotary flooded compressor, an oil separator is fitted as part of the system to minimize oil carry-over. The remaining aerosols in the delivery line are of the order of 0.1 to 0.5 micron. A sub-micron filter is necessary to deal with these. If there should be a separator failure or there is poor maintenance, large quantities of oil can be carried over and cause immediate failure of the filter element.

Filter testing

There are occasions when it is worthwhile installing a filter test set in order to assess the efficiency of a filter element. In most cases tests should be performed by the filter suppliers, but equipment is also available for laboratories who wish to perform their own tests. Such a test set is illustrated in Figure 10. This can be used for batch testing or for the checking of an individual element. The unit generates its own aerosol and measures the subsequent concentration of the test aerosol by a laser photometer.

Check list for compressed air filters

1. Determine the type of compressor lubricant used before selecting a filter. Some filter components such as gaskets, seals and transparent bowls are not compatible with certain synthetic compressor lubricants. Deterioration of these components in contact with such lubricants may cause the filter to burst causing injury.

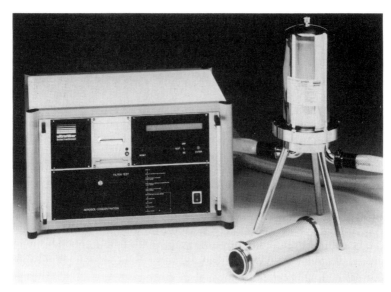

FIGURE 10 – A complete filter test centre to measure filtration efficiency. *(ultrafilter)*

2.   Check that the pressure rating and flow capacity of the filter is greater than the pressure and flow at the point of installation.

3.   Inspect the filter for shipping damage. Replace the whole unit if damage is apparent.

4.   Install filters down stream of aftercoolers and receivers.

5.   On critical lines, where the air supply must not be interrupted, it is good practice to incorporate a good by-pass system complete with standby filter at the filter location.

6.   If the filter does not incorporate an automatic drain facility, install a drain trap or drop leg below the filter and drain frequently. This will prevent liquid from accumulating in the bottom of the filter housing or impairing the efficiency of the filter element. Do not exceed the flow capacity of the drain system on start-up since compressed air lines may accumulate large amounts of water while shut down.

7.   A compressed air filter is a pressure vessel. While in service, filters must be depressurised before any maintenance is attempted. Depressurise filters by slowly opening the drain provided for condensate removal. Failure to do so can result in equipment damage or injury.

8.   Follow the manufacturer's warnings with regard to cleaning the filter, to avoid damage to filter parts. Since some cleaning solvents attack seals, transparent bowls or housings, these should be cleaned only with soap and water.

Standards and legislation

The British Compressed Air Society has published the following documents relevant to the topics discussed in this chapter:

CAC 9407 Compressed Air Condensate. Guide to the legislative control and legal obligations relating to the disposal of industrial effluent.

860900 Methods of Test Part 1 – Filters for Compressed Air. Pressure drop and oil mist removal test procedure for coalescing filters.

860901 Methods of Test Part 2 – Filters for Compressed Air. Absorbent pack filters for hydrocarbon vapours.

For U.K. legislation on condensate discharge refer to:

The Water Resources Act 1991

The Water Industries Act 1991

The Environmental Protection Regulations 1991

STERILE AIR AND GAS FILTERS

The demand for sterile compressed air increases with the adoption of advanced technologies which were unknown a few years ago. The selection of a sterilization filter for a compressed air system can be a difficult task. The production of proteins, vaccines, antibodies, hormones, vitamins and enzymes involves high technology processes which require aseptic and sterile supplies of gases or liquids throughout the manufacturing cycle. The production and packaging of many dairy and food products such as beer, yoghurt, cream and cheese use compressed air or carbon dioxide. The nature of these products makes them susceptible to contamination by micro-organisms held in the compressed air or gas.

Any product that can be contaminated by airborne bacteria must be protected. In the case of food and chemicals produced by fermentation, bacteria would cause serious defects and rejection of the product.

In the fermentation and pharmaceutical industries, compressed air and other gases are used through every stage of the production process from the refining of the raw material to the manufacturing and packaging. Compressed air may be used as a source of energy in a process or as a biomedical aid. Air motors are used for explosion-free mixing of powders, instrumentation and cylinders for the batching of materials.

Process air can be used for aeration of liquids, seed fermentation and laboratory applications. Mixing air into the preparatory chemicals or the final product means that they must be as clean and sterile as the material it serves, so there can be no risk of fouling by solids, liquids or micro-organisms.

Micro-organisms are extremely small and include bacteria, viruses, yeasts, fungus spores and bacteriophages. Bacteria can be from 0.2 micron to 4 micron, viruses less than 0.3 micron and bacteriophages down to 0.04 micron. The filters that have to cope with these organisms have to have a performance rather better than these dimensions. The presence of these living organisms can be a serious problem in process industries, because they are able to multiply in the right conditions.

When selecting compressed air sterilization filters, the follow conditions must be satisfied:

- The filter must not allow penetration of any type of micro-organism that could cause contamination.

- It must be able to operate reliably for long periods.
- The materials of construction must be inert so as not to support biological growth.
- It must be easy to install and maintain.
- It must be capable of being tested for integrity.
- It must be capable of being steam sterilized repeatedly.
- The filter must be as small as possible consistent with efficiency so as to reduce problems of installation.

Modern air sterilization filters use pleated hydrophobic binder-free borosilicate microfibres which have an efficiency of better than 99.9999% at 0.01 micron and remain in service for up to 12 months. The element has been constructed using an amalgamation of filter media (Figure 1); the average diameter of the fibre is about 0.5 micron. The polluted air strikes the outer layer of relatively coarse medium (2 micron). This traps the dirt particles before it reaches the microfibre. The borosilicate collects the particles down to 0.01 micron. Any aerosols present are here converted into liquid. In removing micro-

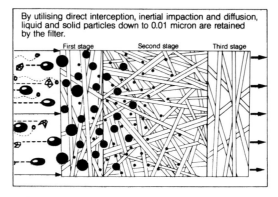

FIGURE 1 – Binder-free microfibre media. *(ultrafilter)*

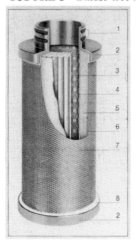

1. Double O-ring seal
2. Stainless steel endcap
3. Internal stainless steel support sleeve
4. Internal supporting medium
5. Pleated filtermedium
6. External supporting medium
7. External stainless steel support sleeve
8. High-temperature resistant silicone bonding material

FIGURE 2 – Construction of a borosilicate filter element. *(Zander)*

organisms, three main mechanisms are present – direct interception, inertial impaction and Brownian motion. The second layer of 2 micron medium is designed to hold the microfibre in place. These two media combined hold back the solid particles. In a dry air stream they will hold back bacteria. As long as the bacteria have no means of multiplying in the medium, they can be held. A filter medium containing a binder material which could act as a nutrient would be unsuitable. The complete construction can be seen in Figure 2.

Membrane cartridge filters are extremely flexible and high in tensile strength. Cartridge construction based on a multi-layer combination of filter media support an irrigation mesh. Nylon and polypropylene polymers that have previously been used in coarser grades are now used as membrane filters. Layers of nylon microporous membrane and polypropylene pre-filter are pleated together and supported by an inner core. The end caps are melt sealed in polypropylene. The membrane is of thin nylon having a controlled pore size. Its construction ensures that it cannot release fibres down stream; it can be repeatedly autoclaved.

Hollow fibre filter

Hollow fibres were originally developed for dialysis and have now been adapted for compressed air. The filter is arranged with a smaller spread of hole sizes, but with a larger number of membranes. This gives a greater retention ability. Due to a closer pore distribution than conventional membrane filters, the number of pores per unit area of filter is greater, which extends the service life. A closer pore distribution means that the largest pore size is much reduced. A conventional membrane filter has a pore size at least 0.3 micron larger than a hollow fibre filter.

The development of a membrane filter element with a rated pore size of 0.1 micron and a reduced pore distribution means the difference between the retention of viruses and bacteria. The construction of the hollow fibre filter means that it is economical to manufacture smaller elements for laboratory use.

Housings

Sterile filter cartridges must be fitted into a pressure holding housing. The method of sealing is by single or double O-ring seals which allow sealing and movement during steam sterilization or shock loading without breaking the seal. Compressed air sterilization filter housings need to be designed to protect the filter cartridge from stress. Stainless steel housings are the usual choice of material; they should be electrolytically polished inside and out, and have no grooves or corners in which micro-organisms can collect. They should incorporate an automatic condensate drain. For economy, sterile filters can be made of galvanised steel housings with stainless steel parts in contact with the compressed air or with a corrosion resistant coating. Aluminium housings with powder coated protection are also available.

Filter selection

Filters are sized according to the flow rate versus pressure drop information supplied by the manufacturer. A sterile filter has an initial pressure drop of about 0.1 bar although the rating can be reduced if necessary to meet the requirements of the process; a pressure drop

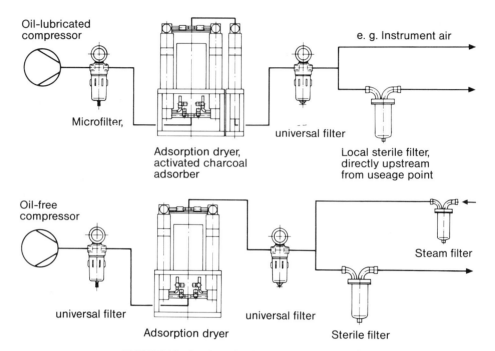

FIGURE 3 – Typical circuits for sterile air. *(Zander)*

Process measurement, monitoring and control systems Compressed air measuring instruments Air storage equipment Paint-spraying technology Fluidics Analysers Pre-filters for adsorption dryers Pneumatic conveying systems	Series X microfilters		Residual oil content* max. up to 0.01 mg/m³ Removal of fluid and solid particles down to 0.01 micron.
			Technically oil-free and dry Pressure dew point – 40° C
Process air Pharmaceutical industry Food industry Breweries Hospital services Packaging industry Breathing air Galvanics Film-laboratories Fluidics circuits	Microfilters and activated charcoal stage		Residual oil content* max. up to 0.005 mg/m³ Removal of fluid and solid particles down to 0.01 micron; free of oil-vapor and odor.
			Technically oil free, odorless, absolutely dry, pressure dew point – 40° C
*Details of residual oil contents are based on a filtration temperature of 20° C at 7 bar			

V	Pre-filter. Efficiency 99.99% at 3 micron	X	Efficiency 99.9999% at 0.01 micron. Oil retention 0.01 mg/m³
AD	Adsorption dryer	A	Activated carbon filter for removal of oil vapour
Z	Efficiency 99.99% at 1 micron. Oil retention 0.5 mg/m³		

FIGURE 4 – Installations for a variety of applications. *(Zander)*

of about 0.2 bar is indicative of dirt collecting in the filter and it must be replaced. A pre-filter should normally be incorporated in the circuit to reduce the load on the sterile filter. Coalescing type pre-filters can sustain a high pressure drop but it is normal to replace the element between 0.3 and 0.6 bar.

A complete filter assembly for producing sterile air will contain a number of separate elements. If the sterile supply is to be taken from a larger system with a variety of applications, an oil lubricated compressor may be the source of air; in this case an activated carbon filter will need to be incorporated. If all the air in the process needs to be sterile, then an oil-free compressor is probably the better choice. Figures 3 and 4 show some of the possible choices.

A decision has to be made about the criterion for replacement of the filter element. It is usual to replace the element after the recommended number of sterilization cycles, unless the pressure drop appears to be excessive. Provided that an adequate level of pre-filtering is incorporated, pressure drop should not normally be the criterion.

Steam sterilization

Routine sterilization is a requirement for any system using a sterile filter. It is often carried out between batches or in a continuous process plant at prescribed intervals. In batch

FIGURE 5 – Steam filter for sterilization cycle. *(Zander)*

operation provided that the filter is kept pressurised and a small air bleed is allowed to flow continuously, several batches can pass before sterilization is carried out. The steam used must be saturated and free from additives or contamination, which implies its own sterile filter, see Figure 5. Steam sterilization is shown in the lower diagram of Figure 3.

A well designed system should last for 12 months before cartridge replacement is

essential that any filter used has been tested for integrity to ensure that it will fulfil its designed duty. The best way of testing for integrity is by a cloud of test particles in the critical range 0.1 to 0.3 micron. The test is based on creating an aerosol of Dioctyl Phthalate (DOP) in the critical particle range, challenging the filter with this and measuring any penetration with an aerosol photometer, downstream of the filter. DOP has been replaced by other substances because of possible health hazards and so corn oil or other oils are used giving the same type of particle spread when atomized. This type of validation test is carried out by the manufacturers and is not normally a procedure used by the end user.

AIR DRYERS

Air dryers remove or modify the water content of air. Moisture is a general term applying to the liquid water aerosol content of air and its water vapour content. No single device can remove all the moisture but a combination of devices can be used to reduce it to an acceptably low level. The ability of air to hold water vapour is dependent on temperature, see Figure 1, so cooling the air allows some of the vapour content to condense, which can then be removed. For every 11°C rise in temperature the water vapour holding capacity of the air doubles.

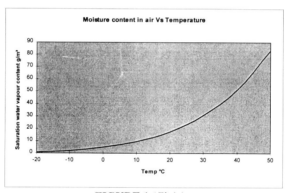

FIGURE 1 *(Flair)*

Methods used for air drying in industry are varied and many types of equipment have been used for the purpose (see Figures 2 and 3). The conditions of the application and the intended service of the air greatly affects the method required to dry it and the choice of equipment to perform the function.

Air is dried in industry for reasons varying from room air conditioning for comfort and warehouse air conditioning to prevent corrosion, to compressed air services for pneumatic tools and instrumentation, where moisture could cause equipment malfunction, product spoilage and corrosion. Examples of the advantages of dry compressed air are:

- Supply to air tools making it possible to lubricate them more effectively. Risk of icing in the exhaust is reduced.

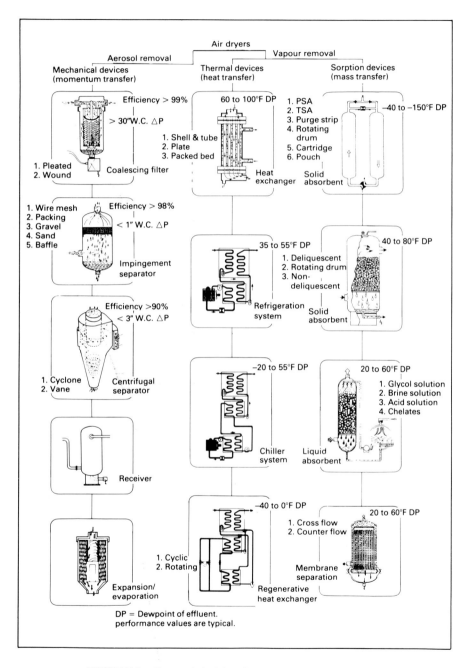

FIGURE 2 – Types of air driers for aerosol and vapour removal.

Dry compressed air at the absolutely lowest possible operation costs

Ready to install, heat-regenerated by any energy source

Electronic control system ensures safe cost-effective operation of the dryer

No loss of valuable compressed air

Pressure dewpoints from -25°C to -70°C

For more information. Telephone: (+44) 01827 58234 Fax: (+44) 01827 310166

ultrafilter
international
ultrafilter and more

TABLE 1 – Characteristics of air dryers; the benefits and limitations of each type of air dryer are unique

	Benefits	*Limitations*
Mechanical devices	Aerosol removal.	Cannot remove water vapour.
Coalescing filters	Highest efficiency.	High pressure loss.
Impingement separators	Low pressure loss, non-fouling.	Low efficiency.
Receivers	Large droplet separation.	Low efficiency, large size.
Expansion/evaporation	Aerosol transformation.	Easily overloaded.
Thermal devices	Water vapour removal.	Cannot remove water aerosols.
Heat exchangers	Cool air, reduces vapour content of air.	Increase relative humidity,
Refrigeration systems	High energy efficiency.	Limited to 32°F dewpoint.
Chiller systems	Reduce vapour content of air.	Require extra transfer loop.
Regenerative exchangers	High-efficiency vapour removal.	Increased system complexity.
Sorption devices	Water vapour removal.	Low aerosol removal.
Adsorption systems	Highest water vapour removal.	Complex valving generally.
Solid absorbent systems	Bulk vapour removal, simplicity.	High effluent dewpoint, corrosive drains.
Liquid absorbent systems	Bulk vapour removal.	High effluent dewpoint, corrosive fluids, complexity.
Membrane separation	Bulk vapour removal only.	High effluent dewpoint, high purge consumption

The principle of the coalescing filter is described in the chapter on Compressed Air Filtration. Coalescing filters are the most efficient aerosol separators, capable of removing in excess of 99% of the entrained water. Impingement separators are less effective, removing 98%. Centrifugal separators are least effective with about 90% removal.

Receivers reduce the aerosol content by settling, due to the low velocity of the air flowing through. They are effective at removing the large droplets.

Expansion and evaporation devices do not actually remove any water, but cause evaporation of the liquid aerosols after impact on a solid surface.

Thermal devices used for drying are: heat exchangers, refrigeration systems, chillers and regenerative heat exchangers.

Counter-current heat exchangers reheat cooled, dried air, causing a reduction in its relative humidity, pre-chilling the incoming air while simultaneously reducing the energy consumption of the system.

Over-pressure

Over-pressure occurs when air is compressed beyond the required working pressure, cooled and expanded to the required pressure. Because of the higher pressure, the volume is reduced and as the air is cooled, its ability to hold water is also reduced. On expansion, the air after the condensate is removed can reach the required dewpoint. The disadvantage of this system is the high energy usage making it useful only for small air flows, although the mechanism works naturally in many compressed air systems.

Refrigeration systems

The ability of air to hold water is dependent on temperature. The traditional method of removing water is by an aftercooler with refrigeration dryers. An aftercooler is a heat exchanger used to cool compressed air after leaving the compressor. Cooling reduces the temperature and precipitates the water into droplets which can be drained off. It also reduces the volume of the air but this represents no power loss (other than that required to run the cooler). As a general rule, air-cooled aftercoolers are used with air-cooled compressors and water cooled aftercoolers with water cooled compressors.

Refrigeration dryers use the simplest drying method; the air passes through an aftercooler but the coolant is maintained at a much lower and constant temperature. The coolant is produced by a closed refrigeration circuit incorporating a compressor and condenser, adjusted to cool the air to between 2°C and 5°C; the limitation of this type of dryer is the freezing point of water, however it is not practical to reduce the temperature completely down to 0°C.

Cooling is done in two stages; first in a pre-cooler which cools with the cool air leaving the dryer, followed by the main cooler using refrigerated coolant. If the water condensed in these coolers is about 2°C, the dryness is adequate for most purposes. In fact drying the air to a dewpoint about 10°C less than the prevailing ambient temperature is usually

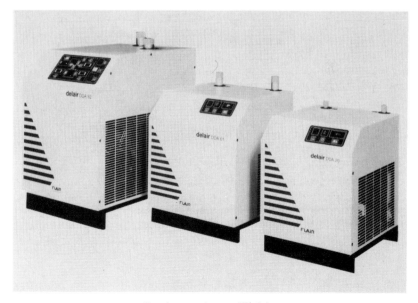

Regrigerant dryers. *(Flair)*

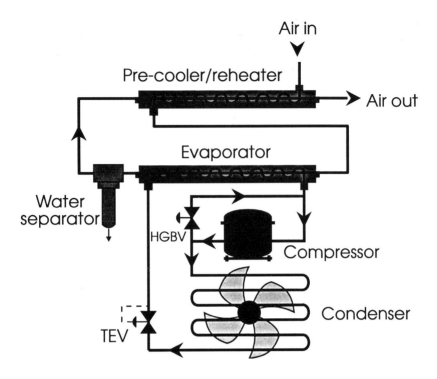

FIGURE 4 – Two stage cooler. *(Flair)*

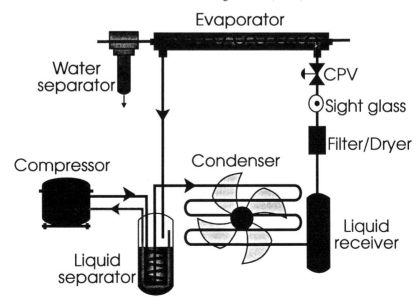

FIGURE 5 – Single stage cooler with separator. *(Flair)*

sufficient to eliminate risk of water condensate in the system. Figure 4 illustrates the system. Note that an internal compressor provides the energy to pump refrigerant round the circuit and represents a small power loss. The thermal expansion valve (TEV) controls the flow of refrigerant to meet demand. As the cool refrigerant passes through the evaporator, heat exchange from the warm compressed air evaporates the refrigerant to a gaseous state, which then re-enters the compressor.

At times of low demand little or no warm air passes through the evaporator, hence the refrigerant is not warmed and remains in the liquid state. If liquid were allowed to enter the compressor, damage could occur. To prevent this, the hot gas by-pass valve (HGBV) is provided which taps warm gas from the outlet of the compressor to evaporate any liquid droplets remaining. Although this system should be fail-safe, HGBVs are notoriously unreliable, and a malfunction could be expensive. To overcome this an alternative system has been developed, as shown in Figure 5.

In this system, at times of low demand, the gas containing liquid is passed directly to a liquid separator, where the liquid settles at the bottom and the warmer gas is passed to the compressor. The hot gas leaving the compressor is passed through the separator housing , evaporating the settled liquid as well as pre-cooling the refrigerant gas prior to its entering the condenser. The liquid separator has the advantage of simplicity without the need for extra sensors and valves and reduces the servicing costs.

TABLE 2 – Dewpoint recommendation for dryers in various compressed air applications

Application	Recommended dewpoint °C
Inside works air	10 to –10
Paint spraying	10 to –25
Instrument air	10 to 40
Air motors	10 to 40
Sand blasting	5 to 0
Pneumatic tools	5 to –25
Packaging	5 to –25
Plastics manufacturing	5 to –40
Pneumatic conveying	5 to –60
Pneumatic powder coating	5 to –60
Optical cleaning	–17 to –33
Bullet-proof glass	–18 to –40
Pressurised cable	–20 to –35
Electronic component drying	–20 to –40
Outside works air	–20 to –40
Chemical and pharmaceutical factories	–25 to –40
Wind tunnels	–40 to –65
Super dry electronic assembly	–40 to –100
Infra-red spectrometry	–50 to –65
Space research	–65 to –100
Drying vessels for nuclear installations	–65 to –100

Lower dewpoints than this are required, for example in systems which have pipes outside in cold weather, for some chemical processes and for instrumentation.

Chillers and regenerative heat exchangers are modifications of the normal refrigeration process. In a chiller system, a brine or glycol solution is circulated between the refrigerant evaporator and the air cooler. Because heat transfer is more efficient between the freon of the refrigerator and the liquid coolant, the evaporator can be made smaller and less expensive; improved temperature control is also possible in a chiller. In refrigeration systems with regenerative heat exchangers, the air is cooled below the freezing point of the condensate, and the heat exchanger is either rotated or cycled to permit removal of the ice by defrosting. Such systems need complex controls to function properly.

Table 2 gives dewpoint recommendations for dryers for various applications.

Sorption systems

These reduce the water vapour content and the relative humidity. They dry air at constant temperature, based on the mass transfer of water vapour into or onto absorbents or adsorbents. They can be either regenerative or non-regenerative. Regenerative systems can be continuous or alternating. The primary types are;

- Adsorption; non-regenerative adsorbent cartridges and pouches
- Regenerative adsorbent systems
- Thermally regenerative; temperature swing adsorbers (alternating beds)
- Continuous rotating bed
- Heatless regenerative types
- Pressure swing adsorbers (alternating beds)
- Purge stripping
- Solid absorbent types
- Deliquescent
- Non-deliquescent
- Liquid absorbents and membrane separation.

Solid adsorbents remove water vapour by physical forces which attract water molecules to the solid surface of the adsorbent. Solid adsorbent particles are highly porous on a microscopic scale, providing a large area for adsorption. The particles are of the order of 1 to 10 mm diameter, to permit high mass transfer rates and allow some aerosol separation by impingement, but at the air velocities permitted for an air adsorption bed, the aerosol removal efficiency for droplets smaller than 10 micron is very low. Common adsorbents used for the removal of water vapour include adsorbent clays, activated aluminium oxide (Al_2O_3), silica gel (SiO_2), molecular sieves (Na, AlO_2) and natural zeolites. Each type of desiccant has its own characteristics, but most are able to produce low effluent dewpoints depending on the adsorbent system and the mode of regeneration.

Adsorption dryers

The adsorption drier consists of two drying columns filled with desiccant (silica gel or alumina). The air passes through one column where the desiccant adsorbs the water vapour in the air until it is saturated. The airflow is then automatically switched to the second

Adsorption Tips

Adsorption Dryer

Series KM

Make the most of the experience of a leading worldwide specialist in compressed air purification.

The compact KM dryer range, for example, dries compressed air for flow rates from 7 m^3/h to 85 m^3/h, remarkably safely with minimum space requirements.

For flow rates up to 30,000 m^3/h, patented processes guarantee the high efficiency of the ZANDER systems at dewpoints of -80°C and pressures of up to 400 bar.

ZANDER UK Ltd.
P. O. Box 2585
Tamworth
Staffordshire B 77 4 QZ
Telephone (0 18 27) 26 00 56
Fax   (0 18 27) 26 11 96

ZANDER ®

FILTRATION
ADSORPTION
CONDENSATION
WASTE WATER
TECHNOLOGY

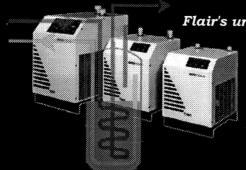

column, while the first column is regenerated, by removing the water from it. Two different methods are used to regenerate the desiccant: in the first method, a separate flow of dry heated air is blown through it; in the second method, no heat is needed but a proportion of the already dried air from the active column is passed through it. The adsorption process is exothermic and the desorption process is endothermic, so it is advantageous to situate the two processes close together to take advantage of heat transfer between the two columns. Both methods waste a proportion of the air to satisfy the purge requirements. The heatless method naturally wastes more, the proportion depending on the inlet temperature and the required dewpoint; the nominal purge air consumption (based on say, a dewpoint of -20°C and an inlet temperature of 35°C) will be about 15% of the rated capacity), but for higher inlet temperatures and a lower required dewpoint, the proportion may be as much as 25%. With a good control system, the energy loss can be minimised. The heat regenerated type uses about 1.5% of the expanded dry air, but requires energy input in the form of heat.

The decision as to which type to choose depends on a calculation of the total energy input. The heatless type is normally the most expensive in running costs because of the wasted air. To save energy, some of the heat of compression may be used to supply regeneration heat if the system can be so designed.

Heatless dryers

An example of the operation of a heatless dryer is shown in Figures 6 and 7. The first stage in drying the air is the removal of all liquid contaminants, so that the desiccant dryer can

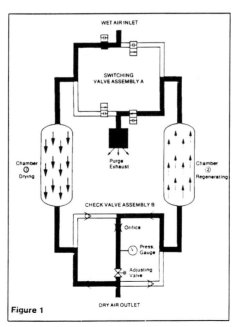

FIGURE 6 – Operation of heatless dryer. *(Flair)*

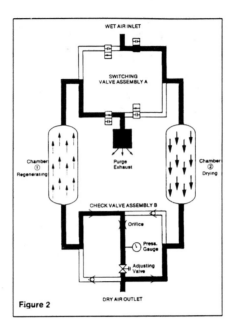

Figure 2

FIGURE 7 *(Flair)*

operate at maximum efficiency by removing the vapour only. In Figure 6, wet incoming air passes through a switching valve assembly A, and is directed downwards through chamber 1 where it contacts the desiccant. Air drying takes place by adsorption, which occurs when the moisture level in the air stream is greater than the moisture level in the desiccant. Water vapour in the air stream condenses as a film on the desiccant. Adsorption is an exothermic process, which means that as drying takes place, heat is released and is stored in the desiccant bed for use in the next cycle. Dry air exits chamber 1 and passes through check valve assembly B and continues to the application main.

While gas is being dried in chamber 1, the desiccant in chamber 2, which has been wetted in the previous cycle is being regenerated. At the start of its regeneration, chamber 2 is depressurised from the operating pressure to atmospheric, allowing flow in an upward direction through valve assembly A; the air goes out through the purge exhaust. A portion of the dry outlet air from chamber 1 passes through valve assembly B and upward through the chamber being regenerated. This dry purge, with the aid of the heat generated during the previous drying period removes the moisture from the desiccant. The purge carries this moisture through valve assembly A and through the purge exhaust. The regenerating purge flow is in the counter direction to that of the drying part of the cycle to ensure desiccant regeneration.

When the regeneration cycle is complete, the chamber being regenerated is re-pressurised to full operating pressure by closing the purge exhaust. Purge gas continues to flow into the chamber until it pressurises to system pressure. Inlet gas is then switched over to chamber 2 (Figure 7) through valve assembly A for drying. Repressurisation before

Heatless dryers. *(Flair)*

Heat regenerated dryer. *(ultrafilter)*

switch over ensures that the desiccant will not be jolted by the inrushing gas flow. A drying period of about 3 minutes provides an effluent dewpoint of -40 °C or better. A full three minutes is used for the drying period. However during the regeneration process, which is occurring in the offstream chamber simultaneously with drying in the onstream chamber, approximately 30 seconds is allowed for chamber pressurisation before going back on stream at switch over. Thus a complete cycle consists of two periods of equal length, one drying, the other depressurisation, regeneration and repressurisation.

Energy efficiency of dryers

The water load on the dryer is determined by the specific flow rate and temperature and is not usually the same as the water load that is present on a continuous basis. For example in the winter, the aftercooler water temperature is lower, decreasing the load on the dryer. Dryers are designed to meet peak conditions. If these peak conditions are not realised, a less severe load is placed on the dryer and energy is wasted in regenerating a desiccant that has not been saturated. It is generally accepted that desiccant dryers may only experience 20 to 30 % of the peak design load; since most dryers work on a fixed cycle, 70 to 80% of the regeneration energy can be wasted, so a system which incorporates the ability to match regeneration energy to moisture load has the capability of saving significant amounts of energy with a resultant cost saving.

Moisture measurement devices

The key to an energy efficient dryer is an accurate, reliable and repeatable moisture measurement device or hygrometer. There are three types of hygrometer for this application, Figure 8:
- in-bed temperature sensing,
- outlet moisture sensing, and
- in-bed moisture sensing.

With in-bed temperature sensing a thermo-switch detects the heat front which passes through the desiccant bed as the heating elements drive off the adsorbed moisture. When

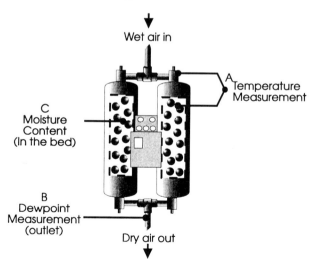

FIGURE 8 – Types of control system the regenerative dryness. *(Flair)*

the heat front meets the thermo-switch, the heating is stopped, with a saving in power. This method is not the most efficient, because at low moisture levels, 50% of the energy is used to heat the bed and the chamber and only the remaining 50% is used to drive off water. This technique can only be used in heat activated dryers and not the heatless dryers described above.

Outlet moisture sensors require a hygrometer to be placed at the outlet of the drying chamber to anticipate the breakthrough of the moisture front before it exits the dessiccant. Three types of hygrometers are used for dryer outlet moisture sensing:

- Chemical cells which exhibit a change in electrical resistance as the moisture increases. These are easily contaminated requiring regular recalibration and frequent replacement.
- Plated wire capacitance probes suffer from the same problem of contamination as the chemical cell.
- Photo-optic sensors incorporate a light source which directs light from a miror on to a light sensor. Changes in reflected light are caused by condensation on the mirror surface. A small heater is used to clear the mirror betwen readings. Any contamination by oil or dessiccant will render the device useless.

All these methods of moisture sensing must be able to measure moisture content of the order of 10 parts per million at -40°C, with very little error.

In-bed moisture sensing employs a capacitance probe embedded in the desiccant and determines the humidity by measuring the dialectric strength of the desiccant. The adsorption of water vapour by the desiccant changes its capacitance.

In a heat regenerated dryer, the probe controls the drying portion of the cycle. The probe senses the moisture front as it passes through the chamber. The air still has to pass through further desiccant after the probe so good dewpoint performance is assured. It can be positioned in the bed to sense moisture well before an outlet sensor could and ensures an accurate and consistent outlet dewpoint of between -45°C and -42°C. Fail-safe operation of this type of sensor is secured by setting a high limit to the system; the dryer is put into a fixed cycle when the high limit is exceeded. A capacitance probe suffers from the same problems of contamination as the other types described above but can be coated with a contaminant resistant material.

In a heatless dryer, the in-bed probe controls the regeneration portion of the cycle. Due to the short time cycle of this type of dryer, there is insufficient time for a well defined moisture front to be formed: a saturated or perfectly dry bed is counter-productive. The heat of adsorption must be kept within the desiccant bed. This is accomplished by moving a pocket of moisture up and down the chambers, through drying and regeneration cycles without allowing any to escape. The probe controls this movement during the regeneration portion of the cycle, and is the only type which will function in a heatless dryer.

This automatic moisture load control (AMLOC) ensures a reliable measure of the moisture content in the bed and limits the regeneration energy whether it derives from the electric heat or purge air. An in-bed capacitance probe measures the order of 200 parts per million, compared with an outlet probe measuring 5 to 7 parts per million, which makes it an industrial measuring device rather than a laboratory instrument. The savings in energy

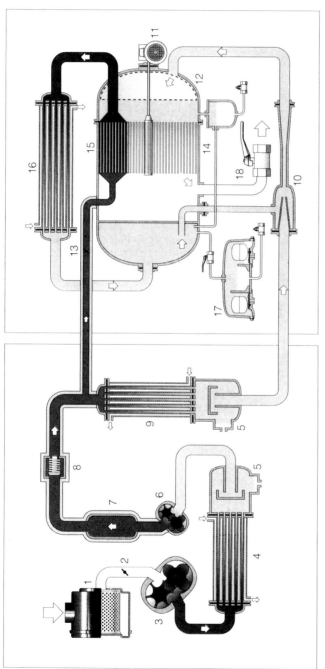

FIGURE 9 – Operation of rotary dryer. (*Atlas Copco*)

compressor

1 Air intake filter
2 Inlet valve
3 Low pressure element
4 Intercooler
5 Water separator
6 High pressure element
7 Pulsation damper
8 Non return valve
9 Aftercooler

dryer

10 Ejector
11 Drive motor
12 Built-in water separator
13 Throttle valve
14 Drying section
15 Regeneration section
16 Regeneration cooler
17 Drain with safety drain
18 Dry air out

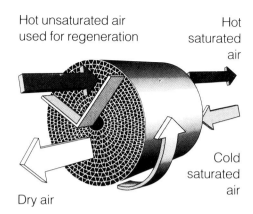

Hot unsaturated air
used for regeneration

Hot
saturated
air

Cold
saturated
air

Dry air

FIGURE 10 – Air flow through dryer section. *(Atlas Copco)*

will depend on the usage – the lower the flow rate the greater the savings, so in a situation where the dryer is not fully utilised during a shift, the savings are greater. A modern dryer, in addition to AMLOC will also incorporate a microprocessor control unit to monitor the dryer condition to identify cycle times and maintenance requirements.

Rotary adsorption dryer

This new development, Figures 9 and 10, is claimed to be particularly efficient in that no external energy is needed, other than a small amount (0.12 kW in this example) to drive the rotating drum and the power needed for the aftercooler; nor is there any wasted air. This dryer is only suitable where the air from the compressor is completely oil-free. It is currently supplied matched to a range of rotary screw compressors (capacities from 6 to 120 m³/min).

The pressure vessel is divided into two sectors, one for drying and one for regenerating. Through these sectors, there rotates a glass fibre based corrugated paper drum impregnated with silica gel. Hot air leaving the compressor is divided into two streams before reaching the aftercooler. The main stream passes through the aftercooler into the drying sector where the moisture is removed from it through adsorption. The hot air branched before the aftercooler passes through the regeneration sector where it removes moisture from the desiccant by evaporation. This air then passes through the regeneration aftercooler. Moisture separated from the air is collected and drained by means of an automatic drain system. The cooling can be either by water or by airblast. The pressure dewpoint can be brought down to -40°C if the temperature of the cooling medium is 10°C.

Absorption dryers

The drying of air by absorption is a true chemical process in that moisture is absorbed into the medium which can be either solid or deliquescent. Solid material reacts with the water vapour without turning it to liquid, whereas in a deliquescent dryer the water vapour is absorbed into the salt until it liquifies. Regeneration is impractical, so the drying medium must be changed regularly. Unfortunately, the absorption media are usually highly corrosive. Deliquescent media carry the danger that they become entrained in the air and carry on down the line. The absorption media also have the characteristic that above a temperature of 30°C , the crystals bake together which leads to a high pressure drop. A dewpoint of about 15°C is the best that absorption dryers can achieve.

Non-deliquescent absorbents form solid hydrates when combined with water vapour; they are not dissolved in the process. These are commonly used in desiccant pouches in the same way as solid absorbents.

Liquid absorbents include water solutions of glycol, salts, acids and other water absorbing chemicals, which are cascaded through a mass transfer tower to effect water vapor absorption from the air. Packing or perforated plates are commonly used in such towers to enhance the rate of mass transfer. The liquid absorbent is drained from the bottom of the tower to a reboiler where it is regenerated; it is then recirculated to the top of the tower.

Solid and liquid absorbents are useful in removing large amounts of water vapor from air of high humidity, but they are not capable of reducing the vapour contents to low levels such as can be achieved with refrigerant or adsorption dryers.

Membrane dryers are continuously regenerating absorption devices.

Wet air passes along one side of a membrane, while a dry purge stream is passed along the other side. Water vapour is absorbed into the semi-permeable membrane and is passed through it from one absorption site to another, from the high pressure side to the low pressure side. The effluent quality is dependent upon the permeability of the membrane, the amount of the membrane surface available for transmission of water vapour and the amount of purge air. Commercial systems produce effluent dewpoints comparable with other absorption systems. As with other such systems they are able to remove large quantities of water but are not suitable for very low dewpoints.

Air heating

A similar effect to drying can be achieved for a limited period of time and for a limited length of pipe by inserting a heating unit in the line. A heater does not extract moisture, it merely reduces the relative humidity so that the vapour does not condense out at the point of use. Because expansion through an orifice causes a drop in temperature, there is a possibility of condensation if the temperature is low before the orifice. An air dryer helps to prevent that. It can be used upstream of a paint sprayer to prevent water droplets ruining the finish and for medical and dental applications where a low temperature would be uncomfortable.

COMPRESSED AIR DISTRIBUTION

Mains systems

Alternatives available are a dead end main or a ring main. A dead end main has the advantage of distributing the compressed air over the shortest possible run, reducing pressure drop and economising on pipe cost. Its chief disadvantage is its lack of flexibility, although the service supplied by a compressor can be extended by the use of a split flow main, taking two separate mains from the compressor. Typical dead end mains are shown in Figure 1.

FIGURE 1 – Types of dead-end main.

A ring main has the advantage that, being in the form of a closed circuit, the velocity of the air in the main will be reduced and the pressure drop will be less. The ring main itself may be of appreciable volume and capable of acting as a form of reservoir, to assist in maintaining pressure at any point when there is a sudden draw-off. Also, if the system is extended by the addition of a single branch, this branch can draw its supply from both ends of the main, reducing velocity and minimising pressure drop. A ring main offers maximum flexibility with functional advantages and is generally the preferred choice, but the longer length makes it essential to take care over reducing leaks. A typical layout is shown in Figure 2.

Regardless of the type of main, the most important consideration is that the pipeline shall be of suitable size to pass the maximum demand without significant pressure drop, which represents a power loss. A low pressure drop means that the operating range of the compressor can be reduced.

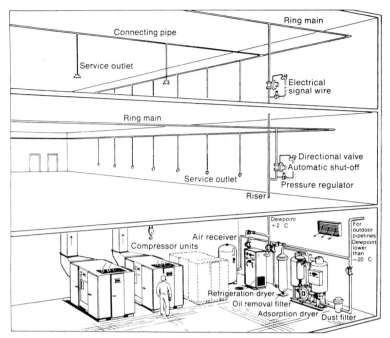

FIGURE 2 – Principle piping layout for compressed air distribution. *(Atlas Copco)*

Mains sizing

It might be thought that the ideal is to choose the largest possible pipe size, to minimize the pressure loss. Apart from the extra cost and weight of oversized pipes, they represent a volume which has to be filled by the compressor and wasted on shut down. A reasonable compromise has to be found, taking into account the acceptable pressure loss at the point of use. A variety of empirical rules have been suggested, but the most satisfactory method of establishing sizes is to calculate the flow rates throughout the system, using the data provided in the chapter on Pressure Drop Calculations.

The critical condition is when all the items of equipment are functioning simultaneously and when the correct pressure to operate them is available. In a ring main the flow will come from two directions. The pressure available at the extremes of the system will vary according to the demand. If a motor or tool has to generate its maximum power and needs a certain pressure to do so, that pressure may not be available if all other devices are acting simultaneously.

Most compressors are capable of generating a pressure at their delivery port greater than that required to operate standard workshop or construction equipment. Most tools give an adequate performance at 6 bar, while compressors are available with a delivery pressure of 8 to 10 bar. A pressure drop along the distribution line of 2 bar might not be noticed by the operator, since his equipment would function satisfactorily, however such a situation would be very wasteful of energy.

Some of the empirical rules suggested for determination of pipe sizes are:
1. Limit the velocity of the air to a prescribed value. 15 m/s is suggested as a limit. The British Compressed Air Society recommend 6 m/s to avoid carry over of moisture past the drain legs. A reasonable compromise is 6 m/s for the main lines and 15 m/s for branch lines that do not exceed 15 m in length.

    A criterion based on applying a velocity limitation results in an equation of the form:

$$q = 0.0007854 \, v \, (p+1) \, d^2$$

where   q is the flow rate in l/s FAD
v is the prescribed velocity in m/s
p is the gauge pressure in bar
d is the internal diameter in mm

For v = 15 m/s, this becomes

$$q = (p+1) \, d^2/85$$

For v = 6 m/s, it becomes

$$q = (p+1) \, d^2/212$$

Table 1 for steel pipes and Table 2 for ABS pipes are calculated on the 6 m/s criterion.

TABLE 1 – Maximum recommended flow through main lines

Nominal bore (mm)	Rate of air-flow at 7 bar (litres/second)
6	1
8	3
10	5
15	10
20	17
25	25
32	50
40	65
50	100
65	180
80	240
100	410
125	610
150	900

TABLE 2A – Maximum recommended flow through ABS pipes

10 BAR PIPE								
Applied pressure	0.6 bar	1.0 bar	1.6 bar	2.5 bar	4.0 bar	6.0 bar	7.0 bar	10.0 bar
Pipe size (o.d.) mm								
12	0.60	0.95	1.52	2.41	4.02	6.25	7.5	11.1
16	1.35	2.35	4.0	6.4	10.9	16.9	19.7	28.8
20	1.80	3.3	5.5	8.6	14.0	21.8	25.7	38.5
25	3.25	6	10.0	16.5	27.5	42.5	50.3	74.5
32	5.5	10	16.5	27.5	45.5	71.5	85	127.5
50	20	36	60	97	163	250	294	425
63	36	64	107	176	290	450	528	760
75	55	105	175	285	475	750	875	1270
90	95	170	280	455	745	1145	1360	1965
110	165	280	465	740	1240	1930	2280	3310

Note: Flow in litres per second

TABLE 2B

12.5 BAR PIPE								
Applied pressure	0.6 bar	1.0 bar	1.6 bar	2.5 bar	4.0 bar	6.0 bar	7.0 bar	10.0 bar
Pipe size (o.d.) mm								
12	0.45	0.8	1.30	2.10	3.50	5.55	6.6	9.80
16	1.05	1.8	2.85	4.65	7.75	12.20	14.6	21.60
20	1.35	2.3	3.7	6.0	9.95	15.75	18.8	27.85
25	2.2	4.0	6.8	11.3	19.2	31.0	36.5	54
32	4.5	9.0	15.5	25.3	41.8	63.5	74.5	107
50	16.5	27.5	44	71	122	193	229	337
63	30	50	83	132	229	366	434	646
75	45	80	137	220	360	565	665	975
90	75	135	225	350	570	885	1040	1525
110	135	225	380	600	1000	1570	1870	2710

2.   Limit the pressure drop to the point of use to a prescribed value, *eg* 0.1 bar or 0.5 bar or as a proportion of the supply pressure, say 5%. This is a rather better method, since it takes into account the length of the pipe run and the pressure needed to

operate the various kinds of equipment. One way of using this method is to take the pipe friction factor f from the tables in the chapter on Pipe Flow Calculations and apply the Darcy equation:

$$\frac{\Delta P}{P} = \frac{\rho_0 f L v^2}{2D}$$

ρ_0 is the density of free air = 1.2 kg/m^3.

The procedure in solving for the correct pipe diameter is:
- Choose a diameter D and obtain a value of f/D from the tables.
- From the calculated maximum flow rate, determine the flow velocity, v.
- Calculate the value of $\Delta P/P$. If the pressure drop is satisfactory, the chosen value of D is correct. Otherwise select another value of D and repeat the calculation. The value of L is to be taken as the effective length of the pipe run, replacing all the bends and fittings by their equivalent length. This method can be tedious, but with the use of a suitable computer program, the burden of calculation can be much reduced.

One advantage of this method is that it is able to take account of the benefits of using smooth bore pipes, such as aluminium or copper, which have a lower pressure loss per unit length than the more common steel pipes.

A further method which combines the two previous methods is to limit the flow velocity to 6 m/s in the ring main and then check that the pressure drop is lower than 0.1 bar (or whatever criterion has been decided), and that the final pressure is high enough to operate the equipment. The diameter can be increased if necessary.

Basic geometry

Overhead location of an air main is the best choice, whenever this is feasible, particularly with ring mains. This enables the mains to cover the working area in a direct way. This is generally cheaper to install, minimizes the number of bends and makes maintenance easier. In the case of a dead end main there can be a case for a low level main, where each building is served by its own compressor. The main run is run at a bench height of 0.45m to 0.6m, carried on floor mounted stanchions at a spacing of 3 to 4m. The airlines should be marked in accordance with BS1710 to identify the various services and to distinguish between high and low pressure.

Mains fall

It is important that an air main shall have a fall in the direction of the airflow with moisture traps at suitable points along the run. A fall of about 1 in 100 or 1 in 120 is enough to ensure that water separating out is carried along to a low point. An automatic trap is to be preferred as these cannot become overfilled and inoperative.

Branch mains or individual take-off points should emerge in the form of an upward loop. This will help to contain any condensate in the main and minimise the carry-over. A filter and separator should be fitted in the main at a high point between the compressor and first take-off point, to reduce the amount of moisture entering the main and branch lines.

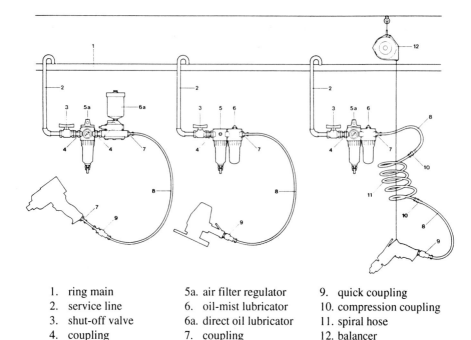

1.	ring main	5a.	air filter regulator	9.	quick coupling
2.	service line	6.	oil-mist lubricator	10.	compression coupling
3.	shut-off valve	6a.	direct oil lubricator	11.	spiral hose
4.	coupling	7.	coupling	12.	balancer
5.	air filter	8.	air hose		

FIGURE 3 – Recommended air supply system for pneumatic tools. *(Atlas Copco)*

Each branch and take-off point should have its own isolating stop valve and drain or drain cock at the blind end. Filters, separators and lubricators should be in the individual lines feeding tools or appliances requiring protection and lubrication. See Figure 3 for some typical installations.

The choice of these devices will depend on the type of tool or appliance concerned. Tools normally require a moisture separator and reducing valve in the feed line. For paint spraying, a moisture separator and reducing valve unit is required, enabling the pressure to be controlled at the take off point. Blow guns should also be used with a pressure reducing valve to control the force of the jet.

When the main involves any appreciable length of run, flange connections and expansion joints should be used throughout the system to make maintenance easier. The run should be above ground so that leaks can be identified and rectified.

Pressure drop is affected mainly by restrictions in the flow caused by changes in diameter or direction. The interior of all pipes should be smooth and clean; steel pipes should be galvanised. Sharp elbows, tees and similar fittings should be avoided in favour of large sweeps. All fittings should be full bore. Similar considerations apply to valves and hose connections.

Adequate pipe support is necessary for ring mains. Table 3 can be used for support spacing.

TABLE 3 – Intervals between pipe supports

Steel pipework				ABS pipework	
Pipe size nominal bore mm	Max. intervals for vertical runs m	Max. intervals for horizontal runs m		Pipe outside diameter mm	Support spacing at 20°C m
8	1.25	1.00		16	1.10
10	1.25	1.00		20	1.25
15	1.75	1.25		25	1.50
20	2.70	1.75		32	1.75
25	2.50	1.75		50	2.0
32	3.00	2.50		63	2.3
40	3.00	2.50		75	2.6
50	3.00	2.75		90	2.9
65	3.50	3.00		110	3.2
100	3.50	3.00			
150	4.25	3.50			
200	4.50	3.50			
250	5.18	4.25			
300	5.48	4.87			
over 300	5.48	4.87			

For each 10°C temperature rise, reduce support spacing by 10%.

These support centres can be doubled for vertical pipes.

Outdoor air lines

Outdoor air lines present special problems, notably the need to protect the outside of pipes with an anti-corrosive paint or similar treatment and the desirability of burying the lines at frost-free depth to prevent freezing of the condensate and consequent blocking of the pipes at low temperature. Underground installations can be costly, particularly as inspection wells are also necessary to gain access to condensate traps and valves.

Overground installation, with the pipes suspended on posts or supported on plinths, is generally cheaper and simpler. Quick-coupling pipes are recommended; these have some expansion/ contraction ability to counteract thermal movement and some misalignment. With overground pipes particular attention must be given to condensate separation, where there is a risk of freezing. It is also recommended that outdoor pipes shall be of welded steel construction , hot galvanised and fitted with flange joints for water traps and valves.

Leakage losses

Even a small leakage hole can reduce the efficiency and waste power. Table 4 gives some examples. The entire system should be checked for leakage at regular intervals. It is commonly said that 5% of the compressor capacity should be allowed for leakage, but many systems are worse than that; anyone who has been in a empty workshop at the end of a shift will be aware of the sound of leaking air lines. The aim of the maintenance engineer should be no leakage at all. The common leakage points are hose connections, couplings and valves.

TABLE 4 – Power wasted through leakage

Hole diameter (mm)	Leakage at (l/s)	Power required by compressor (kW)
0.4	0.19	0.06
1	1.18	0.37
1.6	3.07	0.97
3	10.95	3.36
6	49.1	15.0
10	122.0	37.0

Two standard methods of determining leakage are:

1. Determine the volume V (litres) of the air main downstream of the receiver isolating valve, including all branch and drop lines. Pump up the system to normal operating pressure P_1, and then shut down the compressor. Close the receiver valve and make certain that all tools and equipment are isolated. Check the time for the system to leak to a lower pressure P_2, approximately 75% of P_1. P_1-P_2 is the pressure drop. The leakage in litres per second of free air is given by:

$$V \frac{(P_1 - P_2)}{t}$$

2. The second method uses a compressor of known capacity as a measuring device. This machine should be run at full load with all equipment isolated. When the system is charged to operating pressure, the compressor will unload. Due to air leaks, the air pressure will fall, causing the compressor to come on load. The duration of the period in seconds for which the compressor is on load (T) and off load (t) should be recorded at least four times.

If the compressor capacity is Q litres/s, the air delivered is Q x T litres. The total running time is T + t, so the average leakage is:

$$\frac{QT}{T+t} \text{ litres/s}$$

Branch lines are of relatively small diameter compared with the mains and usually supply only a single piece of equipment or a sub-system. The air consumption is usually more accurately determined than for the mains and often a higher pressure drop can be tolerated. Recommendations for sizing are given in Table 5 for steel and Table 6 for smooth bore copper or nylon. Many branch lines terminate in a flexible hose to connect to hand tools.

Refer to the chapter on Pipe Flow Calculations for advice on hose diameters. Usually when using portable tools the size of the hose connection will determine the hose size. It is not advisable to insert a reducing coupling to allow for a smaller bore hose. Heavy duty reinforced rubber hose is preferred for use on construction sites or in mines, to withstand rough treatment. Reinforced polymers may be used where there is a modest degree of

TABLE 5 – Maximum recommended flow through branch lines of steel pipe
Maximum recommended air flow (litres/second Free Air)†
through medium series steel pipe for branch mains not exceeding 30 metres length
(See BS 1387)

Applied gauge pressure bar	NOMINAL STANDARD PIPE SIZE (NOMINAL BORE) – MILLIMETRES										
	6 mm	8 mm	10 mm	15 mm	20 mm	25 mm	32 mm	40 mm	50 mm	65 mm	80 mm
0.4	0.3	0.6	1.4	2.6	4	7	15	25	45	69	120
0.63	0.4	0.9	1.9	3.5	5	10	20	30	60	90	160
1.0	0.5	1.2	2.8	4.9	7	14	28	45	80	130	230
1.6	0.8	1.7	3.8	7.1	11	20	40	60	120	185	330
2.5	1.1	2.5	5.5	10.2	15	28	57	85	170	265	470
4.0	1.7	3.7	8.3	15.4	23	44	89	135	260	410	725
6.3	2.5	5.7	12.6	23.4	35	65	133	200	390	620	1085
8.0	3.1	7.1	15.8	29.3	44	83	168	255	490	780	1375
10.0	3.9	8.8	19.5	36.2	54	102	208	315	605	965	1695
12.5	4.8	10.9	24.1	44.8	67	127	258	390	755	1195	2110
16.0	6.1	13.8	30.6	56.8	85	160	327	495	955	1515	2665
20.0	7.6	17.1	38.0	70.6	105	199	406	615	1185	1880	3315

General Notes:
† The flow values are based on a pressure drop as follows:
10% of applied pressure per 30 metres of pipe of 6-15 mm nominal bore.
5% of applied pressure per 30 metres of pipe of 20-80 mm nominal bore.

TABLE 6 – Maximum recommended air flow (litres/second)† through copper or nylon tubing and corresponding port thread (ISO 1174 and BS 5409)

Applied gauge pressure bar	NOMINAL TUBING SIZE (OUTSIDE DIAMETER) – MILLIMETRES									
	4 mm G1/16	5 mm G1/8	6 mm G1/8	8 mm G1/4	10 mm G1/4	12 mm G3/8	16 mm G1/2	18 mm G1/2	22 mm G3/4	28 mm G1
1.6	0.05	0.12	0.19	0.48	0.9	1.8	4.2	6.5	10.0	18.0
2.5	0.08	0.18	0.26	0.70	1.3	2.8	7.2	8.8	14.0	30.0
4.0	0.14	0.28	0.45	1.2	2.2	4.5	12.0	15.0	24.0	50.0
6.3	0.22	0.48	0.72	2.0	3.8	7.6	20.0	25.0	40.0	80.0
8.0	0.28	0.62	0.95	2.6	4.8	9.5	26.0	31.0	52.0	110.0
10.0	0.36	0.81	1.3	3.5	6.2	12.0	34.0	41.0	68.0	140.0
12.5	0.46	1.1	1.7	4.3	8.0	17.0	44.0	53.0	85.0	180.0

Notes: † The flow rates are selected to provide an approximate pressure drop with either nylon or copper tubing as follows:
15% of applied pressure per 30 metres of tubing in sizes 4-16 mm inclusive.
10% of applied pressure per 30 metres of tubing in sizes 18-28 mm inclusive.

rough treatment. For fixed and semi-fixed automation applications up to about 7 bar, un-reinforced nylon may be used. For further information on choice of materials, refer to the chapter on Pipe Materials and Fittings.

Calculation of air demand for typical installations

At the planning stage of an installation it is worthwhile preparing a list of all the equipment and its likely frequency of use. This can be used to establish the maximum and likely demand and will assist in determining the correct choice of compressor and the pipe sizes.

One would expect to design the installation on the basis of average consumption multiplied by a factor of 1.5, but this will vary with local circumstances. A calculation should be done on the maximum number of tools that need to be in operation simultaneously. The provision of a generously sized air receiver will assist in dealing with variable demand. If there is a large, intermittent demand at a station distant from the compressor, a further receiver close to the demand point may be worth considering. Refer to the chapter on Air Receivers for further information.

CONDENSATE TREATMENT AND DRAIN VALVES

Water droplets coalesce in receivers, aftercoolers and filters and collect in the lowest part of the system and means must be incorporated to drain off this water. It is far from being clean water. Usually it is mixed with oil from the lubrication system of the compressor and other debris. Most compressor oils contain a detergent which produces an emulsified discharge which does not readily settle out into the constituent parts. This is colloquially known as "mayonnaise" and has that kind of consistency. Unless removed it may adversely affect operation of the circuit valves and other equipment.

The environmental consequences of condensate discharge from compressed air systems is becoming increasingly important, so that instead of merely dealing with the removal of the contaminants that arise from the compression of air and the lubrication of equipment, consideration has to be given to the disposal of those contaminants. It was considered acceptable a few years ago for the contents of the drain traps to be discharged into the sewers with little more than a simple oil trap. Nowadays increasingly stringent regulations are being applied by the Water Authorities and the Environmental Agency (which has taken over the responsibilities of the National Rivers Authority in the U.K.) to limit the amount of oil that is permissible for discharge.

Fortunately industry has made equipment available which can satisfy the most stringent requirements.

Controls on discharge into the environment

The discharge of trade effluent, which includes oil, into the public sewers, soakaways and water courses is governed in the U.K. by:

Public Health Act 1936
The Environmental Protection Act 1990
The Water Resources Act 1991
The Water Industry Act 1991
The Environmental Protection Regulations 1991

In the European Community generally, reference should be made to Directive 76/464/EEC on pollution caused by dangerous substances discharged into the aquatic environment.

Consent for discharge will depend on its situation and the nature of the discharge. It is controlled by the Environmental Agency in the U.K. Discharge into the foul sewer finally ends up at treatment works where it is the responsibility of the water authority. Each Consent is decided on an *ad hoc* basis by the appropriate body. Discharge levels can be between 5 and 500 mg/l of hydrocarbons (approximately the same as parts per million). There may be a more general description such as there must be no visible sheen on the water, which occurs at about 5 mg/l. Most Authorities are at present likely to specify the maximum permissible hydrocarbon content of water discharge to be about 20 mg/l, with the possibility that some may require a significantly lower value. It is well within the capability of modern oil separation techniques to produce a level of 5 mg/l, so that it can be expected that more and more authorities will insist on that level.

Typically, compressor condensate contains 5000 mg/l, which is unacceptable for direct discharge. It is not permissible to discharge untreated condensate into surface water sewers or into the ground.

A useful summary of the issues involved can be found in "Compressed Air Condensate" published by the British Compressed Air Society.

Bio-degradable lubricants

Synthetic compressor lubricants are increasingly being used and so it may be thought that as such a lubricant is bio-degradable, it is permissible to discharge it directly into the sewers.

The view in the U.K. of the Environmental Agency (which has taken over the environmental protection function of the National Rivers Authority and the water authorities) is that the discharge of bio-degradable oils is unacceptable into drains, soakaways and watercourses on the following grounds:

- Oil floats on water, forming a film which can cause environmental damage to wildlife habitat, and although the film will degrade eventually, damage will be done during the time delay.
- A bio-degradable substance in the aquatic environment exerts an oxygen demand on the water reducing the oxygen available for other organisms leading at worst to anaerobic conditions.
- There is a philosophical objection to the discharge of waste material, when recycling is a better environmental option.

Some of the modern compressor lubricants are claimed to be particularly environmentally friendly. One recently introduced synthetic lubricant is claimed to be superior to washing-up liquid in its bio-degradability and therefore permissible for discharge into the drains. Such a claim should be treated with caution and specific advice taken from the appropriate discharge authority. There is also the possibility that there may be toxic elements present in the lubricant which would make it unacceptable for that reason.

Responsibility of the user

It is the duty of the factory operator to ensure that the discharge of compressed air drains does not cause pollution. It should no longer be acceptable for factory drain traps to discharge directly into a foul or surface drain. An automatic drain should always be ducted away to a central station either for separation into its constituents or for industrial disposal, preferably the former. One still sees condensate traps discharging directly over a concrete floor, which is presumably expected to absorb the waste. In fact it will find its way into the underlying soil and eventually into water courses.

It has been estimated that 10% of the air which is wasted in industrial installations is down to blocked or jammed open condensate drains, so it makes economic as well as environmental sense to install efficient separators.

Water drain traps

Manual drain taps can be used provided that care is taken regularly to open the tap and not to leave it open, but a more satisfactory method is the installation of an automatic drain trap, of which several types are available.

A ball float valve, Figure 1, is one of the simplest and most common types of drain; it maintains a positive shut-off, opening only when enough water has collected to lift the float. Another version is shown in Figure 2; the trap must be installed with the inlet (6) at the top so that the float can rise and fall vertically. After the water has discharged, the float

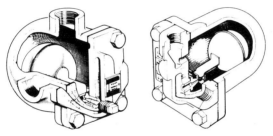

FIGURE 1 – Ball float type compressed air drain traps. *(Spirax Sarco)*

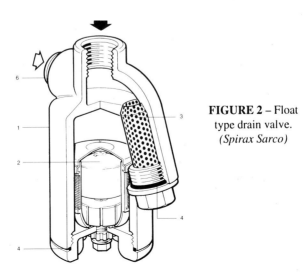

FIGURE 2 – Float type drain valve. *(Spirax Sarco)*

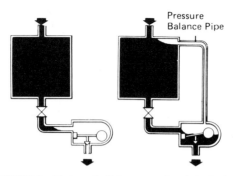

FIGURE 3 – Fitting a ball float trap. An air lock shown on
the left is cured by fitting a balance pipe.

drops and shuts off the valve, so no air is wasted; a strainer (3) is incorporated for safety; the drain can (and in most cases should) be piped away. A simple version of this type of valve can be fitted into the base of a filter bowl of an FRL (filter–regulator–lubricator) assembly, where it is not likely to have to discharge much water because most of it will have been previously removed. Where there is the possibility of a large quantity of water to be discharged from a ball trap, an air lock may occur, which can be cured by fitting a balance pipe, as shown in Figure 3. It is always safer to include a balance pipe, even if the volume of water is considered to be small.

When the moisture collects in the form of an emulsion, this can result in sluggish operation or may block the trap altogether. Regular cleaning out of the trap is essential and should always be done if it appears to have ceased to function. One way of handling heavily contaminated water is to use an aerodynamic type of drain, Figure 4. This blasts open when a specific quantity has collected and this action means it can handle heavily contaminated

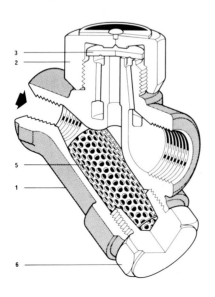

FIGURE 4 – Aerodynamic
type drain trap.
(Spirax Sarco)

FIGURE 5 – Automatic electronic drain trap working to preset timing intervals. *(Spirax Sarco)*

water. The disc (3) across which the air passes can be adjusted according to the degree of contamination; this disc and its seating face require examination and can be serviced by lapping the mating surfaces. Note the screen (5) which should be inspected and cleaned regularly. It is usually fitted horizontally in one of the distribution mains. No trap can handle more than a limited amount of contamination, and so the cause of the contamination should be sought and remedied before considering the use of one of these traps. It may be that a trap in an air receiver, for example, is not working correctly.

Another type of drain incorporates an electronic timer which opens the drain for an adjustable period of time at an adjustable interval. The chosen times can be set by knobs on the face of the trap, see Figure 5 . A disadvantage of this type of drain is that the times have to be set by experiment; also there is a periodic discharge of air whether or not any condensate has collected.

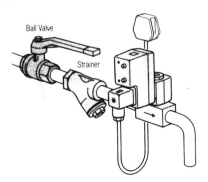

FIGURE 6 – Installation of electronic drain trap. *(Spirax Sarco)*

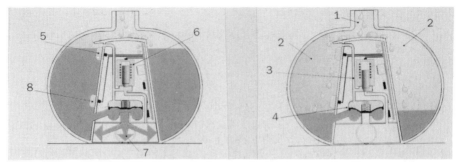

FIGURE 7 – Electronically controlled drain trap. *(ultrafilter)*

Another type of electronically operated drain is illustrated in Figure 7. This incorporates additional safety features which cater for an occasional malfunction. In operation, the condensate collects in the drain through the inlet (1) and accumulates in the tank (2). The diaphragm valve (4) is kept closed by system pressure until the condensate reaches the upper sensor. The solenoid is energised which pulls up the diaphragm valve venting the volume above it. The condensate is driven out by the system pressure until the level reaches the lower sensor, which causes the solenoid to de-energise returning the valve to the closed position. If the unit malfunctions, such as in freezing conditions, a self monitoring program identifies the problem on an LED display, while it continues to function with set timing intervals so as to limit air loss. These units can be equipped with an optional heater if freezing conditions are likely to be experienced.

Oil/water separators

As indicated above, the Environmental or other Agency will probably require that the condensate be treated to separate the oil and water so that the water may be discharged in

FIGURE 8 – A selection of condensate drains suitable for
10m³/min to 1000 m³/min. *(ultrafilter)*

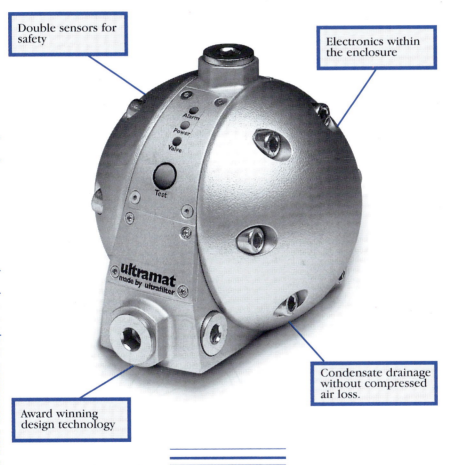

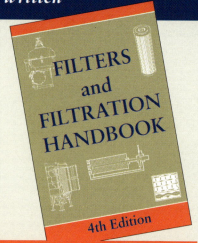

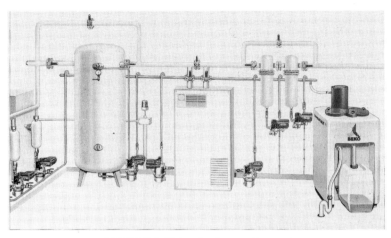

FIGURE 9 *(Beko Condensate Systems)*

the normal way and the oil sent for disposal or recycling. One other option that might be considered is for the whole of the condensate to be collected by a disposal company (at say 10p per litre), but this is likely to prove more expensive than installing a separator.

Even allowing the condensate to settle for hours in a settling tank does not guarantee that the water underneath the oil is safe for discharge. This is because modern compressor oils have a tendency to emulsify and settlement alone is insufficient to produce a liquid free from contamination.

In these circumstances an oil/water separator, should be installed. All the automatic condensate drains in the system, from the air receiver, dryers and filter can be ducted to one separator as shown in the system of Figure 9. Separation is achieved in three stages as illustrated in Figure 10: the first stage relies on gravity separation; the oil forms a layer

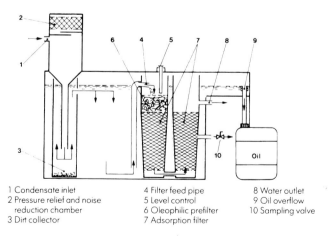

1 Condensate inlet	4 Filter feed pipe	8 Water outlet
2 Pressure relief and noise	5 Level control	9 Oil overflow
reduction chamber	6 Oleophilic prefilter	10 Sampling valve
3 Dirt collector	7 Adsorption filter	

FIGURE 10 – Function of oil–water separator. *(Beko Condensate Systems)*

on the surface of the settlement tank to be collected in a separate oil tank; the water beneath the oil is then filtered to remove solid impurities; and finally passes through an activated carbon filter to remove traces of oil remaining. It should then be pure enough for discharge into the main drainage system.

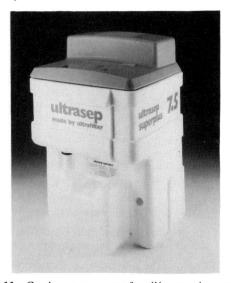

FIGURE 11 – Condensate separator for oil/water mixtures, suitable for a compressor of 40m³/min capacity. *(ultrafilter)*

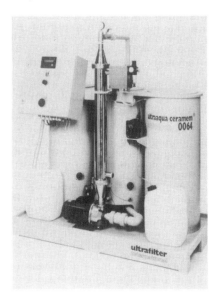

FIGURE 12 – Automatic condensate separator, capable of handling 64 litres/hour of oil/water emulsion with membrane regeneration. *(ultrafilter)*

The oil can be periodically drained off when the container is full, and the filters can be replaced without shut-down of the system. The oil can be sent for processing.

Different separation techniques apply according to whether the oil is present as an emulsion or as separable mixture. A convenient way of deciding this is to take a sample of the condensate in a jar and leave it overnight. If the oil is clearly sitting on top of the water and the water is comparatively clear, the condensate is separable. If it remains mixed and there is little separation, it has to be treated as stable emulsion. An emulsion separator incorporates a membrane filter which has to be replaced at regular service intervals, although some separators include an automatic back-flushing cycle to extend the service intervals. The appropriate oil separator should be chosen in consultation with the separator manufacturer.

Examples of the two types are shown in Figures 11 and 12.

For efficient working the separator must be correctly rated to the compressor capacity and the moisture content of the air. A higher rated separator is required if a dryer is installed.

PIPE MATERIALS AND FITTINGS

Steel is the most common material for main line pipes; ABS copolymer is also used. Copper, stainless steel and various kinds of rubber hose and plastic tubing are suitable for small mains, for branch lines and for connections to tools.

Steel pipes

Steel pipes to BS 1387 are normally specified in the nominal bore sizes given in Table 1. Finishes can be black or galvanised, the latter being preferred for corrosion resistance.

Steel pipes are normally used with threaded connections. Above 65 mm nominal bore, welded fittings to BS 1965 are recommended, although cast iron pipes with integral flanges or steel pipes with welded-on flanges may be used. In sizes smaller than 65 mm malleable iron fittings conforming to BS 143 and BS 1256 may be used for pressures up to 10 bar.

Screwed connections, to BS 21, are restricted to metallic tubes with thick walls.

ABS copolymer pipes

The only plastic material recommended as suitable for air mains is ABS copolymer (Acrylonitrile-Butadiene-Styrene). Provided it is used within its limitations it is a very convenient alternative to steel.

TABLE 1 – Nominal bore size steel tubes to BS 1387

Nominal size	Actual bore					Approximate o.d.	
	Medium			Heavy			
in	in	mm		in	mm	in	mm
1/8	0.226	5.77		0.178	4.52	13/32	10.3
1/4	0.338	8.59		0.290	7.37	17/32	13.5
3/8	0.476	12.09		0.428	10.87	11/16	17.5
1/2	0.623	15.82		0.575	14.60	27/32	21.5
3/4	0.839	21.31		0.791	20.09	1.1/16	27
1	1.060	26.92		0.996	25.30	1.11/32	34
1.1/4	1.401	35.59		1.337	33.96	1.11/16	43
1.1/2	1.633	41.48		1.569	39.85	1.29/32	48
2	2.066	52.48		2.002	50.80	2.3/8	60

The advantages are:

- Corrosion resistant inside and out. No scale or rust can form, so the loss through friction remains low and constant.
- Easy to handle. One-eighth of the weight of steel.
- Easy to joint by cold solvent welding. A correctly made joint cannot leak.
- Self-coloured (light blue) for identification. Meets BS 1710 for compressed air services.
- Non-toxic.
- ABS is available in two pressure ratings NP 10 (up to 10 bar) and NP 12.5 (up to 12.5 bar). The NP 12.5 rating is recommended for airline use.

One limitation on its use is that it softens and loses its strength at increased temperature. It retains its 12.5 bar rating up to 20°C; it should not be used, without reference to the manufacturers, above 50°C, at which temperature its pressure rating is reduced to 8 bar. Installations using ABS must be provided with an aftercooler with a temperature cut-out to ensure that there is no chance of hot air reaching the plastic. Transient pressures up to 10% over the maximum can be tolerated. In installations where there is some carry-over of oil from the compressor, the manufacturers of the tubing should be consulted, as there may be incompatibility between the tube and the oil; synthetic oils should not be used.

Table 2 compares the weight of ABS with alternative steel and copper pipes. Figure 1 shows a selection of fittings used with ABS piping and Figure 2 part of a mains system. Note that screwed fittings may not be used for direct connection to the pipe; only solvent welding is permissible.

ABS pipe may be used underground or in overhead installations.

Copper pipe

Copper tubing for air lines should conform to BS 2871:Part 1. No other earlier standards should be used. It is almost invariably used with compression fittings to BS 2051: Part 2.

TABLE 2 – Comparison between steel, ABS and copper pipes

| ABS (12.5 bar) | | Steel to BS 1387 | | | | Copper to BS 2871 | |
Actual o.d (mm)	Weight kg/m	Nominal bore (in)	Light weight (kg/m)	Medium weight (kg/m)	Heavy weight (kg/m)	o.d.	Weight (kg/m)
16	0.1	$^3/_8$	0.67	0.84	1.02	15	0.28
20	0.13	$^1/_2$	0.95	1.21	1.44	18	0.38
25	0.18	$^3/_4$	1.38	1.56	1.87	22	0.53
32	0.28	1	1.98	2.41	2.94	28	0.68
50	0.69	$1^1/_2$	3.23	3.57	4.38	42	1.37
63	1.09	2	4.08	5.03	6.19	54	1.77
75	1.54	$2^1/_2$	5.71	6.43	7.93	67	2.23
90	2.23	3	6.72	8.37	10.3	76	3.13
100	3.31	4	9.75	12.2	14.5	108	4.47

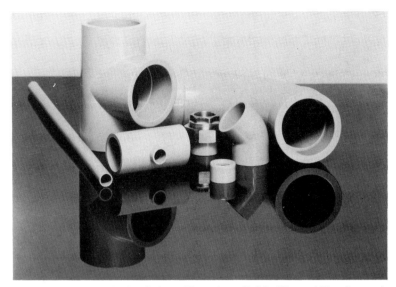

FIGURE 1 – ABS air line fittings. *(Durapipe – S+LP, Glynwed Pipe Systems)*

FIGURE 2 – ABS air line installation. *(Durapipe – S+LP, Glynwed Pipe Systems)*

Copper piping is not normally used for air mains, but because of its ease with which it can be manipulated and joined it is common for branch lines. The cost is higher than steam barrel or ABS. Care should be taken when forming bends; any of the usual bending techniques are permissible, but a smooth ripple-free inner surface is essential to minimise pipe friction.

Steel thin-wall pipes

Thin-wall tubes made of stainless steel or carbon steel are also occasionally used. Compression fittings in stainless steel or brass are used for joining.

TABLE 3A – Bundy tube – inch sizes

o/d (in)	i/d (in)	Minimum bend radius (in)	Max working pressure at 20°C (bar)
3/16	0.131	3/8	340
1/4	0.194	1/2	300
5/16	0.194	3/4	250
3/8	0.319	7/8	195
1/2	0.444	1 3/4	160

TABLE 3B – Bundy tube – metric sizes

o/d (mm)	i/d (mm)	Minimum bend radius (mm)	Max working pressure at 20°C (bar)
4	2.6	10	380
6	4.6	13	300
8	6.6	19	250
10	8.6	22	195
12	10.6	44	160

Bundy tube

Bundy tube is constructed from copper coated steel strip, rolled twice laterally and furnace brazed to produce a double wall tube with a copper bore and a tinned external surface. It has a higher working pressure than copper tube (Table 3) and can be joined with standard compression fittings.

Aluminium tube

This is a recent development, using extruded section anodised aluminium (Figure 3). The advantage of this material is that its smooth interior results in a low friction factor and

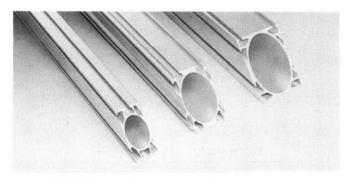

FIGURE 3 – Aluminium extruded pipes. *(Ingersoll Rand)*

consequently a low pressure loss compared with steel pipes (it is quoted as carrying up to 50% more air than steel); see Table 7. Its light weight at about one-third that of steel makes it easier to support. It is non-corrodible by the normal contaminants of an air system. An installation using this material has to have special connections which can be readily attached to the flat side of the extruded section. Because it is only supplied in straight lengths, changes in direction are sharp, which means that the pressure loss at a right angle bend is greater than for a steel swept bend.

Nylon tubes

Nylon tubing is commonly used to connect small hand tools to a branch line and for making up circuits in automation applications. It is available in flexible or semi-rigid forms; and manufactured from nylon 11 or nylon 12 to BS 5409. It can be used for working temperatures up to 80°C, or for short periods up to 120°C (after reference to the manufacturers). It can be obtained in a variety of colours as well as transparent. For use

TABLE 4 – Nylon tube to BS 5409:Part 1

Outside dia (mm)	Inside dia (mm)	Max working pressure at 20°C (bar)	Bend radius on centre line (mm)
4	2.5	26	25
5	3.3	24	30
6	4.0	24	35
8	5.5	22	45
10	6.9	22	60
12	8.5	21	70
14*	11.0	14	80
16	11.9	18	90
22	16.9	16	125
28	21.9	15	160

* 14 mm tube is not to BS 5409.

TABLE 5 – Temperature factors for nylon tube

Working temperature °C	Factor
–40 to 20	1.0
30	0.83
40	0.75
50	0.64
60	0.57
80 (maximum)	0.47

To calculate the working pressure at the given temperature, multiply working pressure at 20°C by the factor in the table.

outside or where it is exposed to daylight over a long period, black tube should be specified as this is ultra-violet stabilised.

Tables 4 and 5 give strength and size information for sizes up to 28 mm o/d, however for pneumatic circuits 12 mm is the usual maximum. Although only metric sizes are quoted here, inch sizes are usually also supplied as an alternative.

Two forms of pipe fittings can be used with nylon tubes: compression fittings as for copper pipes and push-in tube fittings. The latter are available for nylon tubes up to 12 mm o/d (some manufacturers also supply 14 mm). Connections to push-in type fittings are easily made and dismantled so they are very suitable for making up complex circuits which may require to be made and unmade several times. The maximum pressure rating of these fittings is 10 bar.

Nylon spring coils

Spring coils of nylon are also available. Their use obviates the need for trailing hoses or in any application involving a large relative movement between components. See Figure 4 and Table 6.

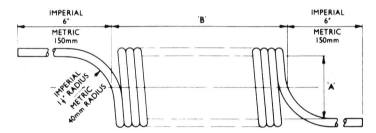

Nylon springs coils are manufactured from a special hard nylon with good mechanical strength and chemical resistance properties, ensuring suitability for a wide range of applications. Each spring coil is supplied with a leader at each end and will extend up to a maximum of 5 m (wound anti-clockwise).

FIGURE 4 – Nylon spring coils. *(Norgren)*

TABLE 6 – Data for nylon spring coils

Tube o/d	'A' bore (mm)	'B' closed length (mm)	Max working pressure at 20 °C (bar)
4	63	140	25
5	63	170	20.6
6	63	205	19.5
8	63	265	19.2
10	82	260	20.6
12	82	305	17.9

Terylene braided PVC hoses

Unreinforced PVC hose is not recommended for use with compressed air but when reinforced with terylene it makes a satisfactory alternative to nylon. End connections are usually swaged or made with O-clips. The working pressure is approximately the same as for nylon tube.

Rubber hose

This is probably the most common material for connecting tools and mobile equipment to a compressor or to a main airline. Rubber hose for airline use should meet the requirements of BS 5118, Types 1 to 3. The construction has a two-ply textile reinforcement in which cords of high tensile textile yarn, fully embedded in rubber, are laid spirally along the hose giving maximum strength and flexibility. For high pressure use, high tensile steel is used in place of the textile. Rubber hose with an outer thermoplastic cover is also available with an added resistance to abrasion and improved weathering properties. Rubber hose has a temperature capability from -30°C to 70°C.

Hose couplings

A variety of different hose couplings are available. For low to medium pressure applications quick action claw couplings are commonly used, Figure 5. Two types of these are found: the 'Universal', American Style with NPT threads as standard and the 'Q Type' European Style with BSP threads as standard.

The Universal Style is also known as 'Air King', 'Chicago Type', 'Lindy' and 'Crowsfoot'. It is interchangeable with the Chicago Pneumatic Type but not the European Type. The seal is by a captive rubber ring in the coupling. Finger-type safety clamps are recommended; these have an extended finger which locks over the security collar of the coupling. With the safety lock pin, the working pressure is 12 bar.

The Q Type meets German standards DIN 3482 and 3483. Twin bolt saddle clamps are recommended; these must match the hose outer diameter. Sealing is by a rubber ring. The working pressure is 7 bar. A safer variation of this is the 'Safeloc', incorporating a security collar and lug extensions, facilitating the use of a safety lock pin; this too has working pressure of 12 bar.

For high pressure use, ground joint (GJ) safety clamps are used. These are threaded connections of a knock-on design. The hose stem has a coned end which seats on the coned face of the spud; the spud has a copper insert giving soft-to-hard metal seating. Finger type clamps are recommended: the smaller size (up to 19mm) have two bolts; larger sizes have four bolts. Working pressures vary from 260 bar in the 12mm size to 70 bar in the 50 mm size.

Swaged and reusable hydraulic type hose couplings are also used, particularly for high pressure use.

Compression fittings

These fittings are usually manufactured from brass bar to BS 2874 and brass stamped components to BS 2872, Figure 6. The dimensions should be in accordance with BS 2051:Part 1 and may be used with copper, aluminium alloy, bundy and nylon tubes. Stainless steel fittings are also available.

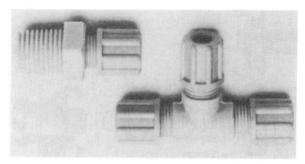

FIGURE 8 – Knurled plastic compression couplings. *(Drallim Controls)*

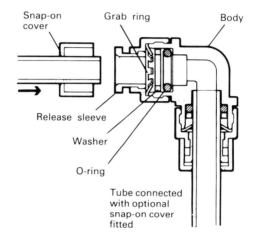

FIGURE 9 – Push-in tube fittings. *(Norgren)*

connections can be made and remade without the need of special tools other than a means of cutting the tube. One type, illustrated in Figure 9, has an acetal co-polymer body; it seals with an O-ring and has a steel grab ring to grip the tube; a special release tool is recommended for disengaging the tube from the grab ring. Another type, made up as a manifold, is illustrated in Figure 10; it has a brass body and a sliding collet with gripping teeth for positive anchorage of the tube. The fitting is disconnected by pulling back the collet and withdrawing the tube. This particular fitting contains a shut-off valve which is held closed by a spring, so that only those connections that have a tube inserted can pass air.

Figure 11 shows another type of push-in fitting, detailing the ease of assembly.

Quick release couplings

These couplings are used for connections which have to be made frequently, *eg* where a variety of tools are employed at one location and need to be plugged in alternately.

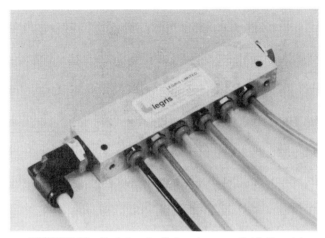

FIGURE 10 – Push-in tube fittings in a manifold. *(Legris)*

Method of assembly

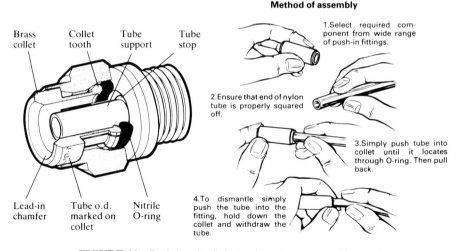

| Brass collet | Collet tooth | Tube support | Tube stop |

1. Select required component from wide range of push-in fittings.

2. Ensure that end of nylon tube is properly squared off.

3. Simply push tube into collet until it locates through O-ring. Then pull back.

4. To dismantle simply push the tube into the fitting, hold down the collet and withdraw the tube.

| Lead-in chamfer | Tube o.d. marked on collet | Nitrile O-ring |

FIGURE 11 – Push-in tube fittings with tube support. *(Norgren)*

Connections can be made and remade under pressure; the ends should be self sealing so that after disconnection the air cannot leak. Some types incorporate a bleed hole so that the down stream pressure is vented before disconnection; this eliminates one of the causes of whipping hoses. Couplings can be single or double cut-off according to whether one or both ends of the line are needed to be sealed. Most of the quick release couplings have a swivel action which reduces strain on the connection and thereby increases the life of the hose. See Figure 12.

They are made from brass, steel or plastic (usually Delrin). Plastic couplings have a pressure limit of 10 bar. Some of the couplings are supplied with a hose stem with a barbed

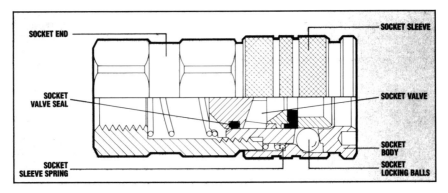

FIGURE 12 – Quick release coupling gives easy one-handed connection
of airlines. *(Hansen Couplings)*

tooth, which are claimed to make a safe connection merely by pushing into the end of the hose. This is not to be recommended, except for diameters below about 6mm. It is preferred to use a worm hose clip.

If one quick release coupling is to be used it should be fitted at the rigid end or next to the compressor, rather than at the tool. The reason for this is the danger of damage to the coupling by the vibration of the tool or merely by handling.

If a self sealing coupling is not used, a shut off valve should be fitted upstream of the hose connection to allow hoses to be changed.

Excess flow valves

One danger inherent in the use of flexible hoses is "hose-whip" which occurs when there is an unexpected downstream break in the line; the hose lashes about uncontrollably which can only be ended by turning off the compressor or isolating the affected part of the system. This can be readily cured by fitting an excess flow shut-off valve which safely stops the main flow. This valve should be employed in all situations where hose rupture due to mechanical damage is a possibility. Figure 13 shows such a valve. It works by cutting off most of the flow when the pressure drop at the valve is excessive. The flow that remains is insufficient to cause hose whip.

FIGURE 13 – Shown is an excess-flow automatic shut-off
valve which stops the main flow of compressed air
if the line is ruptured downstream. It is designed to help
protect personnel and equipment against the hose whip and
uncontrolled escape of air caused by air hose failure. *(Hansen Couplings)*

TABLE 7 – Aluminium pipe – properties

Bore (mm)	External dimension (mm)	Weight kg/m
12	14	0.16
25	28 x 49	0.906
32	36 x 50	1.179
50	60 x 60	1.978
63	68 x 74	2.946

Working pressure is 14 bar. Extrusion length 5m.
Maximum temperature 130°C.

REGULATING VALVES

An important item in a compressed air layout is a regulating valve which ensures that the pressure in the system (or a part of the system) is not exceeded. The first important function of these valves is for safety: the pipes, receiver and other items in the layout have a maximum operating pressure, which if exceeded could cause damage and eventually, if the pressure were allowed to increase indefinitely, result in an explosion. The second function is to provide for a part of the system to operate at a lower pressure than the main supply, *eg* for operating a blow gun, for paint spraying, for pneumatic sensing and for applying a known force to an air cylinder. A well designed valve must maintain a constant pressure despite changes in the primary pressure and the flow rate.

The compressor will normally incorporate its own unloading system which ensures that it operates at a condition which maintains the design system pressure. Further control may be required outside the compressor unit; regulating valves described in this chapter provide that control.

There are various types of regulating valves. Reference should be made to ISO 4126, Safety Valves – Control Requirements, for a full set of definitions, from which the following are derived:

Back pressure regulator. A valve having its inlet port connected to the system so that by automatic adjustment of its outlet flow, the system pressure remains substantially constant. System pressure must remain higher than the preset opening pressure of the valve. Under normal operating conditions the valve may be continuously flowing to waste.

Pressure reducing valve. A valve having its outlet port connected to the system so that with varying inlet pressure or outlet flow, the outlet pressure remains substantially constant. Inlet pressure must remain substantially higher than the selected outlet pressure.

Pressure relief valve. A valve having its inlet port connected to the system which limits the maximum pressure by exhausting fluid when the system pressure exceeds the preset pressure of the valve. Under normal operating conditions the valve remains closed, only operating when unusual system conditions prevail.

Safety valve. A name given to a type of pressure relief valve which protects the system from over-pressurisation in the event of a malfunction of a component forming part

of the system. It opens and exhausts fluid when the system pressure exceeds its
design pressure by a selected amount. The valve is required to have a certified flow
capacity sufficient to prevent the system pressure exceeding its design pressure by
more than a stipulated percentage or amount, usually 10% or 0.3 bar, whichever is
the greater. Under normal conditions a safety valve remains closed and in the event
of its functioning, the cause should be established before the system is put back into
operation.

Safety lock up valve. A valve with its inlet port connected to one part of the system so
that should the inlet pressure fall below a preset amount the valve will close to lock
a minimum pressure in that part of the system connected to the outlet port.

Sequence valve. A valve with its inlet port connected to one part of the system and its
outlet port connected to another part of the system and set so that until the inlet
pressure rises above the preset opening pressure the fluid medium is prevented from
flowing to the outlet port.

Types of pressure relief valves

A direct loaded relief valve is one in which the loading due to the fluid pressure is opposed
by direct mechanical loading such as spring; weight loading is rarely found in compressed
air systems. Direct loaded valves can be either poppet valves in which the sealing member
is directly loaded onto a seat by the spring and the system pressure acts directly on the

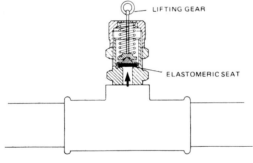

FIGURE 1 – Poppet relief valve with
elastomeric seat. *(Norgren)*

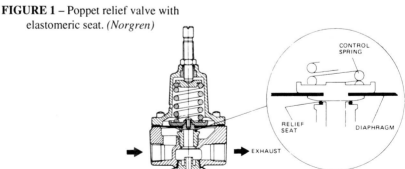

FIGURE 2 – Diaphragm relief valve. *(Norgren)*

seating member (Figure 1), or diaphragm valves in which the spring acts against a diaphragm exposed to the system pressure (Figure 2).

Poppet valves suffer from serious drawbacks. Their performance is inconsistent due to "stiction", they are insensitive because of the small seat areas and they have a limited flow capacity. Diaphragm operated valves eliminate some of the drawbacks of the simple poppet valves: because pressure is sensed over a larger area (that of the diaphragm) they are more sensitive and have a larger flow capacity. However direct operated diaphragm valves still have performance limitations, due to the size of the diaphragm in relation to the spring loading.

Most modern pressure relief valves, at least in the larger sizes, are pilot valves, of which there are two varieties. A 'pilot controlled relief valve' is a valve in which the loading spring is replaced by fluid pressure from a control pilot; the pilot pressure comes from a pressure regulator which is remote from the relief valve and has its own supply which may be independent of the main supply, see Figure 3. A 'pilot operated relief valve', usually a diaphragm type, is one in which a small integral spring-operated valve controls the opening of the main valve; system pressure normally keeps the main valve closed and no external supply is needed.

Pilot controlled relief valves have superior characteristics to those of direct spring systems. Pressure is applied to the control side of the diaphragm from a pilot regulator. A small bleed is necessary to prevent control variations due to temperature fluctuations. Relief valves controlled by pilot regulators do have shortcomings when used for system relief, but they are useful when the control has to come from a separate supply or when the relief pressure needs to be altered from time to time; a pressure gauge can be incorporated in the pilot line for pressure setting purposes. Unfortunately regulators are not fail-safe and if one has to incorporate another relief valve in the pilot line, the whole circuit becomes very complicated.

The integral pilot operated relief valve is now the one most commonly used. This is really two valves in one body: the relief pressure is set by the spring load on the pilot relief

CONVENTIONAL PILOT   **FEEDBACK PILOT**

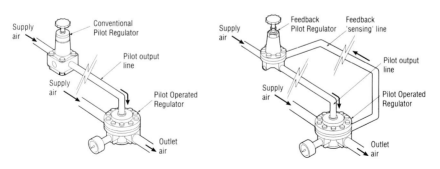

Feedback Pilot Regulators give more sensitive control and quicker reaction to downstream pressure changes.

FIGURE 3 *(Norgren)*

Operation

The unit has a main valve assembly (A) and an integral pilot assembly (B). The main valve assembly (A) incorporates a small orifice to allow air to reach the pilot diaphragm (B).

Primary air enters the inlet port, passes through the restrictor in the main valve assembly (A) and reaches the pilot diaphragm (B). If the pressure under the pilot diaphragm is sufficient to overcome the set spring pressure, the pilot valve will open and air will exhaust through the bonnet holes (C). This will cause the pressure above the main valve assembly (A) to drop creating a pressure differential across the diaphragm of the valve assembly. The main valve will now open and air will exhaust through the relief port (D). The desired relief pressure is set using the adjusting screw (E).

When the excess pressure has been exhausted, the spring force acting on the pilot assembly (B) will close the pilot valve. This will cause the air pressure above the main valve assembly (A) to build up and this, in turn, will close the main valve and prevent further escape of air.

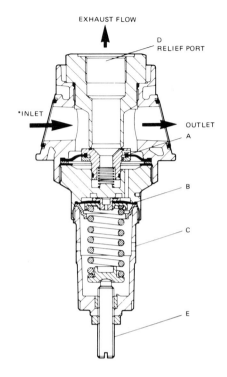

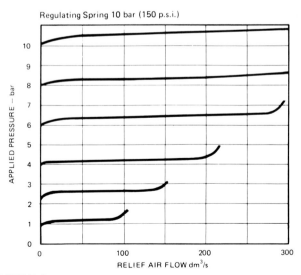

FIGURE 4 – Integral pilot operated relief valve. Above: operation.
Below: operating characteristics. *(Norgren)*

valve; the spring load is easily changed by adjusting a screw on the pilot valve. It can be seen from Figure 4 that, over its operating range, the pressure remains practically constant; such characteristics are greatly superior to a simple directly operated valve.

Installation considerations

Pilot operated relief valves are intended for in-line installation, but can also be used indirectly mounted, Figure 5. A relief valve must be selected to pass the maximum flow rate of the compressor or other source of air, but a valve of this calculated size may not conveniently fit into the main line, which will probably be of a larger bore. To install a small valve in a larger pipe will introduce an excessive pressure drop, so an indirectly mounted valve will be appropriate.

A typical example of excess pressure control is where a relief valve is included between the compressor and receiver to protect the receiver should the control system of the compressor fail. Sizing such a valve is fairly easily done, because the maximum flow rate of the compressor is known, but there may be difficulties in determining the size of a relief valve to install in a branch line. Usually the flow rate required in a branch is a small proportion of the total possible flow, yet if there is a failure of one of the control valves, such as a pressure regulator, the whole of the compressor output may try to pass through the branch. It would be uneconomic either to install a full-flow relief valve at every point in a system where protection was needed or to make every system component capable of resisting the maximum pressure.

There are several ways of dealing with this difficulty: one way would be to ensure that a restriction is inserted in the branch to limit the flow and use the orifice flow to size the relief valve; another way is determine the failure flow of the pressure regulator and size the valve to that. A third way, if the failure flow is unknown, is to take an orifice of one half the supply pipe size of the regulator, double the maximum orifice flow at the supply pressure and use that to size the relief valve. For example if a $G^1/_2$ regulator is used to control a supply pressure of 6 bar, the failure flow is taken as twice the flow through a $^1/_4$ in orifice at 6 bar, *ie* 2 x 45 l/s = 90 l/s. The actual failure flow of the Norgren $G^1/_2$ regulator is 80 l/s, so the rule works in this example. The best method undoubtedly is to

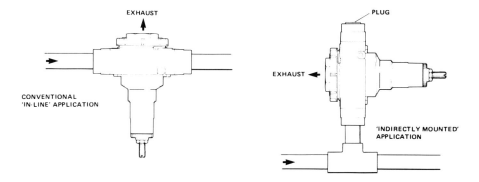

FIGURE 5 – Installation of relief valves. *(Norgren)*

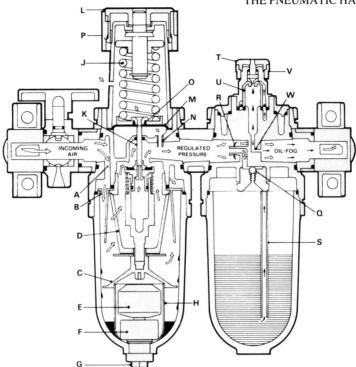

As compressed air passes through the shut-off valve and enters the filter section (A) it passes through the directional louvres (B) forcing it into a whirling flow pattern. Liquid particles and heavy solids are thrown against the inside wall of the bowl by centrifugal force. The liquid then runs to the bottom of the bowl where it is either removed by the automatic drain assembly or by the manual drain. The baffle (C) maintains a 'quiet zone' in the lower portion of the bowl to prevent air turbulence from picking up the liquid and returning it to the air stream. The air then passes through the filter element (D) to remove solid contaminants. The automatic drain assembly dumps the liquid as it collects. As the liquid builds up in the bowl the float (E) rises, causing the piston type automatic-drain assembly (F) to open and release the liquid under pressure. The drain mechanism is protected by a protective screen (H). From the filter, the compressed air goes to the regulator section.

When no load is applied to the adjusting spring (J), the regulator valve (K) is closed. As the knob (L) is turned, it applies a load to the spring which is transmitted to the valve through the flexible diaphragm (M) opening the valve. The air pressure side of the diaphragm is connected to the regulator outlet by syphon tube (N) so that regulated pressure is exerted against the diaphragm. As the regulated pressure increases, the pressure against the diaphragm increases, forcing the diaphragm to compress the adjusting spring until the load exerted by the spring is equal to the load exerted by the regulated pressure. Excess downstream pressure is vented through relief seat (O). The pressure setting can be locked with the snap action lock (P).

As the filtered and regulated air enters the lubricator a portion of it flows through the check valve (Q) to pressurize the bowl. The major portion of the incoming air flows through the lubricator past the flow sensor (R) which allows the lubricator to automatically maintain a constant oil density.

The combination of the pressurized oil reservoir and the pressure differential created at the flow sensor causes the oil to flow up the syphon tube (S). The drip rate can be set by adjusting knob (T) as it flows through the sight-feed dome (U). The setting can be locked with snap action ring (V).

All oil passing through the sight-feed dome to the orifice (W) is converted to constant density fog and continues downstream to the point of application.

FIGURE 6 *(Norgren)*

approach the manufacturers who should be able to quote the failure rates of their valves. If the constructional details are known, it may be possible to determine the equivalent orifice size by examination.

Care should be taken when siting a relief valve. Although it may not operate often, when it does it should not discharge a high pressure stream in a direction that may affect personnel or blow dirt, dust or other objects around; if possible it should be sited at high level away from personnel, or there should be an exhaust pipe leading the flow away. Tamper proof covers are available to cover the case of unauthorised adjustment of the relief setting. Silencers are available to fit the exhaust port.

Pressure reducing valves

In a complex compressed air system there are likely to be several sub-systems or branch lines that need to have a pressure limitation, lower than the main pressure, placed on them. Too high a pressure can result in energy wastage, and many tools and other equipment either work best (or only) at their design pressure.

A pressure reducing valve (commonly known as a pressure regulator) is used in these circumstances. It reduces a high variable primary pressure to a lower, constant secondary pressure; it measures and corrects deviations in pressure. A regulator should also include a bleed so that if the secondary pressure exceeds the setting of the valve, the excess pressure can escape. A regulating valve must possess certain characteristics for reliable operation. The valve must be fitted with a balanced poppet valve otherwise pressure fluctuations will cause instability. The controlled pressure must be practically constant for all flow rates up to the rated maximum.

It is customary to install a combined assembly of filter regulator and lubricator (FRL) to filter out the free water droplets, then control the pressure and finally lubricate the controlled air. This assembly is used to supply regulated air to a branch of the main supply. It is usually built up in modular form and available in this form from a number of manufacturers. Not all elements of the assembly are required in every case. Thus for example if the air has already been filtered and the moisture removed, there is no need for an additional filter. A commonly available modular unit is a combined filter and regulator, the function of which is shown in Figure 6, which also includes a description of a micro-lubricator. A filter regulator of this kind should also incorporate a pressure gauge to assist in setting the spring pre-load. The bowls of the filter and lubricator are often made of a transparent plastic, so that a visual check can be kept on the amount of the moisture and the lubricant. A metal safety guard can be fitted to enclose the polycarbonate bowl and all-metal equipment is also available. It is better, if the installation permits it, to have an automatic drain on the filter bowl, which then allows the fitting of a safer metal bowl.

Electronically-set pressure regulators are available for remote control of the pressure.

PRESSURE GAUGES AND INDICATORS

The SI unit of pressure is the pascal (Pa), which is the special name given to the newton/ m^2 (N/m^2). The pascal is only suitable for small pressures so kilopascal (kPa) and megapascal (MPa) are preferred for higher pressures. The practical unit for pneumatics is the bar, defined as 0.1 MPa, approximately equal to the atmospheric pressure at sea level. The millibar is used to express barometric pressures.

The bar is slightly different from the original metric unit of kgf/cm^2, or kp/cm^2, which is known in Continental Europe as the *technical atmosphere*, (at)

Summarising the relationships:

$$1 \text{ bar} = 1.019716 \text{ kgf/cm}^2 = 14.5038 \text{ lbf/in}^2$$
$$1 \text{ at} = 1 \text{ kgf/cm}^2 = 1 \text{ kp/cm}^2 = 0.980665 \text{ bar} = 14.2234 \text{ lbf/in}^2$$

The *standard atmosphere* (atm) is not used as a unit of pressure, but is in general use as a reference for pressure, equal to 760 mm Hg.

$$1 \text{ atm} = 1.01325 \text{ bar} = 1.033227 \text{ at} = 14.696 \text{ lbf/in}^2$$

In SI units it is not customary to add to the unit a modification to indicate whether absolute pressure or gauge pressure (*ie* pressure above atmospheric) is intended, so the usage has to be determined from the context. In the U.K. it has been customary to use the abbreviation psi for lbf/in^2, and add 'a' or 'g' to give psia or psig. Similarly in Continental Europe, a modifier can be added to the atmosphere, at, to produce ata or atü, for absolute or gauge pressure.

Pressures expressed as the height of liquid columns

Pressure is often measured in terms of the height of a column of liquid, usually mercury (Hg) or water (H_2O). Another unit of pressure is the torr, equal to 1 mm Hg, usually used in vacuum technology.

$$1 \text{ mm Hg} = 1 \text{ torr} = 13.5951 \text{ mm H}_2\text{O} = 1.33322 \text{ mbar} = 0.01934 \text{ lbf/in}^2$$
$$1 \text{ in H}_2\text{O} = 2.49089 \text{ mbar}$$

TABLE 3 – Road breakers and contractors' tools

Ambient temp.	Below 4°C	4°C to 32°C	Above 32°C
Shell	Clavus 15	Clavus 32	Clavus 68
B.P.	Energol HLP32	Energol HLP32	Energol HLP68
Castrol	Hyspin AWS10	Hyspin AWS32	Hyspin AWS68
Esso	Zerice 46	Zerice 46	Zerice 68
Texaco	Capella 15	Capella 32	Capella 68
Mobil	DTE 11	Almo 525	Almo 527
Gulf	Hydrasil 15	Hydrasil 32	Hydrasil 68
Elf	Elfrima 46	Elfrima 46	Elfrima 68

TABLE 4 – Rock drills

Shell	Tonna T32	Torcula 100 Tonna R100	Tonna T220
B.P.	Energol GHL32	Energol GHL68	Energol GHL220
Castrol	RD 32	RD 100	RD 150
Esso	Arox EP 46	Arox EP 46	Arox EP 150
Texaco	Rock drill 32	Rock drill 100	
Mobil	Almo 525	Almo 527	Almo 529
Gulf	Rock drill 32	Rock drill 100	Rock drill 320
	Gulfstone 32	Gulfstone 100	Gulfstone 320
Elf	Perfora 46	Perfora 100	Perfora 320

in the tables are suitable for the purposes indicated, however it should be emphasised that manufacturer's literature should be referred to for specific information. Suppliers other than those listed also have suitable formulations.

There are also some synthetic formulations based on glycol, which have been developed in conjunction with tool manufacturers with the aim of producing a lubricant which absorbs moisture, has a wide temperature range and decomposes without harm to the environment; these tend to be more expensive than mineral oils but have the advantage that because they are able to absorb larger amounts of water, it may be possible to dispense with the water separator. Care should be taken with the use of these lubricants, making sure that they are suitable for the application envisaged.

The lubrication requirements for small industrial tools and for cylinders and valves for automation installations are less severe than for contractor's and mining tools, and for these and similar applications an industrial lubricant meeting the requirements of ISO 3448 VG 32 will generally be satisfactory. See Table 1 for further recommendations. A warning should be given about the use of oils other than those recommended by the tool manufacturer or the supplier of the lubrication equipment. Compounds containing graphite, soaps or fillers should not be used. Certain compounds are not compatible with the transparent bowls of lubricators or the O-rings and seals of the equipment. Attention is drawn to BS 6005 – Specification for Moulded Transparent Polycarbonate Bowls used in Compressed-air Filters and Lubricators.

The problem of moisture condensation is made worse in the exhaust port of a tool, when the pressure drop is so sudden that the fall in temperature is sufficient to cause freezing of the water. This collects as ice in the passages and reduces the flow area, severely reducing the tool's performance and in severe cases stopping it altogether. There are special formulations of synthetic lubricants which cope with this problem, but they tend to be more expensive than mineral oils and to some extent they compromise some of the other desirable properties.

Lubrication methods

There are three methods of lubricating pneumatic tools:

- Integral oil reservoirs and feed devices
- Air-line lubricators
- Manual oiling arrangements

In the chapter on Contractors Tools, the integral type of lubricator is described. The reservoir has to be topped up every shift and is therefore dependent on the human element, which is its main drawback; so it is recommended that, in addition to this kind of built-in lubrication, an air-line lubricator be used.

Manual oil arrangements are seldom satisfactory. They consist of merely pouring oil into the inlet connection and hoping that a sufficient quantity will find its way to the working parts. In practice there is either too much at the beginning or too little at the end. But there is a more fundamental objection to this kind of lubrication. For air lubrication to be satisfactory, the oil must be present in the form of mist particles, so that it can be carried along through the pipes and hoses. Particles larger than about 2 μm tend not be carried along as a mist but are left deposited on the inner surfaces of the distribution system. So an ideal lubricator needs to inject fine droplets in the air stream, which are carried along with the air rather than deposited on the surfaces.

Irrespective of the method of lubrication chosen, an indication of the correct oil feed can be obtained by examining the exhaust air. It should be possible to see a faint oil mist at the exhaust port but no more. On a rock drill or breaker, the exposed portion of the drill steel should be covered with a thin oil film but no more. If the steel is wet with oil or if the outside body of the drill is covered, the amount of oil feed should be reduced.

Lubrication equipment

The best place to fit a lubricator is as close as possible to the equipment being served. Where lubricators are used to provide oil for linear actuators or where the direction of the air flow is reversed, one rule suggested is that the volume of the pipework between the lubricator and the cylinder should not exceed 50% of the volume of free air used by the cylinder per stroke.

There is a minimum air velocity below which a lubricator will not work satisfactorily, and if the velocity is too large there will be excessive pressure drop. The latter condition is hardly likely to occur if the lubricator is fitted into a pipe with the same nominal bore as its thread connections. The size of the lubricator should be matched to the flow rate of the tool being supplied and should have a volume capacity sufficiently large to contain enough oil for a complete shift. It must be regularly refilled; some lubricators provide a visual signal giving a warning when refilling is required; others shut off the air supply altogether.

Three basic types of lubricator design are available:

- Venturi suction with direct injection. This gives a fairly coarse fog where a generous supply is needed.

- Venturi suction with indirect injection giving a fine fog which is easily metered in small doses.

- Pneumatically powered miniature injection pump with a separate line to inject the oil directly into the inlet port of the tool or cylinder. With this system a central lubricant can feed several stations.

Figure 2 shows an air line lubricator of the first type, used in conjunction with rock drills and other mining equipment. It is placed immediately upstream of the tool. The amount of oil feed is varied by changing the feed valve; other similar types are adjustable by means of a screw. Another example is shown in Figure 3.

Figure 4 is a centralised lubrication system providing fine droplets of oil mist. The cabinet houses a filter, pressure regulators, pressure gauges and warning lamps. This system needs an electric supply, but the Micro-fog lubricator itself can be used as a stand-alone unit. The operation of the lubricator is shown in Figure 5. Air introduced into the lubricator pressurises the reservoir and operates a pump supplying oil to liquid level cup; this ensures that oil is drawn from a constant level, eliminating variations in output caused by changing levels in the reservoir itself. Air flow causes oil to be drawn from the cup via

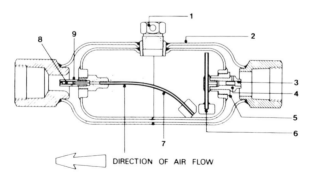

Key:
1. Filler plug.
2. Casing.
3. Components of pendulum.
4. Components of pendulum.
5. Components of pendulum.
6. Components of pendulum.
7. Feed tube.
8. Feed valve.
9. Feed valve body.

DIRECTION OF AIR FLOW

In this design the air stream passes through the space between the outer casing and oil container. When the connected tool is in operation there is a pressure drop at the outlet end. Through a small port live air is admitted into the container, exerting a pressure on the surface of the oil and forcing the oil past the controlling valve. A choice of feed valves is made available with each lubricator. Since the valve outlet is situated in the orifice throat, a venturi effect is induced which also serves to draw oil from the container into the air stream. The weighted flexible tube permits practically all of the contents of the lubricator to be used irrespective of its location, whilst the weighted pendulum ensures the admission of air to the space above the oil without allow a 'leak back' of oil.

FIGURE 2 – Typical line lubricator. *(Compair Holman)*

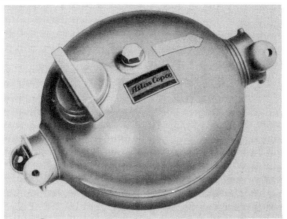

FIGURE 3 – Air line lubricator 1.3 litre capacity. *(Atlas Copco)*

FIGURE 4 – Lubrication system for large complex pneumatic installations. *(Atlas Copco)*

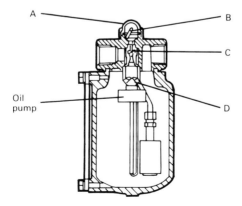

FIGURE 5 – Micro-Fog lubricator. *(Atlas Copco)*

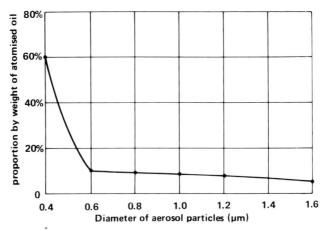

FIGURE 6 – Particle size distribution of the aerosol. *(Atlas Copco)*

the syphon tube B into the sight feed dome A, where it can be seen to drip onto the venturi plug C. The action of the air at the venturi plug causes a fine mist in the upper part of the reservoir, from where it is delivered to the distribution system. This system produces a particularly fine oil mist as shown in Figure 6.

Another type of centralised lubrication system is exemplified by the oil injector pump in Figure 7, which is able to supply oil in pulses direct to the tool. It is actuated by the air pulse generated when the tool is started. Oil is supplied from the main dispenser, either into the air line or to a lubrication point in the equipment. This equipment can be installed to supply oil to a single tool or to an unlimited number of lubrication points; it is suitable for fixed installations only.

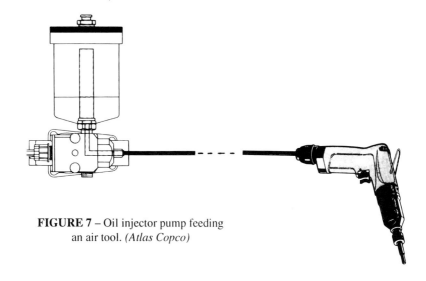

FIGURE 7 – Oil injector pump feeding an air tool. *(Atlas Copco)*

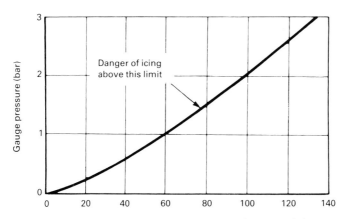

FIGURE 8 – Icing limits for adiabatic discharge of air.

Anti-freeze injectors

In winter conditions, ice formation can be a serious problem, although it occurs also when the ambient temperature is above freezing. Figure 8 shows that, when the exhaust pressure is high, icing can occur at surprisingly high air temperatures. The incidence of ice formation seems to be related to the presence of moisture in the air as much as to temperature. Measures such as injecting glycol with the oil or rinsing out equipment with alcohol should on no account be used. The best method is the use of dry air but it has to be admitted that this may not be a practical solution, on cost grounds alone, particularly if the problem only occurs during winter. If the problem is only in the equipment exhaust, a moisture trap immediately upstream (where the temperature is likely to be closest to ambient) should first be tried.

Another technique, where freezing in the distribution system is the problem, is the installation of an alcohol dispenser immediately downstream of the receiver, where the air is warm (but not above 40°C). The alcohol evaporates and condenses in the water traps, preventing the water which is there from freezing. If the amount of alcohol added does not exceed 4 – 8 drops (0.01 – 0.02 cm per m of free air), then practically no alcohol will get to the tools where it would wash away the lubricant. A conventional air line lubricator, adjusted to give the recommended injection rate, should be adequate to provide the required degree of protection, or a special proprietary injector can be obtained.

Supplier	Lubricant	Remarks
Demag	AES 82	Needs special injector (10 litre capacity) mounted on compressor
Atlas Copco	AIR-OIL	As above (2 -5 litre capacity)
Kilfrost Ltd	Kilfrost 400	No special lubricator. Use in standard lubricators.

There are also available special anti-frost synthetic lubricants based on glycol with inhibitors, intended to prevent icing in the tools. The following products are available.

These lubricants give protection down to -60°C. They possess some other desirable properties, such as being bio-degradable and emulsifying. They are capable of absorbing up to four times their own weight of water. Particular care should be taken if used in a confined area with poor ventilation as there may be health hazards with prolonged exposure. They are more expensive and may not be satisfactory in all circumstances, so they should only be used after reference to the tool manufacturer.

COMMISSIONING AND SAFETY

Pressure testing of the system

After the installation of the main and branch lines and immediately before connection of the compressor, the whole system should be subjected to a hydraulic test (making sure that the water has an added anti-corrosion agent) to 1.5 times the working air pressure, and all leaks remedied. The date of the test should be recorded and attached to the pipe system where it connects to the air receiver. This test should be repeated, for convenience at the time the air receiver is similarly checked, and when substantial changes are made to the layout of the system. After test, the whole system must be thoroughly blown out with low pressure air.

The reader is referred to the BCAS publication "Guide to the selection and installation of compressed air services" which gives useful information and data. It should be used to complement the information in this chapter. There are also a number of publications of the Health and Safety Executive which give information specific to industries using compressed air. A list of these and other safety publications are given at the end of the chapter.

General safety guidelines

- Compressed air can be dangerous unless straightforward precautions are taken. These are mostly common sense, but nonetheless worth listing. In a place where compressed air is used, consideration should be given to placing these or similar guidelines in a prominent place.
- Only approved pressure vessels built to the correct standards should be used. Refer to the chapter on Air Receivers.
- It is essential that a non-return valve and shut-off valve are fitted into the delivery line when the compressor is to be coupled in parallel with another compressor or connected to an existing air supply system; in such cases a safety valve must be provided upstream of the valve, unless one is already fitted to the compressor.
- Distribution pipework and hoses must be of the correct size and be suitable for the expected maximum working pressure. The tightness of compression fittings must be checked regularly.

- Do not use frayed or deteriorated hoses; always store hoses properly and away from heat sources or direct sunlight. A hose failure can cause serious injury.
- Use only the correct type and size of hose end fittings and connections; use only clamps or robust construction made specially for compressed air and rated at the maximum pressure in the system.
- Use eye protection if using compressed air for cleaning down equipment. Do so with extreme caution. Take care never to blow dirt at people or into machinery. Always use a blow gun never an open hose.
- When blowing through a hose or airline, ensure that the open end is held securely; a free end will whip and cause injury; open the supply cock carefully and ensure that any ejected particles will be restrained. A blocked hose can become a compressed air gun.
- Never apply compressed air to the skin or direct it at a person; even air at a pressure of 1 bar can cause injury. Never use compressed air to clean clothing.
- Do not use air directly from a compressor for breathing purposes, unless the system has been specifically designed for such a purpose and breathing air filters and regulators are fitted – see chapter on Breathing Air Filtration.

Precautions during commissioning

- Before installing air line filters, pressure regulators and lubricators, low pressure air should be blown through the system to remove any foreign matter.
- Before removing any blanking flanges and plates from the compressor, it is important to make sure that the pressure in the compressor is released.
- If an isolating or non-return valve is fitted into the compressor discharge, it is essential to check that an adequate safety valve is fitted between this valve and the compressor and that the isolating valve is open.
- Before starting up the compressor plant it is essential to carry out the instructions issued by the supplier in respect of the initial charge of lubricating oil (if any), the setting of the lubricator feed (if a force feed lubricator is used), and the cooling water supply (if applicable).
- Before starting any machinery, all protective guards should be in position and secure.
- On the initial start, the directions of rotation of the compressor must be checked; severe damage may be caused if the compressor is allowed to turn in the wrong direction.
- Before connecting any equipment to an airline lubricator, the pipe or hose should be allowed to discharge freely to atmosphere until visible traces of lubricant are obtained; then equipment can be connected and the lubricator adjustment set. Initially, oil from the lubricator does little more than line the pipes.
- During the first few hours of operation, the performance of the plant should be monitored carefully, in particular the operation of any automatic drains from the intercoolers and aftercoolers.

Precautions during running

- It is always good practice to keep a log of oil consumption, temperature and pressure, interstage and final temperatures and pressures and water inlet and outlet temperatures, so that any deviation from normal running may be noticed quickly. When an abnormality is noticed, the compressor must be stopped and the matter investigated. Power consumption should also be recorded.
- Suppliers' recommendations regarding filters and elements should be observed during and after the initial running-in period.
- The system should be checked regularly for leaks (see the chapter on Compressed Air Distribution).
- Never operate any part of a compressed air system – compressor, pipework or pneumatic appliance – at a higher pressure than for which it was designed or rated. If any equipment has a design pressure less than the maximum output pressure of the compressor, it shall be protected by suitable fail-safe means against over pressure.
- Shut off the adjacent upstream isolating valve and release the air pressure before disconnecting a hose or line, unless there is an automatic valve to give protection at the upstream joint.

Precautions during operation

- Keep doors shut on silenced plant. On other plant, check whether it should be run with doors open or shut.
- Check that all pressures and temperatures are correct; refer to the operating instructions.
- Stop the plant if warning lights show or if gauges register outside the normal limits; untrained personnel must not attempt adjustments. Call in a plant fitter to investigate.
- Do not make adjustments inside the canopy when the machine is running, unless appropriate precautions are taken.
- Do not remove guards.
- Do not use the machine in a fire hazard area unless it is suitably designed; do not operate in the vicinity of toxic fumes.
- Ensure that all pneumatic control equipment and air line accessories are operated below the maximum rated pressure and temperature.

Precautions during maintenance and overhaul

- Ensure that all air pressure is completely released from the system and that it is isolated from any other air systems. In multi-compressor installations, repairs to any of the compressors shall not be carried out whilst the other compressors are working, without first closing the isolator valve and venting and disconnecting the delivery pipe of the compressor to be repaired.
- Ensure that a machine cannot be started inadvertently; isolate the unit and lock the isolator in the safe position; place warning notices on isolator.

- Ensure that any door that opens upwards is securely fastened when open, and that no door can slam shut.

When working on plant:

- Use correct lifting gear of adequate capacity.
- Examine condition of lifting gear before lifting plant.
- Use correct tools for the job.
- When using a chemical or solvent cleaner, follow the manufacturer's instructions. Be sure that any fluid used cannot cause any chemical reactions or explosion in combination with high pressure. Do not use halogenated hydrocarbons on equipment where aluminium or galvanised parts come into contact with solvent or coating material. Trichloroethane and methylene chloride react violently with such parts, causing corrosion and very high pressures if entrapped.
- Do not weld or perform any other operation involving heat near the electric or oil systems. Oil and fuel systems must be purged completely, for example by steam cleaning before any welding repairs are carried out.
- Do not weld or in any way modify any pressure vessel.

Precautions before clearing the machines for use

- Check that the sump is filled with the correct grade of oil. Before using synthetic lubricants ensure that they comply with the compressors manufacturer's instructions and are compatible with all downstream equipment. The long intervals between oil changes permitted by synthetic lubricant manufacturers may lead to neglect of essential servicing of compressors, including filter changes, unless the maintenance schedule is strictly followed.
- On water cooled units check that the coolant is flowing.
- Check direction of rotation.
- Check that the operating pressures, temperatures and speed are correct and that the controls and shut down devices work correctly.
- Every six months examine the discharge pipe and discharge pulsation damper for carbon deposits; if excessive, the deposits should be removed.
- Isolating valves which should be of the self-exhausting type, should be designed for locking in the "off" position. When servicing pneumatically-operated machines, lock the isolating valves so that air pressure cannot be applied inadvertently while the machine is being worked on.
- Maintain filters in accordance with the supplier's instructions, as a reduction in air pressure caused by choked filter elements could result in a malfunction.
- Inspect hoses, flexible lines and plastic pressure-containing components at regular intervals and replace if signs of cracking, crazing, or any form of mechanical damage are evident. If sight glasses or filter bowls are becoming obscured due to internal contamination or scratching, these should be cleaned or replaced.

Portable compressor safety

In addition to the general rules listed above, the following apply to portable compressors:

Towing the compressor
- Check the towbar, braking system and coupling. Check that the wheels are secure that tyres are in good condition and correctly inflated.
- Connect lighting cables to towing vehicle if applicable. Check correct operation of lights.
- On two-wheel plant, raise propstand and jockey wheel fully and lock.
- Ensure breakaway cable is secured to towing vehicle.
- Release the hand brake.

Before starting the compressor
Check that:
- The machine is secure with brakes applied and level within 15 degrees.
- The machine is clean internally.
- All air pressure is released from the machine.
- All hoses and tubing are in good condition, secure and not rubbing.
- No fuel or coolant leaks.
- All fasteners are tight.
- Oil and coolant levels are correct; top up only with correct oils.
- All electrical leads secure and in good condition.
- Fan belt tension correct.
- All guards in place and secure.
- Engine exhaust system in good condition, and no combustible material lying on or against it.
- Start and stop procedures must be understood and carried out. Before starting close the air discharge cocks. Refer to the operating instructions.

During operation
- Do not fill with fuel while the machine is running. Keep fuel away from hot pipes.
- Ensure that the engine exhaust is freely vented to atmosphere.
- Do not remove the oil filler cap or the radiator cap.
- Do no adjustments inside the canopy when the machine is running, other than where instructed.
- Do not disconnect the battery.

After stopping
- Ensure that all air pressure is released from the system.
- Allow radiator to cool before removing the radiator cap. Allow engine to cool before adding coolant.
- Remove starting key if fitted and close panel doors.

Maintenance and overhaul

Before starting work on any machine:

- Ensure that all air pressure is released from the system.
- Stand the plant on a level ground.
- Apply the brakes. Securely fix propstand and jockey wheel on a two wheeled plant.
- Chock wheels securely if jacking-up or if working on the brakes.
- Support towbar on a two-wheel plant.
- Support axles securely if working underneath or removing a wheel. DO NOT RELY ON JACKS. Support tow bar on a four-wheel plant.
- Ensure that the machine cannot be started inadvertently.
- Disconnect battery.

When working on the plant:

- Use lifting gear of adequate capacity.
- Examine the condition of the lifting bale before lifting machine by it.
- Use the correct tools for the job.

Maintenance and repair and tools

- Disconnect the machine from the air supply before doing any work on it.
- For dismantling, hold the machine firmly in a vice or fixture.
- Use the correct tools for dismantling and assembling.
- When using a solvent or chemical cleaner, follow the manufacturer's instructions.
- Ensure that any components that require replacement are those recommended by the manufacturer.
- Before clearing the machine for use, make sure that it has been assembled correctly with all fasteners correctly tightened.

Use of machines and tools

- Use only approved tools. Make sure that the tool shank is the correct size for the machine.
- Do not use blunt tools or those which are worn excessively or damaged in any way on the shank or stem. **A TOOL WHICH BREAKS IN USE CAN CAUSE SEVERE INJURY**.
- Do not use frozen tools. In freezing conditions, store tools undercover, preferably in a warm building. Freezing temperatures cam make hardened steels brittle and cause breakage.
- Lubricate the machine as instructed. Use an approved lubricant.
- Blow through the air supply hose with compressed air before connecting to the machine.
- Connect the hose to the machine before turning on the supply. Ensure that the machine controls are in the "Off" position.
- Position the machine correctly and hold it firmly before operating the controls.
- Persons operating, assisting or working near the machine should position themselves so that they will not be struck or lose their balance if the machine slips or if the drill steel or tool breaks or sticks.

- Do not operate the machine from an insecure footing or staging.
- Where dust is a hazard in rock drilling, use the appropriate dust suppression methods, either dust collectors in quarrying or wet drilling underground.
- **EXPOSURE TO EXCESSIVE NOISE CAN DAMAGE YOUR HEARING.** Wear ear muffs or plugs when operating noisy equipment. The use of silenced equipment, even to EEC standards is no guarantee that the hearing will not be damaged with continued use.
- Noise reduced mufflers fitted to machines lessen the health hazard and reduce environmental noise. Detachable mufflers must be fitted correctly and replaced when damaged. On no account modify the exhaust ports.
- When using vibrating tools continuously and particularly in cold weather, wear suitable protective gloves and keep the hands warm.
- Use vibration reduced equipment where available. Avoid holding the tool bit to guide the work without the use of gloves. Do not start work until the hands are warm. These precautions will help to reduce the incidence of vibration induced white fingers (sometimes known as occupational Raynaud's disease).
- Always turn off the compressed air supply and release the air pressure in the supply hose before changing the tool, removing the oil filler plug or disconnecting the hose.
- When handling lubricants regularly, wear suitable gloves of impervious material. Clothing contaminated by lubricants should be changed.
- Should any lubricant be ingested, seek medical advice immediately.
- When using pneumatic tools, ensure that the area is adequately ventilated. Concentration of oil mist in the air can be dangerous.
- Do not operate a machine at full power without the drill rod or breaker steel securely retrained in the chuck and the steel or bit in contact with the ground.

Safety publications

- For Safety Regulations on the use and manufacture of pressure vessels refer to the chapter on Air Receivers

The following publications should be referred to for detailed advice on safety:

- The Factories Act 1961. Sections 36 and 37 deal with air receivers and are of particular interest to users of compressed air equipment.
- Control of Pollution Act 1974. Section 68 deals with noise from plant and machinery
- Health and Safety at Work Act 1974. Section 6 deals with the general duties of manufacturers and suppliers of equipment. Under Section 15 of the Act, the Secretary of State may make Health and Safety Regulations and a number of these have already been made, in particular:
- Noise at Work Regulations SI 1989/1790
- Control of Substances Hazardous to Health Regulations SI 1994/3246.
- Protecting Hearing at Work, published by CBI.

The following books published by Health and Safety Executive and obtainable from HSE Books contain useful material for those involved in the use of pneumatic equipment:

- Safety in pressure testing GS 4.
- Safety in the use of abrasive wheels HS(G) 17.
- Compressed Air Safety HS(G) 39.
- Noise at work HS(G) 56 Noise Guides 1 to 8.
- Hand Arm Vibration HS(G) 88, gives useful advice on the purchase of equipment and means of reducing vibration.
- The assessment of pressure vessels operating at low temperature HS(G) 93.
- Control of Noise in Quarries HS(G) 109.
- Noise from pneumatic systems PM 57.
- Occupational exposure limits EH40. To be used with the Control of Substances Hazardous to Health Regulations. These limits are updated every year and so the latest issue should be used.

Other miscellaneous publications:

- Compressed Air Safety. Guidelines for the safe use of compressed air in quarries, published by British Quarrying and Slag Federation.
- BS 4575-2 Code of practice for pneumatic equipment and systems.
- BS 6244 Code of Practice for stationary air compressors.
- Recommendations for the Proper Use of Hand-held or Hand-operated Pneumatic Tools. Pneurop Publication, available from BCAS.
- A guide to compressor noise reduction BCAS.

SYSTEM MAINTENANCE

Maintenance requirements can be considered under three headings
- Tools and appliances.
- Systems.
- Compressors.

They are equally important, but require different maintenance intervals. The life of a pneumatic tool is likely to be lower than that of a compressor. Often a percussive tool, for example, will have a commercial life no longer than that of the contract on which it is intended to be used, perhaps a year or 18 months, whereas a compressor may survive for many years, passing from contract to contract. This is bound to be reflected in the attention that is given to planned maintenance. However if it is important that tools should operate with the minimum of trouble, the proper degree of maintenance must be established for each of them.

For underground drilling, where maintenance facilities are limited and where drilling performance in terms of overall cost per metre of drilled tunnel is a crucial factor in mining costs, the fitter responsible for ensuring that the tools are kept at peak performance is known as the "drill doctor", which indicates the importance of his work.

Air tools

A programme for air tool maintenance should cover the following points:
- Ensuring that the air supply is maintained at the required pressure with the required delivery.
- Regular checking and refilling of lubricators to ensure adequate lubrication. This may require attention more than once per shift, so the operator should be supplied with equipment and lubricant for this purpose.
- Checking tools for damage.
- Checking that the tools are handled in a proper manner, that they are cleaned and lubricated after use.
- Checking that the proper tool bit or attachment is kept sharp and in good condition.

The most common cause of early troubles with workshop tools is lack of lubrication. The next is abuse, such as tools being dropped or dragged around by the hose, or generally

TABLE 1 – Air tool maintenance points.

Check of fault	Likely cause	Action
Dirt of gummy deposits on motor	Insufficient or wrong oil; dirty air	Check lubricator Check line filter
High blade wear on motor	Insufficient oil; normal wear	Reduce periods between scheduled maintenance
Bearing play	Excessive load; normal wear	Check tool capacity for job; replace bearing
Bearing rough	Dirty air Lack of lubrication Normal wear	Check filter Check lubricator Replace bearing
Governor action sluggish	Air leakage	Check governor bushings for wear
Loss of power	Low air pressure Air leakage Governor fault	Check line pressure & filter Check and rectify Check governor condition; adjust action if necessary
Vibration	Breakage	Replace faulty part(s)
Loss of speed	Low air pressure Air leakage Governor fault	Check line pressure and filter Check and rectify Check

misused. The design of the work area itself and the allocation of racks for holding tools when not in use can be a form of preventive maintenance. The use of incorrectly sharpened tool bits in the case of a chipping hammer or drill, or a badly dressed wheel in the case of a grinder, can result in an unnecessarily severe reaction on the tool and a correspondingly heavy force having to be applied by the operator; these will reduce the life of the tool.

Scheduled maintenance should preferably be the main responsibility of a nominated individual or section, with the operator bearing his share of that responsibility; he must be aware of the necessity for proper lubrication and simple care of the tools in his charge. Records of each tool should be kept to ensure that they are withdrawn from service at regular intervals for stripping, inspection and replacement of worn or damaged parts. The manufacturer's schedules of maintenance should be strictly adhered to, with a copy placed in every workshop. This applies to all kinds of tool. A record for each tool referenced against the operator will form a guide as to which individuals are responsible for the greater damage and misuse, or which operations are badly matched to the task they have to perform.

In the case of major repairs, the tools may require to be returned to the manufacturer. With good preventive maintenance and efficient scheduled maintenance, a major breakdown is unlikely until the tool nears the end of its natural life. The value of a tool is represented by its initial cost less the amortisation rate, and if the major repair costs exceed the value of the tool, it would be better to replace it with a new one. This is particularly true of workshop tools.

Premium quality service through premium quality people

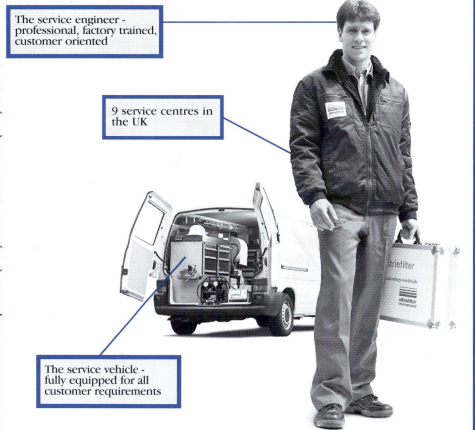

The service engineer - professional, factory trained, customer oriented

9 service centres in the UK

The service vehicle - fully equipped for all customer requirements

For more information. Telephone: (+44) 01827 58234 Fax: (+44) 01827 310166

ultrafilter
international
ultrafilter and more

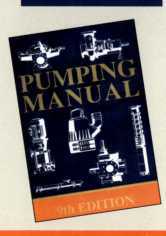

Appliance maintenance

Overhaul of air-operated appliances requires the following precautions:
- All air pressures to be completely released from the system.
- It should be impossible to start the appliance inadvertently.
- The correct tools must be available.
- Adequate lifting gear should be available and checked to be in good condition.

After maintenance and before clearing the machine for use, check:
- The operating pressures, speed and working temperature are correct.
- The controls and shut-down devices work correctly.
- That the sump or oil container is filled with the correct grade of oil, where appropriate.
- That there is a full flow of coolant, where appropriate.
- Condition of exhaust or discharge pipe and pulsation chamber. Clean off any deposits.
- Clean the outside of the appliance.
- Record the work done and replacement parts.

Air cylinder maintenance

Air cylinders require little in the way of maintenance if properly used, except for periodic replacement of expendable parts, such as seals, boots and gaiters. The items to be checked include piston seals, end cover seals, rod seals and bearings.

TABLE 2 – Trouble-shooting – air cylinders.

Fault	Cause	Action
Loss of thrust*	1. Leaking piston seals	Replace seals
	2. Corrosion on bore	Refinish bore as necessary to eliminate wear on seals
	3. Dirt trapped in seals	Replace seals. Clean cylinder. Check condition of rod wiper seal. Fit external gaiter if necessary for better protection.
	4. Excessive friction	Check condition of hose, seals, wear rings and rod bearing
	5. Low pressure	Check supply pressure at inlet
Leakage past rod	Rod seals faulty	Replace seals unless compression type seals are fitted in an adjustable gland. In that case, tighten gland.
Loss of cushioning	1. Cushion valve blocked	Remove and clean
	2. Cushion seals failed	Replace cushion seal(s)

* Loss of thrust can also result from substantially increasing the speed of working with no fault present in the cylinder or system.

The servicing interval will depend on the frequency of operation, the total number of operations and the working conditions. Scraper rings and, in poor environmental conditions, external gaiters together with adequate lubrication can protect the inside of the cylinder against abrasive atmospheric contaminants; the lubricant must be resistant to water washing. Modern cylinders are pre-coated during assembly with a lubricating compound which adheres to the surface for a considerable time. This helps to extend the servicing period but should not be used as an excuse not to supply lubricant in the air.

Seal wear and seal life depend very largely on the condition of the bore and the rod. Corrosion, pitting, abrasion or damage to these surfaces can result in high seal wear and premature failure. Intermittent use, with long periods of idleness during which the seals may dry out is also damaging. Most seal failures can be traced to bad initial fitting, the wrong type of seal or the wrong material specification.

Piston seal leak is not always easy to detect, since the first evidence may appear as leakage past the exhaust ports of the control valve, when a valve seal, rather than a piston seal may be suspect. A simple check is to disconnect the exhaust lines from the cylinder to the valve. If there is leakage through that line, the piston seals are faulty; if there is no leakage, the valve is probably faulty.

Tie rod construction makes cylinder disassembly easy. A point to watch with spring return cylinders is that one end cover will be under spring load and will require careful backing off until the spring pressure is relieved. On long stroke spring return cylinders, it may be advantageous to extend the lengths of the side rods to accommodate the full uncompressed length of the spring with the cover still on the side rods.

Replacement of the piston seals may not require disassembly of the piston, depending on the type of seal fitted, and the manner in which they are retained. Precautions should be taken when refitting seals to make sure that the correct tools are used to avoid stretching, nicking or otherwise damaging them. Always replace static end seals with new ones when they are disturbed. A thorough lubrication of the internal surfaces of the cylinder and the static seals is to be recommended.

System maintenance

Scheduled periods for the inspection of drains, moisture traps and separators should be based on experience with the components and the environmental conditions and the efficiency and type of compressor. Manufacturers recommendations should be used as an initial guide.

Filters and filter separators at individual tool stations must receive regular attention, particularly if they are of the manual drain type. Where practicable the automatic or semi-automatic type should be installed; the semi-automatic drain operates when the air supply is removed. The level of contaminants will normally be visible in the filter bowl if it is a transparent polycarbonate type or in a sight glass if of metal construction.

A tell-tale operated by pressure drop through the filter operates when the contaminant level is excessive, but it should be drained well before that.

All types should be cleaned or replaced in accordance with the manufacturer's instructions. Some can be cleaned by blowing through, others need replacement elements. If a filter should be damaged during servicing it must be replaced. Note that filter bowls

TABLE 3 – Maintenance points – pipework.

Type of line	Possible cause of trouble	Action
Rigid line system	1. Dead weight of piping	Support with pipe clips, closer spaced on horizontal runs
	2. Expansion and contraction	(a) use clips which allow lateral movement of pipes (b) incorporate bends to accommodate movement where there is a large difference in ambient temperatures
	3. Internal pressure	Provide adequate support as necessary to prevent movement and flexing
	4. Leakage	(a) all pipe joints to be properly made (b) replace faulty valves or fittings (c) if caused by damage, check environmental conditions; resite or protect vulnerable fittings
	5. Water entrainment	Check that suitable water traps are provided and that their location is satisfactory. Establish suitable draining period and check period for manual traps. All loop bends should be upward facing, not downward
Flexible lines	1. Leakage	(a) check for wear or deterioration at end couplings (b) resite or protect hoses subjected to bad environmental conditions (c) lay down standard procedure for manipulation of flexible hoses (d) consider possible use of automatic close-coil flexible lines
	2. Excessive pressure drop	(a) check hose for bore condition (b) check that suitable size of hose is being used.

made of polycarbonate should only be washed in warm water with a detergent, not with a solvent.

Clogged filters are the most common reason for pressure drop at the point of use. Air leaks or constrictions in the flexible lines are less obvious causes of pressure drop; the former can usually be heard, with the tool connected but switched off. Kinks or damage that has not produced puncturing show in the form of variable performance as the tool position is changed.

Less obvious sources of pressure drop may occur further back in the system. With adequate line-sizing and all connections in good condition, the pressure drop between receiver and tool station should not exceed 0.4 bar. Additional losses at the tool itself should not exceed a further 0.4 bar. About 10% of the system pressure is acceptable as an overall pressure loss.

Regardless of the pressure loss as installed and originally worked, any further loss noticed is a cause for concern and should be immediately attended to. Do not increase the setting of the regulator pressure to compensate for the pressure loss, otherwise when the system faults are corrected, the pressure may be too high. Loss of pressure not only reduces the tool working efficiency, but may be expensive in terms of air wastage if due to a leak. Leakage may develop at joints, at hose connections or in traps where the valve has been jammed open by contaminants. Excessive line pressure can be as bad as lack of line pressure. The latter reduces the working efficiency of the tool, often proportionally more than just by the pressure reduction, while the former can result in accelerated wear and over-stressing of parts, with no increase in performance.

Lubricators at tool stations need regular attention to ensure that an adequate supply of oil is maintained. Only the specified oil should be used, not only because an oil of different viscosity may affect the lubrication rate, but also because an unapproved oil may not have the qualities necessary for a tool lubricant. Refer to the chapter on Lubricants.

Pipe work should require minimum attention, usually only inspection for possible leaks. Installation faults needing correction are:

- Sagging or displaced pipes requiring proper support.
- Evidence of whipping at bends or vibration due to inadequate support during pressure surges.
- Distortion of runs caused by expansion or contraction movement, which can be cured by incorporating bends or expansion joints.

Compressor maintenance

While a general indication can be given of the principles of compressor maintenance (see Tables 4 to 8), it is better to rely on the manufacturer's schedules where available. Basic requirements include:

- cleanliness (regular cleaning of intake filters);
- attention to drain points on the compressor, intercooler and receiver;
- ensuring an adequate supply of lubricant;
- correct tensioning of intercooler and main driving belts;
- routine oil changes in accordance with the manufacturer's recommendations;
- cleaning or replacement of intake filters;
- inspection and cleaning of valves on a reciprocating compressor;
- inspection and replacement of vanes on a rotary vane compressor.

Compressor manufacturers normally specify comprehensive maintenance requirements and run courses for skilled personnel responsible for that work. They will also issue recommended lists of spares to be carried and have service kits available with installation instructions. It is always advisable to buy approved components from the same source as the original purchase.

TABLE 4 – Maintenance Points - reciprocating compressors

Component	Interval†	Action
Air filter (inlet)	Fortnightly	Clean as necessary
Cooling system	Continuous As necessary	Check water temperatures Clean water side on water-cooled systems
Lubrication	Periodically†	1. Check cylinder oil feed rate 2. Check oil levels 3. Change crankcase oil
Bearings	Periodically†	Check wear. Adjust for wear or replace as necessary
Drains	Hourly† Daily†	Intercooler drains Other drains or all automatic drains
Valves	Periodically†	Check for condition and replace plates or springs as necessary
Safety valves	Periodically†	Clean and inspect
Piston rings	Annually	Inspect for condition and replace as necessary
Packing glands	Periodically†	Adjust if applicable; otherwise inspect for condition and replace if necessary

†Suitable intervals can only be established by experience or on manufacturer's recommandations

TABLE 5 – Oil Changes - reciprocating compressors

Compressor type	Working conditions	Change interval
All stationary	Running in	After 100 hours
All portable	Running in	After 50 hours
Stationery	Clean atmosphere Dirty atmosphere	6 months or 2000 hours 3 months or 1000 hours†
Portable	Average Dirty atmosphere Very dirty atmosphere	1 month or 500 hours 2 weeks or 250 hours† 1 week or 100 hours†

†Suitable intervals can only be established by experience or on manufacturer's recommendations

TABLE 6 – Trouble-shooting - reciprocating compressors

Fault or symptom	Possible causes	Action
'Knocking' or heavy vibration	Drive or coupling elements loose Unstable oil-relief valve causing hydraulic hammer Excessive working clearances on crosshead or bearings Cylinder clearance volume too small	Check and rectify Check valve springs (a) Check for wear (b) Check for misalignment Check for accumulation or water, oil or other deposits
Abnormal air temperature	Faulty cooling system Defective air filter Excessive cylinder wear	Check and clean Replace Check and replace
Abnormal air pressure	Leakage Loss of compression efficiency Seal failure	Find and correct leakage Check cylinder and ring wear Replace if faulty Check compression seals
Rumbling noise	Air leakage past piston rings	Replace
Valve chatter	Defective valve springs	Replace

TABLE 7 – Trouble-shooting - rotary compressors

Fault	Possible cause	Remedy
Compressor fails to build up pressure	Regulator needs adjustment Unloader valve stuck Faulty blow down valve Choked intake cleaner (indicator shows red)	Adjust to give normal pressure Inspect and release Overhaul or renew Renew element
Compressor fails to offload (safety valve blows)	Unloader/compressor joint leak Unloader valve stuck open Unloader diaphragm punctured Faulty safety valve	Renew Inspect and release Renew Test and adjust
Plant operates at incorrect speed (portable compressors)	Plant operating at incorrect pressure Speed control linkage worn or needs adjustment	Adjust pressure regulator Renew or adjust
Air delivery temperature too high	Exterior of oil cooler clogged Insufficient oil in cooling system Incorrect grade oil Oil filter clogged Faulty thermal bypass valve	Clean Fill to correct level Drain and refill Renew Check temp. in cooler & adjust
Excessive oil consumption (oil carryover with air)	Incorrect oil grade Minimum air pressure too low Too much oil Air/oil separator faulty	Drain and refill Inspect and adjust minimum pressure valve Drain to correct level Clean or renew
Emission of oil from air inlet after machine has stopped	Oil control valve stuck open	Examine and clean. Check for lacquer around valve

TABLE 8 – Maintenance Points - rotary compressors

Component	Inspection	Further action
Electric motor	Clean motor; inspect for wear; check bearings; check clearances; check general condition	As necessary
Couplings (where appropriate)	Clean and examine	Replace if necessary
Casing	Open and examine for corrosion and wear	
Journal bearings	Examine for wear	Replace, reline, adjust with shims (as necessary)
Shaft seals	Check condition of clearances	Replace if necessary
Rotor	Examine for corrosion, pitting and wear; check clearances; check balance	As necessary
Gear case (where appropriate)	Examine for corrosion and wear; check damper	As necessary
Other parts	Examine for corrosion, erosion erosion and wear	As necessary
Governor/Regulator	Clean and inspect; check for wear; check adjustment	Replace worn parts; adjust for proper operation
Auxiliary drives	Check	Replace worn parts
Instrumentation	Check for operation and correct reading	Recalibrate or adjust as necessary

SECTION 5

Applications

TOOL CLASSIFICATION AND PERFORMANCE

Tools operated by compressed air can adopt a variety of different forms as in Table 1. They can be broadly classified according to the method of operation – percussive, rotary or combination (a combination tool is one in which percussion and rotation is provided pneumatically in the same tool, as in some rock drills and impact wrenches). A further broad classification is according to the place of use – the usual distinctions being industrial (or workshop) tools for factory operations, contractors tools for use on construction sites, and mining tools for quarry and underground use. Although the general principles of design are the same, the practical realisation of those principles can be very different according to the needs of the user.

Industrial tools form part of the total factory environment and as such they may be automatically operated and controlled. In a factory, tools which rely for their motive power on compressed air may nevertheless for convenience be controlled electrically.

In other situations, for example in mining and quarrying operations, the percussion action of a tool may be generated pneumatically whilst the rotary action may be hydraulic. The optimum configuration uses the ideal properties of each medium.

Performance information

Table 2 provides data for typical air consumption of tools. There is much variation between manufacturers. Efficiency is always improving, so it is wise to obtain up-to-date values direct from manufacturers. Air consumption figures are always available but other performance figures for tools are not so easily obtained.

Torque values for air motors, screwdrivers and nutrunners should be supplied and are easily checked by a conventional dynamometer, but the energy output of percussive tools is rarely provided, and for good reason. It is only in recent years that reliable techniques for the measurement of blow energy in percussive tools have become available, these are now embodied in BS 5344.

The technique described in that standard requires the use of strain gauges attached to the drill or chisel bit of the tool. When used with the appropriate analysis equipment, the energy content of each shock wave can be determined. Before this standard was

TABLE 1 – Tool classification

Portable rotary tools for removing material	
Drill	A rotary tool driving an output spindle generally through a gear box. The output spindle is normally fitted with a chuck or Morse taper or other socket, making the tool suitable for drilling, reaming, tube expanding and for boring metal, wood and other materials.
Straight drill	A drill with the rotary tool in line with the handle.
Piston grip drill	A drill where the handle of the tool is at an angle to the motor and the chuck axis.
Angle drill	A drill with the rotary cutting tool at an angle to the motor axis.
Grinder	A rotary tool driving an output spindle adapted to carry an abrasive device.
Straight grinder	A grinder where the handle, motor and spindle are in line.
Vertical grinder	A grinder where the handle or handles are at right angles to the motor and spindle axis which are in line or parallel.
Angle grinder	A grinder where the output shaft is usually driven through bevel gears so that the output spindle is at an angle (usually at a right angle) to the motor axis.
Straight die grinder	A small straight grinder for use with collets and mounted points.
Angle die grinder	A die grinder where the output shaft is driven through bevel gears so that the output spindle is at an angle (generally 90°) to the main axis of the tool.
Shear	A rotary pneumatic tool having a cutter reciprocating relative to a fixed cutter and used for cutting sheet metal by means of a shearing action.
Nibbler	A pneumatic tool for cutting sheet metal by means of material removal by reciprocating a punch through a fixed die.
Portable percussive tools	
Chipping/caulking hammer	A tool for chipping, caulking, trimming or fettling castings, welds, etc.
Scaler/scaling hammer	A tool for removing rust, scale, paint, *etc* through one or several reciprocating work pieces.
Needle scaler	A tool fitted with reciprocating metal needles for rust or scale removal.
Bush hammer/scabbler	A tool for surfacing stone *etc.*
Stone hammer	A tool for carving and working stone for sculpture purposes.
Sand rammer	A tool for ramming sand in foundry moulds by means of a butt attached to an extension of the piston.
Backfill rammer/tamper	A tool for consolidating earth *etc.*
Paving rammer	A tool for levelling paving-stones.
Tie tamper	A tool for tamping ballast beneath the sleepers of railway track using a special ancillary tamping tool.

TABLE 1 – Tool classification *(continued)*

Concrete breaker/pavement breaker/road breaker	A tool for breaking up concrete, rock, brickwork *etc.*
Sheet pile driver	A tool for driving sheet piles.
Pile driver	A tool for driving steel or wooden poles.
Sheet pile and pile extractor	A tool for extracting piles and sheet piles.
Spade	A tool fitted with a spade for digging clay, loam or peat.
Pick hammer/pick	A tool for light demolition or mine work.

Percussive tools with rotation	
Rock drill	A tool for drilling holes in rocks, concrete, *etc.*
Blower-type rock drill (blowing rock drill)	A rock drill fitted with a device for blowing out drilling chips with compressed air; usually the air passes through a hole down the centre of the drill steel.
Wet type rock drill (with water flushing head)	A rock drill fitted with a device for washing out drilling chips with pressurised water. *N.B.* Some rock drills may combine both air blowing and wet flushing.
Dry suction rock drill	A rock drill fitted with a device for the removal of drilling chips by suction.
Feed leg/air leg	A telescopic leg on which a rock drill can be mounted producing the thrust required for penetration and hole drilling.
Sinker or jackhammer	A hand-held rock drill used to drill vertical holes in rock. Its prime use is for sinking shafts and for quarry work.
Plug drill	A light hand-held drill for small holes in rock. Used in conjunction with a plug and feathers for splitting larger pieces of rock.
Stoping drill/stoper	A rock drill for drilling holes vertically upwards in a mine stope. Consists of a drill mounted on the end of a pneumatic cylinder.

Fixed pneumatic tools (rotary)	
Air motor	A pneumatic motor for driving.
Drilling unit	A pneumatic drilling tool with a feeding (and rectracting) device to be used as a component of a special machine tool.
Drilling unit with manual feed	A drilling tool with manual feed through a rack or any other means.
Drilling unit with automatic feed	A drilling tool with a feed controlled by a built-in powered feed device with adjustable stroke.
Tapping unit	A tapping tool with a built-in powered device with adjustable stroke.
Grinding unit	A straight or angle drive grinder designed for mounting on special grinding machine-tools or as a rapid grinding spindle on lathes.

Supported percussive tools	
Pile driver	A tool for driving steel or wood piles. Often this is an adaption of a road breaker.

TABLE 1 – Tool classification *(continued)*

Pile extractor	A tool for extracting piles.
Rock drill drifter	A percussive tool of heavy construction with rotating chuck for drilling holes in rock, used with a suitable support.
Carriage rock drill	A drifter, slide mounted on a carriage or cradle. The carriage can be mounted on a wagon drill, crawler, hydraulic boom, *etc.* The feed of the drifter along the carriage can be by chain, screw or cylinder.
Down-the-hole hammer	A hammer which is placed at the end of a drill rod and entirely enters the hole as it is being drilled. Rotation is effected by an independent motor turning the drill rod to which the hammer is attached.
Stone breaker	A heavy percussive machine mounted on a tractor for breaking stone, concrete, *etc.*

Pneumatic machines for assembly work
Portable rotary machines

Screwdriver	A rotary, reversible or non-reversible tool driving a spindle fitted with a screwdriver blade.
Straight screwdriver	A screwdriver where the axis of the bit is in line with the handle.
Pistol grip screwdriver	A screwdriver where the handle of the tool is at an angle to the motor and bit axis.
Stall-type screwdriver	A screwdriver without a clutch.
Clutch-type screwdriver	A screwdriver fitted with a clutch which may be adjustable.
Automatic clutch-type screwdriver	A screwdriver fitted with a clutch ensuring drive disengagement as a set torque is reached.
Automatic clutch-type screwdriver with air flow cut-off	A screwdriver fitted with an automatic clutch which actuates motor cut-off once a preset torque is reached.
Automatic clutch-type screwdriver with push start and air flow cut-off	A screwdriver fitted with an automatic clutch with axial push start by pressure on the bit, and automatic cut-off as a required torque is reached.
Angle screwdriver	A screwdriver with or without a clutch where the axis of the output spindle is at an angle to the motor axis.
Screwdriver with automatic feed of fasteners	A screwdriver with automatic feed of the screws *etc.*
Studrunner	A reversible nutrunner fitted with a special chuck for stud running.
Nutrunner	A rotary reversible or non-reversible machine driving a socket adapter.
Stall type nutrunner	A nutrunner where the only means for setting the tightening torque is by air pressure regulations.
Torque nutrunner	A nutrunner where the setting of the tightening torque is achieved by means of a clutch or any other means.

TABLE 1 – Tool classification (*continued*)

Angle drive nutrunner (wrench)	A rotary machine incorporating a socket adapter, the output axis of which is offset or at an angle to the rotor axis.
Two-speed nutrunner	A nutrunner incorporating a speed reduction gear or second motor, with an automatic tripping device reducing the speed once the torque has reached a certain value.
Ratchet wrench	An angle drive wrench progressively rotating a socket by means of a ratchet and pawl coupling.
Impact wrench	A rotary machine fitted with a multi-vane or oscillating motor driving a hammer which periodically strikes an anvil to which a socket is attached.
Straight impact wrench	An impact wrench with the motor-axis, handle and output spindle in line.
Angle drive impact wrench	An impact wrench with the output spindle at an angle to the motor axis.
Torque controlled impact wrench	An impact wrench in which a device automatically cuts off the motor when a preset torque is achieved.
Portable percussive tools	
Riveting hammer	A percussive machine for forming rivet heads.
One-shot riveter	A riveting hammer which delivers a single blow for every depression of the throttle actuator.
Yoke riveter	A combination of a riveting hammer and a holder-on, one on each side of a yoke.
Holder-on	A hammer or piston acting as a counterset on a rivet, the other end of which is being riveted.
Squeeze (compression) riveter	A linear piston machine without percussion which forms rivets by squeezing.
Nailer (stapler)	A tool for driving nails (staples) with one or more strokes. The nail (or staple) feed is often automatic.
Pneumatic machines for mechanical handling	
Hoist	A device for lifting and lowering loads.
Winch or capstan	A pulling or hoisting appliance, incorporating a cable drum.
Rotary vibrator	A device with a motor rotating an eccentric mass for vibrating containers, *etc.*
Percussive vibrator	A percussive tool designed to generate vibration.
Immersion vibrator (poker/vibrator)	A vibrating tool designed to be immersed in a fluid, usually concrete.
Miscellaneous pneumatic tools	
Sand or shot blasting machine	A system for sand or shot blasting where the carrier medium is compressed air.
Paint spraying gun	A gun by which the paint is atomized.
Air starter	A pneumatic motor or compressor for starting engines.

TABLE 2 – Air consumption of pneumatic tools and appliances

Description of tool or appliance	Litre/s (6 bar)
Rock drills	
Drifter drill (cradle mounted), 75 mm	70
Drifter drill (cradle mounted), 82 mm	82
Drifter drill (cradle mounted), 100 mm	100
Hand hammer drill (jack hammer), 13 kg	25
Hand hammer drill (jack hammer), 16 kg	30
Hand hammer drill (jack hammer), 21 kg	40
Hand hammer drill (jack hammer), 28 kg	50
Plug drill	15
Sinker drill, 75 mm	70
Sinker drill, 82 mm	80
Stoping drill, light	40
Stoping drill, heavy	70
Pneumatic tools (percussive)	
Concrete breaker, 35 kg	35
Concrete breaker, 23 kg	30
Concrete breaker, 14 kg	25
Pile driver	35
Chipping and caulking hammer, light	5 and 7
Chipping and caulking hammer, medium	9
Chipping and caulking hammer, heavy (suitable also for $1/2$" hot rivets)	12
Riveting hammer, 20 mm rivets	12
Riveting hammer, 25 mm rivets	13
Riveting hammer, 32 mm rivets	15
Riveting hammer, 35 mm rivets	17
Auger drill, for coal	40
Deck planer, for wood decks	30
Tube cutter, for tubes up to 65 mm	25
Tube cutter, for tubes up to 60 to 100 mm	30
Tube expander, for tubes up to 60 mm	25
Tube expander, for tubes up to 75 mm	30
Tube expander, for tubes up to 100 mm	35
Pneumatic appliances	
Wrenches (rotary torque), for nuts 6 mm	5
Wrenches (rotary torque), for nuts 10 mm	10
Wrenches (rotary torque), for nuts 12 to 18 mm	15
Wrenches (rotary torque), for nuts 20 to 25 mm	20
Impact wrench, for nuts up to 18 mm	15
Impact wrench, for nuts up to 30 mm	25
Spray gun, small at 4 bar pressure	1
Spray gun, medium at 4 bar pressure	4
Spray gun, large at 4 bar pressure	10
Air gun or duster	4
Sump pump, 6 to 20 l/s	20 to 60

TABLE 2 – Air consumption of pneumatic tools and appliances *(continued)*

Description of tool or appliance	Litre/s (6 bar)
Air motors, up to 0.75 kW	20 (per kW)
Air motors, 0.75 to 4.0 kW	18 (per kW)
Air motors, over 4.0 kW	16 (per kW)
Pick, light	12
Pick, medium	13
Pick, heavy	18
Spader, light	15
Spader, medium	20
Tie tamper	15
Rammer, foundry, bench type	7
Rammer, foundry, floor type, medium	10
Rammer, foundry, floor type, heavy	14
Rammer, back fill trench type	17
Spike driver	30
Scaling hammer, valveless, for surface work	3
Scaling hammer, for large boiler tubes	12
Paint scraper	4
Stone tool, for lettering and light carving	3
Stone tool, for medium dressing	5
Stone tool, for roughing and bushing	7
Weld flux chipper	8
Pneumatic tools (rotary)	
Drilling machine, for 6 mm holes in steel	10
Drilling machine, for 9 mm holes in steel	12
Drilling machine, for 18 mm holes in steel	15
Drilling machine, for 25 mm holes in steel	20
Drilling machine, for 30 mm holes in steel	25
Drilling machine, for 38 mm holes in steel	30
Drilling machine, for 50 mm holes in steel	40
Drilling machine, for 75 mm holes in steel	50
(For wood boring, take next size smaller.	
For reaming and tapping in steel, take next size larger).	
Corner drills, as above	
Grinders, with 18 mm wheels	5
Grinders, with 50 mm wheels	8
Grinders, with 100 mm wheels	20
Grinders, with 150 mm wheels	30
Grinders, with 200 mm wheels	40
Saw, 150 mm	15
Internal vibrator, o.d. 62 mm	30
Internal vibrator, o.d. 75 mm	40
Internal vibrator, o.d. 113 mm	50
Internal vibrator, o.d. 140 mm	60

Note: The air consumption of tools and appliances operated by compressed air is expressed in free air per minute – FAD. The above air consumptions should be taken as a guide and are subject to variations in different circumstances, as is the nature of the work, condition of the tools, intermittent use. The figures given are based upon operation at approximately sea-level. For higher altitudes a slightly higher allowance per tool is necessary.

developed, the performance of a percussive tool was assessed by such unreliable methods as measurement of the permanent deformation of a soft pellet when hammered by the tool bit or by the height of the pressure pulse when the tool output is absorbed by an hydraulic buffer. These methods are now only used for comparison between similar tools, and should not be quoted as an absolute measure of tool performance.

Even when the stress energy content of the shock wave is known, it will not necessarily be helpful to the user of the tool, who is concerned primarily with the amount of material that can be removed or drilled. Another way of defining the power output of a percussive tool is to quote, for example, the drilling speed in a specified kind of rock, or the amount of concrete that can be broken by a concrete breaker in a given time, or the amount of steel that a chipping hammer can remove. None of these is particularly easy to measure under controlled conditions. The prospective purchaser should be aware of information presented in such a way when trying to compare different tools. There is really no alternative to actually trying out a tool in the circumstances of one's own application.

As a general statement, a tool made by a reputable manufacturer will have a power output proportional to the its consumption, and that is probably the best guide that can be given. The weight of the tool is another useful indication of the power, but it should not be assumed that more actual work will be done by the heavier tool.

A further warning should be given when assessing raw performance data. At first sight it might be thought desirable to choose the tool with the highest output, paying no regard to the ease of use. All tools require the operator to apply a feed force and to sustain the vibration present at the handle. Usually a tool with the highest performance is the hardest to handle, but some tools are proportionally worse than others in this respect. Many tools are coming on the market which have been designed for ease of use; feed force, vibration levels and noise have been reduced, although sometimes at the expense of bulk and manoeuvrability.

Performance related to air pressure

Table 2 quotes air consumption for a variety of tools at 6 bar gauge pressure. Most manufacturers use this as a standard at which to measure performance, in spite of 6.3 bar being the recommended ISO standard; most compressed air systems should be capable of supplying this pressure as a minimum. The performance at other pressures is not necessarily proportional to the pressure. Theoretically the power output of percussive tools varies according to (pressure)$^{1.5}$, but tools are designed to work ideally over a limited range of pressures, and they may not work anything like as well at a different pressure.

Some tools, particularly rock drills and down-the-hole machines for quarrying, work at much higher pressures (up to 20 bar). The reason for these high pressures is the need to maximise the drilling speed within the restricted space limitations of the tool. The overall efficiency of power conversion in high pressure tools may be lower than for standard pressure tools, but efficiency is less important than high performance in these applications.

INDUSTRIAL TOOLS

Rotary tools

Most rotary tools employ a vane type motor. This is the simplest and cheapest motor available and it can be made in a small diameter suitable for hand tools. Vane motors have a low inertia which means that they quickly reach maximum speed. Stationary machinery such as hoists and winches tend to employ piston and gear motors which are more robust and can withstand frequent stopping and starting. For high speed die grinders and increasingly for angle grinders, turbine motors are used.

Rotary tools can be divided into:

- grinders and sanders
- drills, tappers and reamers
- screwdrivers, nut runners
- impact wrenches
- impulse wrenches, impulse screwdrivers
- miscellaneous

Grinders and sanders

Figure 1 shows three common kinds of grinder. A die grinder uses a small diameter wheel for access to awkward corners and the bottom of dies. Each wheel has a maximum permissible rotary speed which must not be greater than the free speed of the grinder. Attention must be paid to ensuring that the designated speed on the blotter of the wheel cannot under any circumstances be exceeded. In the case of a die grinder, the free speed can be as high as 100 000 r/min, and special precision collets are needed to mount the grinding points. BS 4390 should be referred to for safety requirements; as well as information on safe speeds, it gives design parameters for the construction of safety guards.

The larger sizes of grinders (for wheels greater than about 50 mm diameter) are equipped with a centrifugal governor, which ensures that the free speed is limited to about 10 000 r/min. Smaller grinders have to rely on the limitation imposed by the maximum air

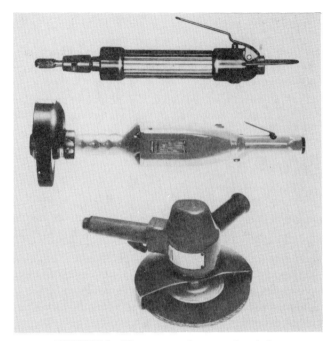

FIGURE 1 – Three types of pneumatic grinders.
A die grinder, straight grinder and vertical grinder. *(Atlas Copco)*

FIGURE 2 – Orbital sander cut away to show turbine motor.
20 000 rev/min. Consumption 0.2m³/min. *(Nitto)*

flow through the inlet passages. Grinders are supplied with safety guards, which must always be used.

Different shaped wheels are available: cylindrical wheels, cup wheels, depressed centre wheels and cut-off wheels. A variety of special shaped points and tungsten carbide burrs are provided for die grinders.

Rotary sanders are essentially the same as vertical grinders but fitted with a sanding disc in place of a grinding wheel. Orbital sanders driven by a turbine are also available. One example using sanding paper and suitable for fine finishing is shown in Figure 2.

Theory of grinding

Grinding is a process that involves material removal using a succession of cutting edges. As the grains penetrate the material being ground they become worn, and the cutting force increases until they fracture and break out, revealing new and sharp grains. This gives a degree of self-sharpening which depends on the matching between the composition of the wheel and the material being ground. The choice of composition and grain size of the wheel is largely a matter of personal experience and choice. Usually a fine finish is not required of a hand grinder, in contrast to stationary grinders used in a machine shop, so a coarse grit for maximum stock removal is to be preferred.

All grinding wheels have a maximum speed which depends on their composition. BS 4390 sets the following limits on speed:

Type of wheel	Diameter (mm)	Thickness (mm)	Max. surface speed (m/s)
Reinforced resin-bonded straight wheels	50 – 200 50 – 200	6.5 max over 6.5	90 80
Reinforced resin-bonded depressed centre and cut-off wheels			80
Resin-bonded (resinoid) all types			48

The power curve of an ungoverned air motor, on which the design of grinding machines is based, peaks at about half the free speed (refer to the chapter on Air Motors), and this represents the condition at which maximum metal removal is possible. On a governed tool, the peak power point varies from 75 to 90% of the free speed. Most operators are able to judge the best speed fairly accurately with experience. The operator has to press down on the handles, and this force added to the dead weight should be sufficient to realise the ideal working speed.

Pneurop have organised extensive tests on hand-held grinders in assessing the vibration experienced by the operator of the tool ("Investigations on hand-held grinding machines", obtainable from BCAS). It appears that most of the vibration derives from unbalance in

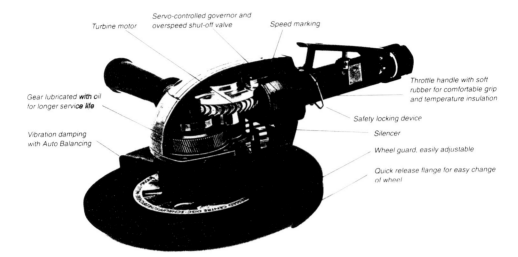

Turbine motor

Servo-controlled governor and overspeed shut-off valve

Speed marking

Throttle handle with soft rubber for comfortable grip and temperature insulation

Gear lubricated with oil for longer service life

Safety locking device

Silencer

Vibration damping with Auto Balancing

Wheel guard, easily adjustable

Quick release flange for easy change of wheel

FIGURE 3 – Angle grinder cut away to show features. *(Atlas Copco)*

the wheel, rather than the cutting operation. This unbalance comes either from geometric changes, *eg* ovality, eccentricity of the centre hole or thickness variation. It may also be induced by uneven wear, or from uneven density in the composition of the wheel; it is possible for a wheel that is initially balanced to develop an unbalance and vice versa. It has been established, contrary to a common belief, that a vibrating tool is not more effective at removing material than a well balanced one. This reinforces the importance of regular dressing of the wheel. The exposure of grinding wheel operators to vibration should be monitored and their time at work restricted.

There are now becoming available hand grinders which incorporate devices which compensate for wheel unbalance. One example is the vertical (angle) grinder of Figure 3 which has a counterbalance mechanism employing bearing balls.

See the chapter on Noise and Vibration for further information.

Screwdrivers and nut runners

A selection of screwdrivers and nut runners is shown in Figure 4. For access to awkward places double angle wrenches are available. Wrenches are provided with square drive to accept standard sockets.

These tools transmit the torque either by direct drive or through a clutch. Direct drive relies on the operator to assess the applied torque, which can be as much as the stall torque of the motor or rather more if account is taken of the inertia of the motor. Clutches are of two broad types – slipping or torque release; the former type slips at the set value of the torque; the latter cuts off the supply of air to the motor. Magnetic clutches are also available. See also Figure 5. These tools have to be reversible, but usually the clutch does not operate when undoing.

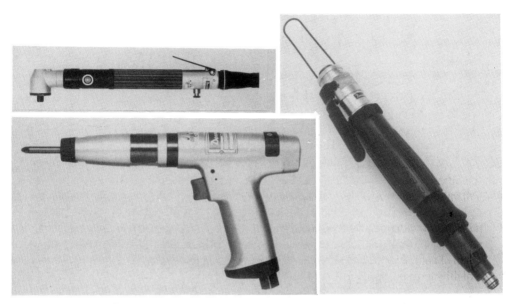

FIGURE 4 – Angle head nut runner (angle wrench), pistol grip screwdriver and straight screwdriver with ergonomic grip. *(Desoutter)*

TORQUE CONTROL POWER SHUT OFF

Torque is transmitted through two ramped clutch members which need only the smallest axial movement to simultaneously disengage the drive & shut off the motor. This clutch ensures excellent torque accuracy & minimal mean-shift on most applications

SPRING TENSION

The clutch faces ride up and slip over each other when the preset torque has been achieved. The spring tension clutch is a good general purpose clutch.

POSITIVE

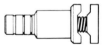

Torque transmitted by the positive clutch is proportional to axial thrust exerted on the screwdriver. This feature makes it ideal for woodworking & self tapping applications where the torque needs to be "driven up" to overcome variable resistances – *ie* knots.

FRICTION

One clutch face is flat so only friction between faces has to be overcome. Excellent clutch for very light duty applications such as self tapping in plastics.

LOW TORQUE

Similar in operation to the spring tension or the one shot clutch, but with the clutch face ramp angle reduced.

DIRECT DRIVE

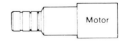

The direct drive between screwdriver bit and motor ensures maximum available torque from the motor is transmitted. Variable torque is achieved by varying tool inlet pressure – stall point.

FIGURE 5 – Types of clutch for screwdriver and nut runners. *(Desoutter)*

Drills, tappers and reamers

These are similar in construction to screw drivers. The larger sizes carry a standard chuck; the smaller sizes employ a collet to grip the drill bit. Tappers can incorporate a reversing mechanism which automatically withdraws the tap at double speed when the cutting operation finishes.

Precision feed drilling

For precision drilling in the aerospace industry and wherever high quality performance is required, a variety of controlled feed mechanisms are available. The following are typical examples:

1. The controlled feed drill has an automatic feed/drill/retract cycle, all operated pneumatically on a hand held tool. The complete cycle is initiated by depressing the trigger. Hydraulic damping controls the feed rate to minimise burring on break-through, and the trigger can be locked to allow simultaneous use of more than one unit.
2. The automatic feed drill is similar, but in this case the cycle is initiated by touch of a button and allows for remote start/stop signals, ideal for robotic or special purpose machine applications.
3. Variable feed drills combine power feed with manual feed. The manual feed allows for rapid advance and retraction, and for such applications as swarf clearance when deep hole drilling.
4. Rack feed drills rely on manual feed and retraction, although automatic stop and start of the motor is possible at any point in the stroke.

Combination of air and electric motors for special applications are also available – it is possible to have air feed combined with an electric drill, and also both feed and drilling performed electrically.

Impact wrenches

These are rotary vane motors with an additional rotary impact mechanism to impart a blow as the resistance increases (Figure 6). Impact wrenches are available to apply a torque of up to 100 000 Nm or more with an output drive of $3^1/_2$ in. The motor first runs freely until the nut offers resistance to the torque, whereupon the impact mechanism comes into play to assist further tightening. It is possible to exert high values of torque with small reaction

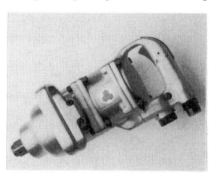

FIGURE 6 – Impact wrench.
1" socket, 2500 Nm torque.
(CompAir Power Tools)

on the operator. Some types have an adjustable torque setting, ensuring that the tension in the bolt is not exceeded. Although the torque setting is unaffected by changes in air pressure, unless the available air pressure is 6 bar or greater the maximum setting of the torque may not be reached.

Various types of torque control are available, according to the manufacturer. Some have a device which cuts off the air supply when the torque setting is reached; some have a built-in timer which shuts off the air after a certain time; some have an adjustable restriction which limits the air supply. The first type is more accurate, but they all need to be calibrated on the joint to be tightened, which makes them more suitable for production runs than for the occasional tightening operation. Devices which work by cutting off the supply pressure rely on an ability to sense the amount of rebound off the tightened nut, followed by triggering a cut-off valve. With this method, the response depends on the stiffness of the joint; some tools work better with a stiff joint, some are better for a flexible joint.

For accurate torque settings, torque bars can be used. These are graduated extension bars, inserted between the wrench and socket. two kinds are available: adjustable and non-adjustable. They have a calibrated flexibility which is able to trigger a shut-off device when the torque is reached

Impulse tools

These represent a new development, which for many purposes are replacing impact wrenches, particularly for smaller sizes of bolt (for a torque range up to about 280 Nm and a driver size up to $^3/_4$ in); see Figure 7. These have a conventional vane motor but instead

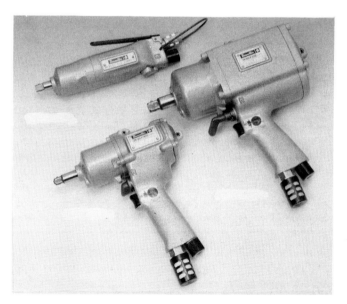

FIGURE 7 – Impulse wrenches: Straight case, torque 19Nm.
Pistol grip torque, 17Nm and 245 Nm. Note the torque setting control. *(Desoutter)*

of metallic impact, an impulse is applied to the output shaft of the tool through a rotating, cam operated, hydraulic piston unit. This results in high speed rundown, much quieter operation, repeatable torque settings and low vibratory reaction on the operator. The absence of metallic impact in the tool means that the service life is likely to be longer than for an impact wrench; a factor of 10 x has been quoted, but this will depend on the particular design.

As the impulse unit rotates and the cams engage every 360°, a pulse of energy is transmitted to the output shaft, increasing the fastener torque each time the cams engage. The torque is set by a control knob on the body of the tool.

These tools also incorporate a shut off mechanism which acts to divert the energy of the tool and restrict the supply of air when the resistance of the bolt reaches the torque setting.

Considerations of joint tightness

Although the use of a calibrated torque wrench to control the degree of tightening of a bolted joint is an advance over relying on the feel of the operator, it is not as reliable as a method which directly measures the tension in the bolt. Friction is the dominating influence in trying to correlate the torque with the bolt tension. The coefficient of friction is not easily determined: it varies with the amount of lubricant present, the condition of the mating surfaces and variation in the geometry of the joint. A high degree of accuracy in the applied torque is unlikely to decrease the inherent scatter in the bolt tension values. Two methods have been developed to overcome this problem.

The first method is known as Angle-controlled Tightening. It is a method of indirect elongation measurement avoiding the practical difficulties inherent in trying to assess the total bolt stretch from head to point. Angle-controlled tightening measures the movement of the nut relative to the clamped component after the mating surfaces are fully seated. The stretch of the bolt and the tension in it can be calculated from the angular movement of the nut and the pitch of the thread.

A plot of torque versus deflection would show a distinct knee in the curve, at the point where the clamping action changed from bringing the surfaces together to stretching the bolt, and it is from this point that the angular movement is measured.

Practical experience in the use of this method indicates that high precision is obtained only if the bolt is tightened beyond its yield point, which is satisfactory in many applications but in others, as for example in joints where the clamping length is short or where the material is brittle, the bolt may be overloaded. The method is unsuitable when bolts experience a working tensile load which adds to the pre-load. Bolts which have been stressed into the plastic region cannot be reused.

A second, more accurate method, known as Gradient-controlled Tightening has been developed. This method, instead of measuring the angle alone, measures both torque and angle and calculates the slope of the torque/angle curve. As soon as this slope reaches the yield point, the air supply is cut off. This technique is only suitable for use on a production line in conjunction with the appropriate electronic analysis equipment; however for less sophisticated applications a portable calibrated torque wrench is available which measures the gradient and can be used to calibrate a power driven wrench and for quality control.

Percussive tools

Percussive industrial tools include:
- chipping hammers
- scaling hammers
- rivetters
- sand rammers

Chipping hammers

Chipping hammers are percussive tools used for a wide range of industrial purposes:
- removal of welding slag and rust
- dressing castings
- chamfering steel plate for weld preparation
- bolt and rivet shearing

Table 1 enables a choice to be made of the correct size of hammer.

TABLE 1

Tool weight kg	Frequency blows/min	Recommended duty
1.5	3500 – 5000	Very light chipping, rust scaling, panel cutting.
2.5	2000 – 3500	Light chipping, weld dressing, bolt and rivet shearing, driving shackle pins, caulking.
4.5	1500 – 2000	Aluminium and light cast iron dressing.
6.0	1500	Heavy cast iron and light steel dressing.
6.5	1500	Heavy steel dressing.

Chipping hammers of a traditional design produce high levels of vibration, and operators who spend most of their working time using one for foundry fettling or some other continuous operation should be monitored for signs of vibration injury (see chapter on Vibration Exposure).

Figure 8 shows an example of a modern vibration reduced chipping hammer. The design of this tool has succeeded in reducing the vibration level when compared with a conventional tool (a claimed reduction from a weighted vibration of 10.5 m/s^2 down to 2.1 m/s^2).

Chipping hammer chisels are made in a wide choice of shapes to suit different purposes, Figure 9. Shanks can be hexagonal or round, according to manufacturer's choice; chisels are not necessarily interchangeable between manufacturers, although BS 673 specifies standard dimensions which most manufacturers follow.

Some of the hammers are provided with a boss on the handles for attachment of an

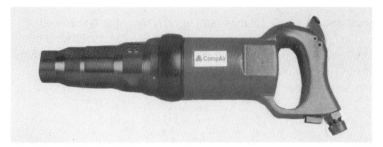

FIGURE 8 – Vibration reduced chipping hammer. *(CompAir Power Tools)*

Description	Shape	Description	Shape
Rivet Cutter		102mm Flat Chisel	
179mm Flat Chisel		3-way Silencer Cutter	
Edging Tool			
Double Edge		25mm Forked Chisel	
Punch			
Spot Weld Splitter		Needle Scaler	
Paint Scraper			

FIGURE 9 – Tool attachments for chipping hammers. *(Desoutter)*

FIGURE 10 – Vibration reduced needle scaler. *(CompAir Power Tools)*

FIGURE 11 – Triple head scaler. *(CompAir Power Tools)*

exhaust hose which can be led away to a silencer. It should be borne in mind, when using a silencer, that it may make the tool rather more awkward to handle and may not significantly reduce the noise level, particularly if most of the noise comes from the actual process (drumming of the panel or ringing of the casting).

Scalers

These are similar to light chipping hammers, Figure 10. They are specially designed to remove scale and rust, and for surface preparation. They are available either with a conventional chisel bit or with a bunch of hard spring steel or beryllium copper needles suitable for negotiating awkward shapes.

Figure 11 illustrates another type of tool designed primarily for surface dressing of stone and concrete.

CONTRACTORS TOOLS

The designation "Contractors Tools" refers to compressed air equipment used on construction and building sites. There is a degree of common usage between the three broad site classifications – industrial, contractors and mining. Thus chipping hammers may be used for metal dressing in a factory and for concrete dressing on a construction site. The needs of the miner and the construction site worker are also similar; a rock drill can be used for drilling rock on a road construction contract and for quarrying and mining. So the distinctions made here are somewhat arbitrary.

Road breakers

A common contractor's tool is the road breaker, paving breaker or concrete breaker. This is sometimes referred to as a road drill, but this is an incorrect usage; the term "drill" is restricted to those tools which either incorporate rotation as part of their mechanism or are separately and externally rotated. A road breaker has no rotation; it is a tool specifically designed to produce a heavy blow to break up road surfaces, and to perform similar allied functions. With the addition of various tool bits and chisels, it can also act as a clay spader, tie tamper, post-hole driver, back-fill rammer and pile driver. See Figure 1.

Modern road breakers are designed with a degree of exhaust muffling to reduce the possibility of noise damage to the operator. Some tools are available in both standard and muffled form, although it is unlikely that the standard form would be able to meet the EC limits for maximum noise level. The standard form may only be sold in those countries outside the European Union where noise regulations do not apply. Some modern tools are designed in such a way that they may only be used in the muffled form – the muffler forms an integral part of the construction which, if removed, would render the tool unusable. A number of manufacturers make tools of this kind. This has the advantage from environmental considerations that the tool may only be operated in the muffled condition.

Most mufflers make use of rigid polyurethane, a material which has demonstrated its ability to withstand the rough usage of a construction site. It can be moulded without expensive equipment, so it can be economically made up in small batches. It is available in a variety of colours.

Figure 2 shows a modern road breaker. It has many features common to other percussive

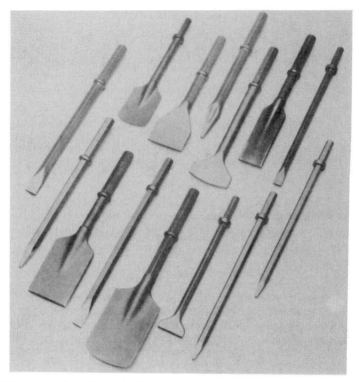

FIGURE 1 – A selection of accessories for use with road breakers. *(Padley and Venables)*
Clockwise from the left: narrow chisel, digger steel (4¹/₄ in), tarmac cutter, 'Easibust' steel, axe
blade, digger steel, narrow chisel (4¹/₄ in), two moil points (4¹/₄ in), wide chisel, clay spade,
narrow chisel, digger steel and moil point.

Accessories have shank sizes to match standard tool chucks. The available shank sizes are:
32 mm (1¹/₄ in) hex x 152 mm (6 in) long
28 mm (1¹/₈ in) hex x 152 mm (6 in) long
25 mm (1 in) hex x 108 mm (4¹/₄ in) long
22 mm (⁷/₈ in) hex x 82 mm (3¹/₄ in) long

tools such as picks and spaders which incorporate a distribution valve. The method of
operation is as follows:

Air is admitted through the main swivel connection. When the trigger is depressed, the
ball valve opens admitting air to the distribution valve situated at the top of the piston. The
valve in this example is a simple plate valve, although more complex shapes are also found.
It is held on its top or bottom seat by differential pressures on the upper and lower surface:
when the valve is on its bottom seat, air is directed to the under surface of the piston; when
the valve is on its top seat, air passes to the top of the piston. When air is first admitted to
the tool, both piston and valve are at their down position, so air passes into the bottom
chamber of the cylinder forcing the piston up. Towards the end of its upwards travel, the
piston compresses the air in the upper cylinder, which causes the valve to move on to the

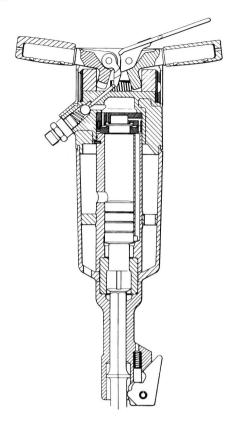

FIGURE 2 – A modern road breaker tool, cut away to show the construction.
The components are held together by long side belts (not shown). Incorporated in the
handle is an oil reservoir. *(CompAir Holman)*

opposite seat. At the same time the air in the bottom chamber exhausts to atmosphere via
the passages in the muffler. The air then acts on the top of the piston forcing it down. When
the piston impacts the anvil block, its energy passes into the chisel and then into the
concrete to be broken. The air exhausts from the upper cylinder and the fall in pressure
allows the valve to move once more onto its bottom seat. The cycle then repeats at a
frequency, of the order of 1000 to 1500 rev/min, depending on the design and the air
pressure.

The shank of the chisel is of hexagonal cross-section; the latch-type retainer keeps the
chisel held in the front head by the collar on the chisel shank; round shanks are also found
in picks where it is not required to turn the tool. A variety of different retainers are made:
a latch type is illustrated and is the most common, but screw and spring types are also
supplied.

Most tools incorporate an oil reservoir which dispenses a small amount of oil into the
working parts of the tool each time the trigger is pressed. It is important to keep this
reservoir topped up at least once during each shift, but this form of lubrication is not as

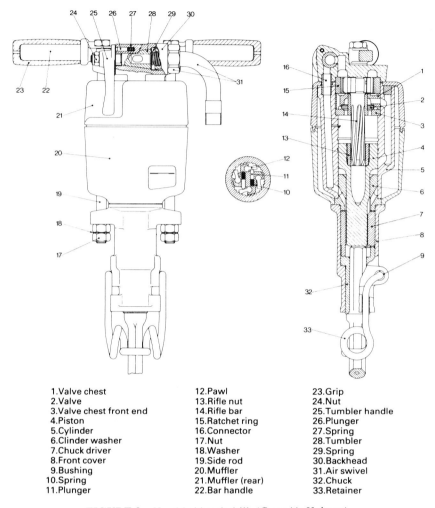

1.Valve chest	12.Pawl	23.Grip
2.Valve	13.Rifle nut	24.Nut
3.Valve chest front end	14.Rifle bar	25.Tumbler handle
4.Piston	15.Ratchet ring	26.Plunger
5.Cylinder	16.Connector	27.Spring
6.Clinder washer	17.Nut	28.Tumbler
7.Chuck driver	18.Washer	29.Spring
8.Front cover	19.Side rod	30.Backhead
9.Bushing	20.Muffler	31.Air swivel
10.Spring	21.Muffler (rear)	32.Chuck
11.Plunger	22.Bar handle	33.Retainer

FIGURE 3 – Hand-held rock drill. *(CompAir Holman)*

satisfactory as a separate lubricator inserted in the air line. The use of a specially formulated oil is recommended for this tool.

This tool also incorporates a vibration suppression device, which can be seen on the handles. The handles can pivot on a pair of pins and the anti-vibration element consists of a lateral spring and a rubber cushion. There is an increasing demand for some form of cushioning to meet health and safety requirements.

A range of contractors' percussive tools is shown in Figure 4.

Rock drills

Figure 3 is a cross-section of a typical modern hand-held rock drill, also illustrated in Figure 4. Superficially it has many similarities to the road breaker discussed above, but

FIGURE 4 – A range of contractors' percussion tools. Left to right: 2 road breakers, a medium duty pick, a light weight rock drill with rotation, a light pick, a light road breaker, a road breaker with a 'D' handle, two road breakers. Note the numbers 108, 111 on the bodies, which are the sound power levels to satisfy EC regulations. *(CompAir Holman)*

there are some major differences. It has the same type of disc valve and the cycle of operations is similar except that, in addition to the reciprocation of the piston, the drill bit is rotated by an internal ratchet mechanism. The piston has an internally rifled bush which matches a similarly rifled bar held in the backhead of the drill; a ratchet mechanism ensures that the piston rotates only on the upward (or return) stroke. On the power (*ie* the down) stroke there is no resistance to the motion of the piston, so that the full pressure is available to produce impact energy. On the return stroke, the piston rotates along the rifling and transfers that rotation to the drill steel by means of straight splines on the trunk of the piston which mesh with similar splines on the chuck.

A rock drill of this kind incorporates a means for flushing out the drilled hole, which can be done either by air or water. Air flushing is either continuous, when air passes directly from the exhaust port or is under operator control by selection of the blowing position on the tumbler handle. In the latter alternative the full flow of the air passes down through the hole in the drill steel to give full pressure clearance. Water flushing, which is necessary underground where the dust has to be suppressed, is achieved by passing water from a separate connection on the backhead down a tube passing through the piston into the drill steel. The water pressure has to be kept at a minimum of about 1 bar.

This particular drill is capable of drilling a 45 mm diameter hole at a penetration speed of 180 mm/min in granite; in limestone and softer rocks , the drilling speed can be up to twice as much.

A rock drill has to stand much harder usage than a breaker. The blow frequency is higher and the level of energy is higher. The extra complication of the rotation mechanism underlines the need for good lubrication at all times, so a separate line oiler should be included in the air line feeding the drill, even though some manufacturers offer a drill with an oil reservoir in the handle for occasional use. The consumption of spare parts is high, particularly of such items as the chuck, the rotation ratchet and the rifle nut. Typically during the life of a drill, it will consume two to three times its first cost in spare parts.

One distinction between a drill and a breaker is the magnitude of the energy in each blow and the shape of the cylinder to produce that energy. A breaker has a long stroke and smaller bore, typically about 45 mm bore with 150 mm stroke, producing a heavy-hitting low-frequency blow. The hand-held rock drill described above has a 64 mm bore and a 48 mm stroke. The energy content of the individual blow is not so important as the total energy per minute, so the drill maximises that energy by increasing the frequency of operation.

Rock drills of this general type incorporating rifle-bar rotation can be as small as a miniature drill weighing 2 to 3 kg, consuming 12 l/s of air, up to the large rig-mounted drills weighing 150 kg and consuming 240 l/s. Even larger drills, with independent rotation (the rotation is achieved with a separate motor directly connected to the drill chuck) are available, weighing 300 kg, consuming 300 l/s. Drills heavier than 30 kg are not suitable for hand use.

Performance estimation of percussive tools

Estimating the performance of percussive tools is not a particularly easy exercise, when compared with their hydraulic or pneumatic counterparts. The difficulty stems partly from an uncertainty about the amount of expansion of the air at the point of use. Modern methods of analysis rely on computer simulation, for which specialist programs are available. A description of these methods is outside the scope of this volume and the interested reader is referred to one of the companies specialising in simulation techniques.

The "broad brush" method of analysis adopted here makes it possible to compare the performance of one tool with another, and enables the influence of the main tool parameters, such as bore, stroke and air pressure to be studied.

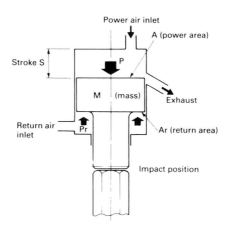

FIGURE 5 – Forces acting on percussive tool.

Figure 5 shows the forces acting on the piston of a percussive tool. The following quantities are derived from a consideration of the acceleration of the piston.

Work done on the power (impact) stroke = PAS

$$\text{Time for power stroke} = \left(\frac{2MS}{PA} \right)^{1/2}$$

$$\text{Terminal velocity} = \left(\frac{2PAS}{M} \right)^{1/2}$$

$$\text{Time for power and return stroke} = (1 + k) \left(\frac{2MS}{PA} \right)^{1/2}$$

where   k is the ratio of the time spent on the return stroke to the time spent on the power stroke.

$$\text{Blow rate} = \frac{1}{1 + k} \left(\frac{PA}{2MS} \right)^{1/2}$$

$$\text{Total work output} = \text{Blow rate} \times \text{work/blow} = \frac{PAS}{1 + k} \left(\frac{PA}{2MS} \right)^{1/2}$$

$$= \frac{(PA)^{3/2}}{1 + k} \left(\frac{S}{M} \right)^{1/2}$$

These equation are applicable when all units are consistent. In terms of the standard engineering units which have been used throughout, *ie* stroke in metre, area in (metre)^2, pressure in bar and mass in kg they become:

Work done on power stroke = 10^5 PAS joule

$$\text{Time for power stroke} = 0.00447 \left(\frac{MS}{PA} \right)^{1/2} s$$

$$\text{Terminal velocity} = 447.0 \left(\frac{PAS}{M} \right)^{1/2} m/s$$

$$\text{Time for power and return stroke} = 0.00447 (1 + k) \left(\frac{MS}{PA} \right)^{1/2} s$$

$$\text{Blow rate} = \frac{224}{1 + k} \left(\frac{PA}{MS} \right)^{1/2} \text{per s}$$

$$\text{Total work output} = 22.4 \times 10^3 \times \frac{(PA)^{3/2}}{1 + k} \left(\frac{S}{M} \right)^{1/2} kW$$

These equations give only an indication of the behaviour of a tool and as such are useful for estimating what is likely to be the result of changing the pressure or the dimensions of an existing tool. They are less useful when attempting to predict the performance of a tool off the drawing board. For such purposes, a more complex simulation, involving a computer analysis is required.

When considering the above expressions note that the pressure P is the mean effective pressure on the power stroke, which will generally be lower than the applied pressure at the inlet. Furthermore the mean pressure on the return stroke will be lower than that on the power stroke. The factor by which the inlet pressure is to be multiplied to get the effective pressure depends on the detail design of the tool, particularly that of the distributor valve. P on the power stroke will be about 0.6 times the inlet pressure and P on the return stroke will be about 0.4 times the inlet pressure. The return pressure and the area below the piston on which it acts will govern the value of k. Typically k will be 2.0 to 2.5.

The air consumption of a tool of this kind is theoretically equal to the total swept volume of the power and return stroke per second of the compressed air. In terms of the usual FAD (free air delivered), the pressure ratio must be taken into account.

$$\text{Air consumption} = \frac{224 \times 10^3}{1 + k} \left(\frac{PA}{MS}\right)^{1/2} PAS \ \text{l/s}$$

This expression should be used with reservation. It takes no account of leakage and other wastage, which is inevitable because there are no seals on the piston. In practice the actual air consumption may be as much as 50% in excess of the calculated value.

Another expression may sometimes be useful. This concerns the external reaction force on the handle or on the support structure of a rig-mounted drill. It is a measure of the force that must be applied by the operator to use the power generated by the tool.

$$\text{Reaction force} = 10^5 \frac{PA}{1 + k} \ \text{N}$$

If the tool is held vertically, as for example the case of a road breaker, the operator applies the difference between this and the dead weight of the tool. Over a sustained period an operator cannot apply more than 100 N, and for short periods more than his body weight. It is this reaction force more than any other factor which limits the usable power of a hand held tool. If a tool were designed with a greater potential power, the operator would find it impossible to handle.

The reaction force calculated here is the mean static force. There is in addition a vibratory element, which varies with the tool design. Apart from vibration induced white fingers, a disease of tool operators, the presence of vibration at the tool handle represents a fatigue-inducing factor which may be more limiting than the static force.

Rotation systems

Another aspect of percussive tool behaviour, capable of theoretical estimation, is the torque produced by a rifle-bar rotation mechanism.

The function of this method of rotation should be understood first. The modern rock drill derives its action from the traditional method of hand drilling with hammer and chisel. The old time miners used to drill in teams: two men would wield hammers, and a third would hold the chisel and rotate it. The rock was broken by chisel action rather than by rotary cutting, and so it is in a modern rock drill; only in relatively soft rock such as coal and shale is it possible to drill by rotary action. It is easier to rotate a drill by hand in an anti-clockwise direction, so that is the direction that a drill turns. Smaller drills that can be used for both rotary and impact drilling are also fitted with clockwise rotation; and the heavier rock drills, using coupled rods that have to be screwed together, include a facility for reversing the direction of rotation to allow for uncoupling of the rods. Some stoper drills, which serve the dual purpose of drilling and acting as torque wrench for roof bolting, are also equipped with a clockwise rotation mechanism.

A rock drill must be forced against the rock by a feed force having the magnitude given in the equation above. The applied force should be just sufficient to keep the bit in contact with the rock, but no more. As the bit rebounds after the rock is broken, there is no resistance to rotation, and so a comparatively small torque is needed just to index the bit to the next cutting position. Air or water removes the chippings so that the chisel action is not cushioned by drilling detritus.

It is for this reason that the rotation occurs on the return stroke of the piston. The angle of the splines on the rifling is just sufficient to index the bit by about 20 to 30 degrees per blow, so that while the blow frequency may be of the order of 2000 per minute, the rotational speed will be about 100 per minute. It is usual in English usage to refer to a drill having a rotation (for example) of 1 in 30, as meaning that the bit rotates by one complete turn for every 30 inches of travel of the piston – a lead of 30 inches. Common values of lead are 30, 35, and 40. In SI units, the lead can be expressed in mm.

The shallowness of the angle of the rifle-bar, used to achieve such a slow rotation, means that friction between the splines and the rifle nut and between the chuck and the piston absorbs a high proportion of the available energy on the return stroke. It is important that the sliding parts are well lubricated to minimise the coefficient of friction and to prevent a common cause of wear – the break-up of the splined surfaces.

An advantage of this form of rotation is the absence of any appreciable reaction torque to be sustained by the operator.

Estimation of the torque provided by rifle bar rotation

Refer to Figure 3. The relationship between rotational torque and force on the piston is as follows:

$$\frac{T}{F} = \frac{(r - \mu\pi d_1)\, d_1 d_2}{2\,(\mu r\,[d_1 + d_2] + \pi d_1 d_2 - \mu^2 \pi d_1^2)}$$

where T = the torque available for rotation of the bit
F = the force applied by the air pressure to the piston on the return stroke
d_1 = mean diameter of the rifled splines

d_2 = mean diameter of the straight splines on the chuck

μ = coefficient of friction

r = lead of rifling

This formula is correct for consistent units.

The force F is the product of the area of the piston A_r (Figure 5) and the air pressure P_r. This formula is accurate for consistent units.

As an example of the use of this formula, take a drill with a bore diameter of 66.7 mm in which d_1 = 19.3 mm, d_2 = 34 mm, r = 0.762 m (30 in). Two values of the coefficient of friction can be taken, μ = 0 and 0.2.

For $\mu = 0$,

$$\frac{T}{F} = \frac{r}{2\pi} = 121.3 \text{ mm}$$

For $\mu = 0.2$

$$\frac{T}{F} = 24.3 \text{ mm}$$

A value of $\mu = 0.2$ is fairly typical for lubricated steel surfaces. It may be a little less between the steel rifle bar and the bronze rifle nut.

The return area, A_r, of the chosen drill is 2303 mm². The maximum operating pressure of the drill can be taken as 6 bar; whilst this pressure is not realised in the return chamber of the drill during normal operation, it is approached under stall conditions (when the calculation is likely to be of most use).

So the stall torque when $\mu = 0$ is given by:

$$T = 121.3 \times 6 \times 2303 \times 10^{-4} = 168 \text{ N m}$$

and when $\mu = 0.2$,

$$T = 33.58 \text{ N m.}$$

The latter figure has been demonstrated to be experimentally correct for the chosen drill. The calculated values indicate two important points: the importance of reducing friction by supplying adequate and correct lubrication, and the torque available for rotating the drill is comparatively small. Should this value be applied, it will stop the operation of the drill. This can easily occur if the drill bit is allowed to jam in the hole through an accumulation of drilling debris. This is the main reason why, in most modern heavy rig-mounted drills, the rotation mechanism is provided independently of the impact mechanism by a separate motor able to supply much greater torque.

Drilling performance of a percussive rock drill

Figure 6 illustrates the principle of rock drilling. When the piston impacts the end of the drill stem, its kinetic energy is transformed into elastic energy which then passes down the rods as a stress wave until the interface between the bit and the rock is met, whereupon the

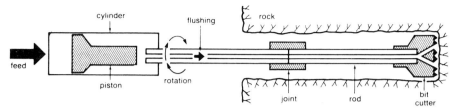

FIGURE 6 – Principle of percussive drilling.

rock fractures. The shape of the stress wave (*ie* the plot of stress magnitude versus time) is an important element in assessing the effectiveness of the drill in breaking rock. The shape can be calculated from considerations of the geometry of the piston and the impact train of drill rod and bit. Its overall magnitude is proportional to the velocity of the piston. The actual calculation of the stress pattern is a complex and time-consuming exercise, which is conveniently solved by a graphical method as described for example in "Down-the-hole drilling using elevated air pressures", M.G. Adamson, Quarry Managers' Journal August 1967. Computer programs are also available to perform the same calculation ("Digital machine computations of the stress waves produced by striker impact in percussive drilling machines", R. Simon, Rock Mechanics, ed. by C. Fairhurst, Pergamon Press).

Figure 7 compares a theoretical calculation with the actual stress pattern measured by a strain gauge.

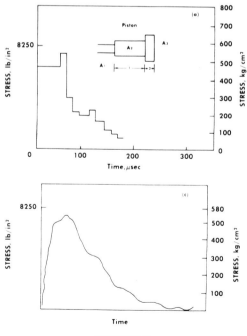

FIGURE 7

The total energy of the stress wave is given by:

$$W = \frac{Ac}{E} \int \sigma^2 \, dt$$

where   W = energy content of the stress wave (J)
        A = cross-section area of the drill rod (m^2)
        c = velocity of sound in the drill (5000 m/s for steel)
        E = modulus of elasticity (2×10^5 MN/m^2 for steel)
        σ = stress in the drill rod (N/m^2)
        t = time (s)

The energy content of the wave can be used to assess the penetration rate of the drill, but it is difficult to relate directly the value of W to the drilling speed, because of the variability of different rocks. Although a number of attempts have been made to define a property of a rock which allows a calculation of its "drillability", none has been entirely successful. There is no substitute for drilling trials in the actual rock.

The stress pattern in the drill stem can also be used to assess the maximum stress that occurs in the elements of the impact train. Most modern drills are limited to an impact velocity of 10 m/s.

The initial stress caused by impact is given by

$$\sigma = \rho \, c \, v$$

where v is the impact velocity of the piston.

If c = 5000 m/s, $\rho = 7.85 \times 10^3$ kg/m^3 (for steel), and v = 10 m/s,

$$\sigma = 7.85 \times 10^3 \times 5000 \times 10 = 393 \text{ MN/m}^2$$

A stress level of this magnitude will be recognised as a very high one for repeated applications and calls for a good quality steel.

A number of specialist manufacturers make drill rods and screwed couplings. A wide variety of different types of thread and couplings are available, the choice is very wide and reference should be made to the manufacturer for recommendations.

Most modern drill bits use a tungsten carbide insert, which can adopt several different forms:

- for small diameter holes a single chisel bit;
- for larger diameter holes a cross or X-shaped bit;
- for the largest sizes, button bits.

Button bits have come into widespread use in recent years, mainly because they require little or no dressing to retain their drilling efficiency, whereas chisel bits require regular sharpening.

MINING AND QUARRYING EQUIPMENT

Many of the tools described under Contractor's Tools are also employed in mining and quarrying, although the emphasis in modern mining is towards the removal and processing of large quantities of rock, for which high powered drilling rigs are required. The principle of rock drilling is the same, whether a hand-held drill or a multi-head drilling rig is used.

Mining is one area where hydraulic drills are offering a real challenge to the former superiority of pneumatic drills because, when the emphasis is on maximising drilling speed, hydraulic power comes into its own. However, a great deal of mining is still done using compressed air for the drilling process, sometimes combined with hydraulic power for support of the drills and for applying feed and rotation forces.

Hand-held rock drills

The small sinker drills discussed in the chapter on Contractor's Tools are occasionally used in mines and quarries, mainly for work such as drilling for secondary blasting. Two other tools, used in considerable numbers are air-leg drills and stopers. These two drills were first used as a stage in the development towards the modern powered support; they take out some of the physical effort that was formerly needed when using a simple sinker drill.

Air-leg or pusher leg drills

An example of a drill of this kind is shown in Figure 1. The drill as shown is basically a standard sinker drill, without the handles and supported by a pneumatic cylinder which, under pressure, is able to support the weight of the rock drill and supply the feed thrust for drilling. The support leg is hinged to the drill body, so there can be any chosen angle between the drill axis and the leg. This drill is used primarily for tunnelling and mine development purposes.

Figure 1 shows the drill in a rather ideal geometric arrangement, where, if the correct pressure is supplied to the leg, the feed force, weight and leg force are in balance. In practice, the geometry changes considerably during the drilling of the round, so that for some of the time the ideal balance cannot be achieved without the operator applying an

FIGURE 1 – Air-leg drill. *(CompAir Holman)*

extra manual load. When the hole has advanced some distance into the rock, the drill rod itself can give some extra support by bearing against the side of the hole. However, friction against the sides of the hole slows down the rotation and reduces the drilling speed. This kind of drilling requires constant attention from the operator, and also requires a great deal of hard physical effort; in modern, large scale mining and tunnelling it has been largely superseded by techniques using support rigs.

The drill shown is of integrated design, which means that all the controls for drilling and for supplying air to the leg are incorporated in the backhead of the drill. Air passes from the backhead control via passages in the hinge pin to the cylinder. The operator is able to select feed or retract and adjust the feed force from the one control tumbler.

Note that two hoses supply the drill: the larger is for delivering the compressed air; the smaller is for the flushing water, which passes down through the centre of the drill and washes out the rock chippings. Water, rather than air, is universally used for flushing in underground drilling. It has been found that air flushing causes harmful dust to be released into the air, which in the past has been responsible for crippling lung diseases such as silicosis.

Stoper drills

An example of a stoper is shown in Figure 2. This is very similar to the air-leg drill, except that it is designed for overhead work in a stope (a chamber formed for excavating ore) or for drilling a raise (a shaft excavated upwards). The support leg is rigidly attached to the body of the drill and controlled in a similar way to the air-leg drill.

A stoper is also employed for roof bolting (a technique for strengthening a roof in a mine stope by bolting it back into sound rock with long bolts). A stoper intended for rock bolting

FIGURE 2 – Stoper drill. *(CompAir Holman)*

can be supplied with a clockwise rotation, rather than the more usual anti-clockwise rotation mechanism so that, as well as drilling the hole, it can also tighten the nuts.

Rig mounted drills

Rig mounted pneumatic drills, operating on very much the same principle as the hand-held drills described above are available for underground and surface work.

In recent years, there has been a tendency to move from pneumatic to hydraulic drilling, since hydraulic operation is much more efficient in terms of energy consumption. However, the hydraulic drill is more of a precision tool requiring a higher level of maintenance and the initial cost is higher. The majority of hard rock drilling is still done by pneumatic drills.

Figure 3 shows three drills suitable for this kind of work. In this type of drill the rotation is obtained by a separate rotation motor, which is usually of the meshing gear type. This makes it possible to generate a much higher torque than with the rifle bar rotation discussed in the chapter on Contractor's Tools. The rotation may be separately controlled in speed and direction to meet a variety of drilling conditions. There is also a separate supply of air or water for flushing the hole.

The drill is capable of drilling 100 mm diameter holes several metres in length. The drilling speed depends upon the type of rock drilled and to some extent on the length of

AIR MOTORS

General characteristics

Air motors should not be looked upon as a substitute for hydraulic or electric motors. They have their own characteristics which make them ideal in certain applications.

They offer a compact and lightweight source of rotary power, reversible and easily adjustable in speed and torque. They possess characteristics similar to those of d.c. series-wound electric motors.

Air motors can be stalled indefinitely and start immediately with maximum torque. They can be designed to produce equal power in either direction of rotation, merely by reversing the supply and exhaust ports, although maximum efficiency is usually obtainable only when the rotation is uni-directional. They can operate at any speed throughout their design range, and are easily geared to produce maximum power at any required shaft speed. They can be run from any available compressed gas, *eg* from natural or a process gas. They can be run at any attitude.

When comparing pneumatic motors with the alternative using hydraulic or electric power, the following should be borne in mind:

- There is no heat build-up when continuously stalled. When the load is reduced to allow the motor to turn, it will resume normal operation. They can tolerate being driven counter to the applied pressure. The air flow through the motor acts as a self cooler.
- Maintenance is low compared with hydraulic motors.
- There is no sparking, so they are safe in explosive atmospheres. In wet conditions, there is no shock hazard.
- When compared with electric motors they have a higher power/weight ratio (for the same output, the weight is about one-third).
- The moment of inertia is lower than electric and hydraulic motors, so they reach maximum speed quickly and brake instantly.
- They can be installed and operated in any position from horizontal to vertical.

Performance characteristics

All pneumatic motors of any design possess similar theoretical performance characteristics, illustrated in Figure 1. The torque is a maximum at zero speed, although in practice

TABLE 1 – Survey of characteristics of the most important types of pneumatic motors *(Atlas Copco)*

Type characteristics	radial piston	Displacement piston link	vane	gear	Dynamic turbine
Maximum working pressure, bar	100	8	8	100	8
Output range, kW	1.5 to 30	1 to 6	0.1 to 18	0.5 to 50	0.01 to 0.2
Maximum shaft-speed, r/min	6000	5000	30 000	15 000	120 000
Specific air consumption, l/kJ = l/(W.s)	15 to 23	20 to 25	25 to 50	30 to 50	30 to 60
Maximum expansion ratio	2:1	1.5:1	1.6:1	1:1	-
Number of cylinders or working spaces per rev.	4 to 6	4	2 to 10	10 to 25	Single-stage
Torque variation during one revolution as percentage of mean value	30 to 15	60 to 40	60 to 2	20 to 10	-
Seal	Piston/ring valve clearance	Piston/ring valve clearance	Gap/positive	Gap/positive	Clearance
Lubrication	Sump and/or air-borne	Air-borne	Air-borne	Air-borne	Only bearing lubrication
Maximum internal relative velocities, m/s	25	20	30	30	70

to limit the maximum speed; it should not affect the peak power. One basic motor can have its characteristics changed by altering the governor weights or spring stiffness.

Table 1 shows the most popular types of motors that are available, with their characteristics.

Vane motors

These motors are similar in concept to vane compressors. Torque is developed by pressure difference on the vanes. There may be between 3 and 10 vanes per motor, but usually 4 or 6. Vanes are usually radial (the centre line of the vane passes through the axis of the motor), but some manufacturers prefer to angle the vanes slightly off the vertical. Vane motors are recommended for high speed operation up to 30 000 r/min. They tend not to be very effective at low speed, because they rely on centrifugal force to seal the periphery of the vanes and without this seal, the pressure cannot be sustained. The air pressure is usually limited to 6 bar in vane motors.

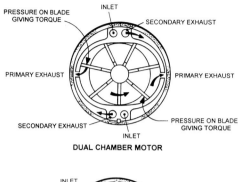

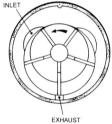

FIGURE 5 – Two configurations for a vane motor. *(Desoutter)*

The more usual vane motor is a single chamber, but it is also possible to make a dual chamber motor. The two are shown in Figure 5. The advantages of a dual chamber are a higher torque at lower operating speeds.

The detailed design of an air motor differs according to whether or not it is intended to be reversible. Reversible operation is achieved by making the inlet port into the exhaust and vice versa; such a design is bound to be a compromise, so rather better efficiencies are available in single direction motors.

The vanes are the vulnerable components of a motor. At very high speed the centrifugal force is so high that frictional forces can cause rapid wear; and at low speed the centrifugal force is too low to ensure sealing contact at all times. If a motor is continuously operated at low speed or under conditions of frequent reversals of direction, the vanes tend to hammer the stator and suffer impact failure. A variety of different materials have been tried for vane manufacture but it has been found that resin-impregnated, fibre-reinforced materials are generally best; carbon reinforced materials are used for heavy duties. The conditions in a motor are quite different from those in a vane compressor, where the speed range is limited, so the same vane materials will not necessarily do. Good air-mist lubrication is essential to reduce friction and wear. The air should be filtered to 64 micron or better. Refer to the manufacturers for recommendations for suitable lubricants. There may be different lubricants specified for the gear-box, the bearings and the oil mist.

Small vane motors are use for hand tools such as drills, grinders and screwdrivers. They

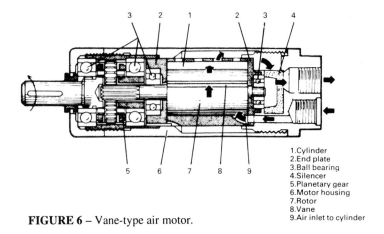

1. Cylinder
2. End plate
3. Ball bearing
4. Silencer
5. Planetary gear
6. Motor housing
7. Rotor
8. Vane
9. Air inlet to cylinder

FIGURE 6 – Vane-type air motor.

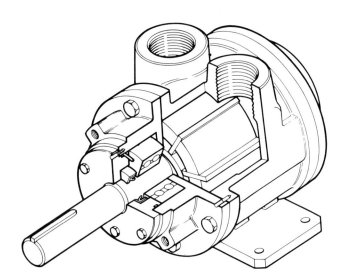

FIGURE 7 – Reversible vane motor for industrial applications. *(Fenner Fluid Power)*

often incorporate a step down gear box in the main casing as in Figure 6. Larger sizes are used for winches, pump drivers, drive motors on pneumatic drill rigs and general industrial applications. Figure 7 shows a large foot-mounted industrial vane motor.

Piston motors

Piston motors operate at lower speeds than vane motors, the limiting factor being the inertia of the reciprocating parts. Free running speed is usually 3000 r/min or less, with the maximum power being developed at 1000 to 2000 r/min.

The majority of piston motors are multi-cylinder units, which may be in-line, vee, H,

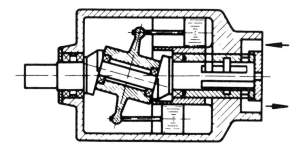

FIGURE 8 – Operating principle of an axial piston motor. *(Atlas Copco)*

flat-four, radial or other variants. In the main, however, the vee is the most popular configuration, with the radial configuration equally popular for three or more cylinders. A radial configuration offers the possibility of weight reduction and is frequently adopted for larger motors. Either an odd or even number of cylinders can used; three-, four-, five- and six-cylinders are commonly found.

The vee-four motor normally has two cylinder banks at 90° to each other. Drop forged two-throw crankshafts are normally used with two connecting rods on each crank pin. Distribution valves may be oscillatory or rotary. Valve timing can be arranged to give equal power in each direction, or may be biased to provide asymmetrical timing and increased speed and power in one direction of rotation.

The typical radial motor has all connecting rods mounted on a common crank as in Figure 8. Power and exhaust strokes are controlled by suitable distribution valves and a two-stage exhaust is often adopted. In this case the inlet valve opens just after top dead centre and remains open for 160°; the inlet valve then closes and a large primary exhaust valve opens quickly to release the air to atmosphere. After a further 40° of rotation, this valve closes and a smaller exhaust valve opens to continue the evacuation of the air.

Reversal of direction of rotation is obtained by interchanging the inlet and secondary exhaust connections, which can be done by means of a control valve some distance away from the motor.

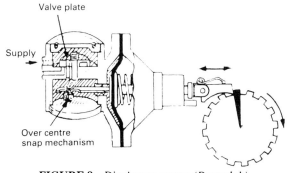

FIGURE 9 – Diaphragm motor. *(Dusterloh)*

FIGURE 10 – Free piston air motor. Suitable for speeds up to 800 rev/min
and a starting torque of 13.5 Nm. *(Dynatork)*

Axial piston motors can have 4, 5 or 6 pistons, but are limited to smaller sizes, less than 2.5 kW. They have the advantage of low inertia, high starting torque and so they reach operating speeds almost instantly. See Figure 9.

Piston motors develop high starting torque, which falls with increasing speed. Both the torque and power are directly dependent on supply air pressure. The torque characteristics can be modified to some extent by detailed design such as the valve timing and valve sizes. They are positive displacement machines, so the air consumption is directly related to speed, within the limitation of the air passages to supply that air. Power output is of the order of 5 to 10 kW/litre of cylinder volume.

Running speeds are generally low to moderate, although small machines may be designed to run at high speeds.

Motors may have output gears to increase the torque and lower the speed.

The overall characteristics of piston motors make them suitable for a wide range of applications. With modification some standard designs can often work satisfactorily down to 0.35 bar. Continuous lubrication is essential and the method of lubrication is dependent upon the design and the need to separate the various lubricants that may be used. A lubricant containing an emulsifying agent may be desirable in a cylinder to prevent water washing, but would not be suitable for splash lubrication in a crank case.

Another variation of piston motors employs a free piston acting on an elliptic cam ring illustrated in Figure 10. The air supply to the cylinder comes from a rotary valve on the main output shaft. Because the design has no need of a connecting rod, the air supply need not be lubricated, which makes it useful for process industries where the presence of oil could not be tolerated. This type of motor is available with either a metal or plastic (acetal) body. It is particularly suitable for low speed applications (up to 800 r/min), and can be driven through a reduction gear box for slow speed process control applications.

Turbine motors

These are very high speed motors capable of generating a small torque only; in the stalled condition the torque is practically negligible. The range of speeds is limited in the range of 50 000 to 80 000 r/min, with the torque a maximum in the mid-range. Applications are limited to those where high speeds with light loads are suitable – high speed pencil grinders and dental drills. Like other air motors a turbine motor can be stalled without damage. Some turbine motors can be fitted with mechanical governors to limit the maximum speed.

Turbine motors operate with clearances throughout, with no sliding or rubbing other than the bearings. Bearing friction is thus the limiting factor governing free speed. Turbines can operate well on dry air – no lubrication other than in the bearings which can be self-lubricating or have their own lubrication system; this can be an advantage in certain applications.

Impulse turbine motors

Originally developed in small powers up to 2.25 kW, very much larger turbine motors are now available up to 70 kW. These are normally single-stage impulse turbines coupled to a planetary gear reduction set. They are competitive with vane and piston motors, being smaller and lighter for the same size.

Gear type motors

These motors are based on the meshing together of gears. The simplest and most common type of gear motor is based on a spur gear pair (Figure 11), although other types are available. Helical gear motors have better sealing characteristics and for that reason are claimed to be more efficient in the use of air, but they are more expensive. A helical gear motor having an axial rather than a radial flow path, designed to the same principles as a screw compressor, is likely to be the most efficient of all, but except for special applications its cost will be prohibitive.

Roots motors (similar to the Roots compressor, described in the Compressor chapter) can be included in this category, but they have an inherently high leakage rate and are not normally competitive as motors.

As can be seen from Figure 11, there is little or no expansion of the air between admission and exhaust, compared with the vane or piston type, and this limits the power that can be extracted from the air. In spite of this the spur gear motor is in practice as efficient as a vane or piston motor and, provided that adequate lubrication is supplied with the air, is very robust and has a long life. One advantage of a gear motor is that there is very little variation between minimum and maximum starting torque, which makes it suitable for such applications as starter motors. It is more compact than the radial piston motor and has better low-speed performance than the vane motor.

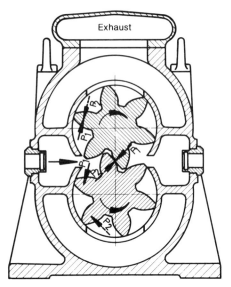

P₁ is the supply pressure,
P₂ is the exhaust pressure.
Power is generated by P₁ acting on the sides
of the teeth.

FIGURE 11 – Mode of operation of gear type motors. The pressure medium enters in the
direction of the arrow.

Motor control systems

The simplest control system is a manual on/off valve. For reversible motors a two way
spool valve is required, which may be operated either by a simple hand control or remotely.
Differential power in forward and reverse directions (as may be required in hoists) can be
achieved in three ways:

- through pressure control valves in the supply lines,
- by a biased power spool valve,
- by a restriction in the lines.

A common requirement on air motors is an automatic brake to prevent creep when the
air is turned off; it is essential for hoists and winches which have to hold the load safely.
Most manufacturers supply the brake as a module, mounted on the output face of the
motor. It consists of two spring-applied shoes pressed against a central hub. When air is
applied to either of the input ports, a shuttle valve sends air to the brake which overcomes
the springs, allowing the motor to turn. See Figure 12.

Gear boxes for air motors

Manufacturers supply gear boxes for industrial motors, either flange or base mounted. One
company, for example, offers gear boxes with ratios from 2.3 : 1 to 85 : 1. Gearing may

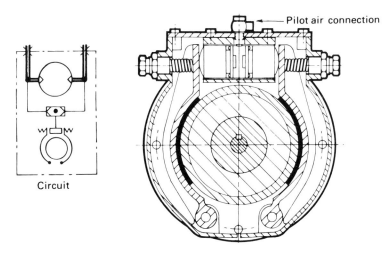

FIGURE 12 – Air motor brake.

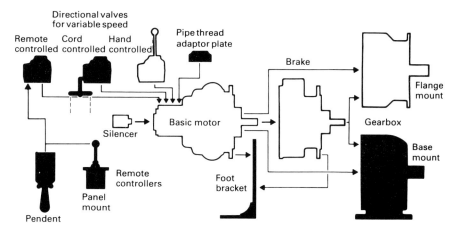

FIGURE 13 – Air motor with a variety of options. Example shows a motor with hand controller valve, gear box, brake and silencer.

be spur, helical, bevel or worm. Figure 13 shows how a basic motor may be supplied with a variety of different controls, brakes and gear boxes. An efficiency of about 85% can be assumed for the gears.

Summary of characteristics of air motors

It will be apparent that air motors are available to meet a variety of applications, and so performance characteristics should be obtained from the manufacturers. From published data on power output, the values of Table 2 have been selected.

The rotational speed at which the maximum power is reached depends on the motor design.

TABLE 2 – Typical power/air consumption values for commercial motors

Motor type	Power (kW)	Relative power (kW per litre/s)	Manufacturer
Vane	1.5	0.039	Atlas Copco
Vane	3.0	0.045	Fenner
Vane	8.0	0.057	Fenner
Radial Piston	2.0	0.055	Atlas Copco
Radial Piston	6.0	0.046	Fenner
Radial Piston	16.0	0.053	Fenner
Free Piston	0.075	0.03	Dynatork
Gear	9.0	0.055	Dusterloh
Gear	66.0	0.059	Dusterloh
Gear (10 bar)	6.5	0.043	Dusterloh
Gear (10 bar)	14.0	0.049	Dusterloh

Motors chosen are a small selection of those available. Pressure 6 bar, except where quoted.

APPLICATIONS FOR AIR MOTORS

Air motors are used in a wide range of applications, such as supplying the motive power for winches, cranes, pumps, dispensing machines, stirrers and agitators. The are to be found in drilling rigs (see the chapter on Mining and Quarrying Equipment), on account of their inherent safety.

They are found in a range of rotary hand tools such as drills, screwdrivers, impact wrenches and grinders where their light weight makes them ideally suitable for continuous manual operation.

Hoists and winches

Lifting and hoisting machines can be powered by either air cylinders or air motors. Pneumatic winches are conventional winching mechanisms powered by air motors, although some types of winching action are also performed by air cylinders.

Air motors or cylinders can be used to advantage where other types of driver are less economic to operate or are excluded because of the conditions involved, for example in high temperatures or in explosive atmospheres. An air motor has the advantage of being simple to control. Motor hoists are equally suitable for fixed installations or for running on an elevated rail.

Cylinder hoists

Cylinder hoists have a more limited field of application than motor hoists when working at extreme or variable heights. Cylinders tend to be more bulky than motors of the same capacity, and need to be able to accommodate the hoist length in the overall length of the cylinder unless double or triple extension cylinders or pulley mechanisms are used.

Cylinder hoists are best suited to fixed installations, although they may be trolley mounted. For particular lifting operations, they have an advantage over air motors in that they can provide greater load stability and a more rigid system.

Air motor hoists

The advantages provided by motor-powered air hoists include:
*   small, light weight compact construction
*   low maintenance costs

- variable lift speeds
- safe if stalled or overloaded
- continuous operation if required
- suitable for use in explosive atmospheres.

The characteristics of a motor hoist or winch are determined by the type of motor fitted. A piston motor would be used when a high starting torque is required; heavy duty or general purpose hoists are usually fitted with piston motors. Vane motors are used where an economical, compact, light weight unit which does not have frequent stop/start cycles is appropriate.

The basic hoist unit can be equipped in different ways for lifting – wire roller or link chain. Control can be by rope or pendant. Rail-mounted units can also incorporate traction drive to provide traverse motion along a horizontal rail and up gradients. An extension of this is the mounting of the hoist on a wire rope suspended from a safety booth above the working area; the hoist runs up and down the wire rope which can also be swung to one side to provide pick-up over a large area. This form of suspension is limited to smaller capacity hoists with a lifting capacity up to 100 kg. The all-up weight of an air hoist can be as low as 1.5% of its lifting capacity.

Table 1 gives some typical performance data for air hoists.

TABLE 1 – Typical performance data for air hoists

Capacity tonne	Lifting speed loaded m/min	Air consumption litres/metre	litre/s
0.25	18.6	120	40
0.5	12.6	180	40
1	6.3	360	40
2.5	3.2	750	40
5	1.6	1500	40

The free lifting speed (*ie* under no load) is approximately twice the speed under load unless governed. The air consumption, expressed as litres/metre when running free, is about 90% of the consumption under load.

Air winches

The air winch is an efficient device where any force has to be applied by a rope or cable, such as heavy hauling in mines, workshops, shipyards, ships and construction sites. They are available for applying a force up to several tonnes in useful speeds up to about 20 m/min. Winches have either vane or piston motors with a large number of vanes or pistons to give good slow speed performance, particularly a high break-out torque. Normally a winch will incorporate a reduction gear box built into the drum and an automatic spring loaded brake which ensures fail-safe holding when the air control lever is put into the neutral or if there is no air supply. As with all motors, it is essential that the air supply is lubricated and filtered. Remote control is usually available as an option.

Figure 2 shows a typical winch.

FIGURE 1 – LLA chain hoists. *(Atlas Copco)*

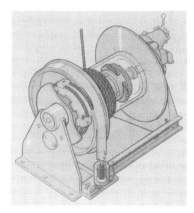

FIGURE 2 – Rope winch. Note the planetary internal gear and the strap type brake. The rope
is locked to the drum by clamp jaws and load-relieving turns of the rope.

Compressed air starters

Heavy engines, such as ship's diesels, can be started by direct injection of air into the
engine cylinders. The air pressure is usually high (up to 40 bar), so special safety
precautions have to be taken, which makes this type of starting system difficult to apply
to the smaller type of mobile engine. The compressed air for this application is held in a

high pressure reservoir and passes via a distributor valve, which is usually mounted on the engine camshaft, to the correct cylinder. A non-return valve in the cylinder head ensures that there is no return flow after the cylinder has fired. The engine can be brought up to speed very quickly and the system is economical in the use of air.

Apart from this specialist application, when compressed air starting is referred to it means the use of an air motor as an alternative to an electric motor, driving a gear on the flywheel rim.

Air motor starters

Any type of air motor can be used as a starter motor – vane, piston, gear or turbine. Most starter motors have been of the vane type, because they are compact units which can be fitted in the space of an electric starter. Indeed some manufacturers offer a pneumatic starter unit as a bolt-on replacement for an electric motor. In small engines, the meshing of the starter pinion is accomplished in the same way as in an electric starter, with a Bendix drive, *ie* at the commencement of the turning of the motor, the pinion is pushed by a coarse thread into engagement with the rim gear. Figure 1 shows a typical arrangement.

FIGURE 3 – Air starters based on gear type motors. Top figure shows a manually operated starter with an external valve. Bottom figure is an automatic starter with a sensing valve. This design is capable of generating a starting torque up to 330 Nm. *(Dusterloh)*

One disadvantage of the vane motor is that its starting torque is less than its slow-speed running torque, but starting a diesel engine requires maximum torque at its breakaway point. Both gear and piston motors have better starting torque characteristics than a vane motor, and theoretically are more suited to this application.

Air starters have an advantage over electric starters in that the motor can be rotated slowly by admitting air through a small pilot bleed until the gears can mesh. Instead of a Bendix drive, the meshing can be accomplished by an air piston and the disengagement by a spring. The operation can be manually controlled, in which case before the full air supply is admitted to the motor, a visual (or audible) check is needed to confirm engagement of the starter gear. Alternatively for standby sets, where it is required that an engine should start without manual intervention, it can be automatically controlled by means of a sensing valve which ensures that the main air valve does not open until there is complete meshing engagement of the pinion.

Air supply for a starter is typically taken from a reservoir charged by the air brake system on a vehicle or by a separate compressor for shipboard installations. For emergency use, the reservoir can be refilled from a high pressure gas bottle.

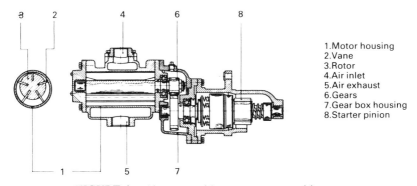

1.Motor housing
2.Vane
3.Rotor
4.Air inlet
5.Air exhaust
6.Gears
7.Gear box housing
8.Starter pinion

FIGURE 4 – Air starter with vane-type motor drive.

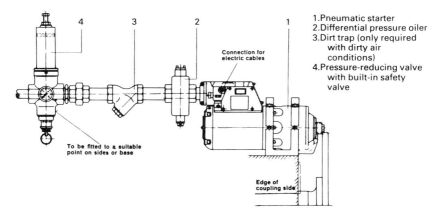

1.Pneumatic starter
2.Differential pressure oiler
3.Dirt trap (only required with dirty air conditions)
4.Pressure-reducing valve with built-in safety valve

Connection for electric cables

To be fitted to a suitable point on sides or base

Edge of coupling side

FIGURE 5 – Remote-controlled air starter, saddle-mounted and ancillary parts in the main air supply line.

Unsilenced compressors and road breakers are not available today (at least in the developed world), but the values in the table are useful in indicating the progress of silencing techniques. It can be seen that the breaker is by far the noisier piece of equipment and swamps the noise from the compressor. It makes very little sense to pay for an expensively silenced compressor and use it to operate a noisy breaker. As a general statement, the noise of a compressor can be reduced to any prescribed level, merely by adding progressively better treatment to its enclosure. A compressor can be as bulky as necessary without affecting its functioning; the only drawback is the expense. The same is not true of a breaker, which has to be both efficient and convenient to use. If a breaker cannot be handled, it will not find favour with the operator; so however well silenced it may be, ergonomic considerations will predominate in the choice of a tool.

When trying to reduce the total noise on a site it is clearly desirable to tackle first the most noisy equipment but there may be a practical limit to the noise reduction possible for certain kinds of equipment. The limit appears to have been nearly reached by a modern pneumatic breaker working on a conventional percussion system. Even the use of an hydraulic or electric breaker will not result in a quieter tool – an hydraulic breaker contrary to expectations is just as noisy as a pneumatic one, because most of the residual noise comes from the vibration of the drill bit.

Noise reduction in pneumatic tools

Most tool manufacturers apply some sort of noise reduction treatment to their tools. Figure 1 shows the proportion of noise from the various sources in a road breaker. It can be seen that the first and largest noise source to tackle is the exhaust noise, followed by the ringing noise from the steel, and then the internal clatter of the working parts. Fortunately exhaust noise is reasonably easy to suppress, at least in theory.

The principle to follow is to reduce the velocity of the jet noise from the exhaust port. The pressure ratio at exhaust is as high as 3 to 1, which implies that the exhaust velocity is sonic and the noise is produced by turbulent mixing of the high velocity jet. The theory

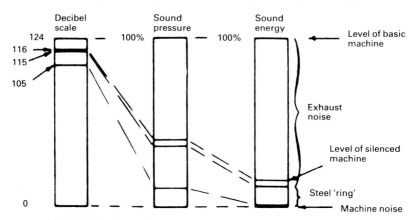

FIGURE 1 – Comparison of the decibel scale with sound pressure
and energy for an experimental silenced rock drill.

on which the silencing of engine exhausts is based uses the assumption that the pressure variations are small. Such an assumption is not applicable to pneumatic tools and does not lead to useful designs. The technique that has proved most successful is diffusion of the exhaust stream by a gradually increasing cross section of the air passages. It has to be admitted that much of the design of exhaust mufflers is empirical, and most mufflers have been developed by trial and error. If the only problem were to reduce the exhaust noise, the solution would be easy – it would consist of a succession of expansion volumes joined by restricted passages. However the noise reduction must be achieved without changing the performance of the tool; so there must be no back pressure to impede the motion of the piston and no restrictions in the flow passages which could be clogged by ice formation. Balancing these factors is not easy.

Most manufacturers have found that a flexible plastic such as polyurethane is the most suitable material for manufacture of a muffler. Solid polyurethane is a very robust material which is capable of withstanding hard usage, and ideally it should form an integral part of the tool construction so that it cannot be removed without the tool ceasing to operate. Refer to the chapter on Contractors Tools for a description of the construction.

The next source of noise, particularly in a road breaker, is the steel "ring". As the impact stress wave passes down the tool stem, part of the energy will be absorbed by the road surface, but a proportion of it is reflected back and forth along the length of the tool. The stem is made of a high quality steel and so has a low internal damping, which ensures that it acts as an efficient radiator of noise. Although the energy from this source is small, it occurs at a single frequency and is subjectively very annoying. There have been attempts to suppress the noise from this source by the addition of damping rings to the stem of the tool. Such devices suffered from having a short life and have largely gone out of fashion.

The clatter from the internal working of the tool is reduced by the presence of the muffler itself, particularly if made of a flexible plastic.

The one source of noise which is very difficult to suppress is that produced by vibration of the material being worked. In the case of a road breaker, noise is produced by the shock waves in the road surface. In the case of a chipping hammer or a riveter, noise is generated by vibration of the casting or other component being worked.

Legislation on road breaker noise

Legislation in the European Community limits the sale of pneumatic breakers to those which meet the prescribed noise levels. The EEC Commission Directives, to which reference should be made, are 84/537/EEC and 85/409/EEC. In the U.K. these Directives are implemented by SI 1985:1968 and by the appropriate legislation in other European countries; they quote the maximum levels of noise and the proper test methods to be used. It should be noted that the marketing or use of breakers and compressors (as well as certain other construction equipment) which emit noise in excess of the permitted levels is a criminal offence. Breakers which satisfy the regulations have to bear an approved mark specifying the sound power level (refer to the chapter on Contractor's Tools for an illustration of this). In order to establish conformity with the Directive, the measurements have to be determined at a Test Station approved by the appropriate authority inside the Member Country of the Community. A breaker approved in one country can be freely

TABLE 2 – Approved bodies for pneumatic noise testing

Organisation	Type Examination for	
	Road breakers	Compressors
A V Technology Avtech House Cheadle Heath Stockport SK3 OXU	x	x
Taywood Engineering 345 Ruislip Road Southall UB1 2QX	x	x

imported into another without further inspection. In the U.K., the authority is the Department of Trade and Industry. The laboratories which are able to perform the tests are commercial bodies (rather than Government Laboratories, as in some countries of the Community) and are inspected by NAMAS (National Measurement Accreditation Service). At the present time only the bodies listed in Table 2 are approved to perform tests. They are in commercial competition, so the charge for performing the tests will vary.

It should be understood that the noise level measured according to the Directive is not necessarily typical of the actual noise that is likely to be experienced by an operator of a tool or by a passer-by. The test is performed for type approval of the breaker alone, and so it is devised in such a way that the steel ring and the radiated noise from the concrete block are suppressed. The quoted value is the sound *power* level emitted by the breaker, in the rather artificial test arrangement shown in Figure 2. In order to assess the actual noise exposure in practice, sound pressure measurements should be taken. The type approval test is useful in comparing one breaker with another, but should not be used to determine the noise environment on a particular site.

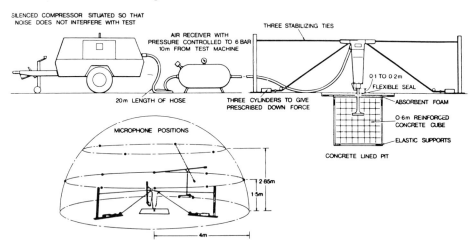

FIGURE 2 – Test arrangement to measure sound power from road breakers according to EEC Test Code. *(CompAir Holman)*

Measurement of noise from tools other than road breakers

So far, road breakers and compressors are the only items of pneumatic equipment subject to legislation. To measure the noise emitted by other items, the only available test procedure is the CAGI-Pneurop Code, which should be studied for further details. This code defines the measurement procedure for all kinds of pneumatic equipment. The readings have to be reported in terms of the sound *pressure* level. The measurement distance from the noise source is 1 m for tools and 7 m for compressors and other large equipment, and the measurement points are situated on the sides of an (assumed) enclosing parallelepiped. This code is useful for assessment of the actual noise experienced by the operator or by the public, since it measures the noise from all sources. It has been shown that when this Code is applied to compressors it gives results (when the sound pressure readings are converted to sound power) that are as accurate as the EEC method. For the reasons given above, this is not the case for road breakers.

TABLE 3 – Permissible sound levels for portable tools

Mass of appliance m in kg	Permissible sound power level in dB(A)/1 pW
m < 20	108
20 ≤ m ≤ 35	111
m > 35 (and devices with an internal combustion engine)	114

Vibration of pneumatic tools

Another health hazard for the user of a percussive tool is vibration. Almost any power tool will generate vibration which, if the level is high enough and the exposure is sustained for long enough, will affect the health of the operator. The main disease caused by vibration is known by a variety of names – Vibration White Fingers, Raynaud's Disease of Occupational Origin or Vibration Syndrome. The Health and Safety Executive prefer the term Hand Arm Vibration Syndrome (HAVS), which includes Vibration White Fingers (VWF). VWF is characterised by intermittent blanching of one or more of the fingers, due to impaired circulation, which gets progressively worse with continued exposure to vibration. There is still much that is unknown about the disease, but the generally accepted view is that it is caused by vibration damage to the peripheral arteries in the fingers; the nerves are also affected. For a discussion on the various methods that are available for the diagnosis of VWF refer to "Hand-Arm Vibration – HS(G)88" published by HSE Books. These methods are not suitable for routine workplace surveillance. Attacks of White Finger are often precipitated by cold. They last about an hour and may be associated with considerable pain as the attack is terminated.

Even if the use of vibrating tools does not result in the disease of White Fingers, the operator may still be adversely affected by the presence of vibration – he is likely to tire earlier and be less effective in his work – so it is sensible to take all reasonable steps to reduce the level of vibration.

Protection of operators from vibration damage

Pneumatic tools which are known to have caused White Fingers include road breakers, riveters, chipping hammers, rock drills and grinders (both hand-held and pedestal). The damage caused by White Fingers is generally considered to be irreversible, so any worker who complains of attacks should be removed from use of vibrating tools and placed in an environment where he is not likely to be subject to cold. Regular checks should be made on those operators who regularly use these tools. Apart from the use of specially designed tools with reduced vibration, the following measures are recommended:

- The tool should be held as lightly as possible consistent with proper control.
- Wearing of gloves to keep warm. Note: there is little evidence that gloves, by themselves, do much to reduce the magnitude of vibration.
- Keep the workshop warm and ensure that operators do not use tools before their hands are properly warm.
- Chisels should be kept sharp and grinding wheels properly dressed.
- Regular periods of rest allow the hands and arms time to recover and circulation to be restored.

Note that there is a recent standard ISO 10819 – Hand Arm Vibration – Methods for the measurement and evaluation of vibration transmission of gloves at the palm of the hand; this can be used to compare gloves, but this is still at a development stage and should be applied with caution.

Acceptable levels of vibration

Several attempts have been made in recent years to assess safe levels of vibration. Any investigation into this subject can only be epidemiological, *ie* vibration injuries and the exposure which causes them can only be assessed after they have occurred; there seems to be no reliable predictive method of determining the chance of an individual sustaining vibration injury.

There are some standards which have been prepared by the International Standards Organisation to which reference should be made. The ISO Standard covering the assessment of human exposure to hand transmitted vibration is ISO 5349 (also BS 6842 and DDENV 25349) which embodies the best current knowledge on vibration exposure. The method of expressing vibration level is by use of a weighted root mean square acceleration which takes into account the whole vibration spectrum. Legislation is not yet in place which specifies maximum vibration levels of a tool, although attention is drawn to the Machinery Directive 89/392/EEC (amended by 91/368/EEC) implemented in the U.K. by The Supply of Machinery (Safety) Regulations 1992 (SI 1992:3073). This Directive requires that the instructions supplied with the tool must state that tests have been done and either the rms acceleration does not exceed 2.5 m/s^2 or if it does, the value must be stated; the test regime under which these measurements are made must be the appropriate one for the tool being tested, see below. Most manufacturers now take 2.5 m/s^2 as a target for vibration levels of their tools.

Figure 3 gives the recommendations of ISO 5349. It should be noted that the sustainable

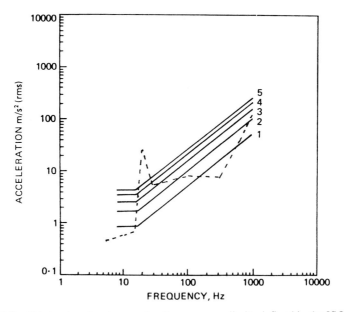

FIGURE 3 – Third-octave hand-arm vibration exposure limits defined in the ISO Standard 5349. The five curves correspond to the five multiplying factors given in Table 4. Superimposed is the vibration level of a typical road breaker. Note the maximum response at the tool operating frequency.

TABLE 4 – Multiplying factors for various exposure durations
(as given in the International Standards Organisations Standard 5349)

Exposure time during 8 hour daily shift	Multiplying factors for various minutes of interruption in vibration exposure per hour				
	Not regularly interrupted < 10 min/hr	Regularly interrupted			
		10 to 20	20 to 30	30 to 40	> 40 min/hr
Up to 30 min	5	–	–	–	–
30 min to 1 hour	4	–	–	–	–
1 hour to 2 hour	3	3	4	5	5
2 hour to 4 hour	2	2	3	4	5
4 hour to 8 hour	1	1	2	3	4

vibration depends on the amount of exposure during a working day and on the frequency of rest periods. Any recommendations must be provisional in the light of present knowledge on the subject, but most authorities accept the general validity of the data. The standard gives much useful information on the precautions which should be taken to prevent vibration injuries.

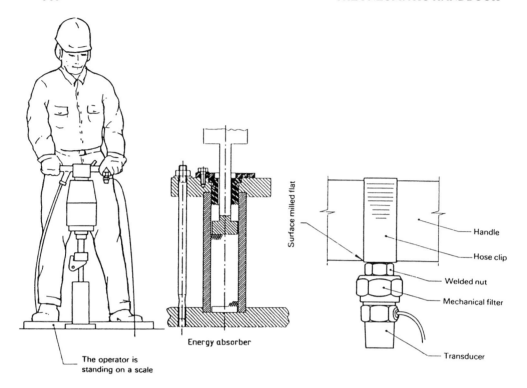

FIGURE 4 – Arrangements for performing vibration tests on a roadbreaker. Note the energy of the impact blow is absorbed by a steel tube 60 mm diameter filled with hardened steel balls of 4 mm diameter.

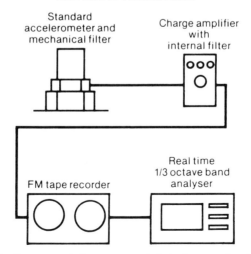

FIGURE 5 – Recommended measurement chain for assessing the vibration of percussive tools.

Measurement of tool vibration

The usefulness of vibration standards necessarily depends on an exact method of determining the vibration of the tools. Standards for the measurement of handle vibrations of percussive tools have been issued for chipping hammers, rivetting hammers, rock drills, rotary hammers, grinding machines, paving breakers, hammers for construction work, impact drills, impact wrenches and orbital sanders, most of which are operated by compressed air. These are to be found in ISO 8662 (BSEN 28662).

There are two fundamental problems in the measurement of vibrations in percussive tools. The first is the establishment of a consistent means of absorbing the energy of the tool. It is not practical to allow the tool to operate in the same way as it would in practice because it would be impossible to ensure consistency; in a road breaker, for example, the variability of the concrete would make comparisons between different testing stations impossible. It might be thought that the energy absorbing method illustrated in Figure 2 would be satisfactory, but this has not been adopted by ISO. Instead the percussive energy is absorbed in a steel tube full of hardened steel balls as illustrated in Figure 4. Tests on this method have shown that the reflected energy from the absorber is of the order of 15 to 20 %, which is typical of a working situation. In ISO 8662-5, detailed dimensions are given for various sizes of tool. The down force, expressed in newtons, to be applied by the operator is to be 15 times the value of the mass of the tool in kilograms; this is in addition to the weight of the tool.

The second problem is the mounting of the accelerometer on the tool handle. Most modern tool handles are covered in resilient handgrips to reduce high frequency vibrations; the attachment of an accelerometer to these is unreliable, so ISO recommend the use of a rigid adaptor clamped to the handles. Mainly through the work done by Pneurop, the correct techniques for vibration measurement have been established and incorporated in ISO 8662. One feature of the vibration spectrum which has to be borne in mind is the high level of shock present which, unless precautions are taken, can seriously affect the accuracy of the readings. Vibration readings of the order of a few metres per second have to be measured in the presence of short-period shocks several thousand times higher.

Pneurop found that the only satisfactory measurement technique is as shown in Figures 4 and 5. A piezo-electric accelerometer is mounted on a mechanical filter which isolates the high shocks (a mechanical filter is a special accelerometer mount, with a rubber insert, which has a flat response well beyond the measurement frequency). The use of so-called shock accelerometers has been found to be ineffective in this application; all the shock accelerometers that have been tried suffer from a phenomenon known as d.c. shift, resulting in false readings. The analysis equipment must be of high quality; an FM recorder should be used, and the amplifier must give indication of signal overload. For further advice on the mounting of accelerometers, refer to ISO 5348 (BS 7129).

The difficulties inherent in the measurement of vibration makes it imperative that any manufacturer or importer of tools must chose a test laboratory familiar with the technique described above. Such a laboratory may be an in-house facility or a commercial company prepared to do the work. There are as yet no laboratories accredited by NAMAS. It appears that the only one currently able to do this work is situated at ISVR Southampton University, although some manufacturers have their own equipment.

Vibration reduced equipment

Some pneumatic tools have been designed with a degree of vibration isolation and their use should be encouraged where they are available. Most manufacturers now supply this kind of equipment.

There are several methods which are used to suppress the vibration of percussive tools; most of the work has been done on the road breaker. Modern tools have comfortably shaped handles, usually made of rubber or plastic or they have resilient grips which help to take out the high frequency "sting". If the grips are removable, they must be regularly inspected and replaced when worn. A rather more elaborate form of the same idea uses spring bushes and hinges to support the handles; an example of these can be found in the chapter on Contractor's Tools.

The predominant low frequency vibration occurs at the operating frequency of the tool and is the hardest to suppress. One method uses spring/mass isolation of the handles; the spring can be a metallic helical spring or it can use the compressed air as a spring. The sprung mass attached to the handles has to be fairly large to give a cut-off frequency at the operating frequency of the tool (about 16 Hz for a road breaker); another method incorporates an internal mass which moves in opposition to the piston, so as to neutralise the external vibration.

These methods can be effective, and a reduction of the order of 90% is realisable, but a word of warning should be given to anyone considering a purchase. Any artificial flexibility introduced into the handle is bound to affect the response of the operator, and while on a test rig the tool may behave very well, it may fail to meet the test of user acceptance. Because of the extra complexity in construction, the tool may be heavier and more expensive. Special tool bits may also be required. The assessment of vibration exposure is a complex matter, depending not only on the magnitude of vibration of the tool handles, but also on other factors such as the grip force. It would make little sense to reduce vibration yet at the same time require the operator to apply a greater force to keep the tool on the work surface.

Other pneumatic tools

Chipping hammers have also been made available with a degree of vibration isolation, see Figure 6. This tool has a piston on which the air pressure constantly acts on its rear surface, so the reaction force on the handle is constant; air is alternately admitted to the front of the piston and exhausted from it; the piston reciprocates under the unbalanced forces. The force in the front chamber acts only on the tool bit, which is not felt by the operator. Shock reflections from the chisel are cushioned by the front collet. This tool is claimed to have a considerable reduction in vibration when compared with a conventional tool.

Hand held grinders also have high vibration levels. As mentioned in the chapter on Industrial Tools, vibration is produced mainly by imbalance of the wheel so it is important to keep the grinding wheel regularly dressed and balanced. In the chapter on Industrial

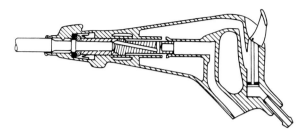

FIGURE 6 – Chipping hammer designed to have minimum vibration. *(Atlas Copco)*

Tools, a grinder is illustrated which incorporates a set of bearing balls which can compensate for imbalance. Fortunately the measurement of acceleration does not the involve the same high shock levels as in percussive tools, so the accelerometer mounting is not as much of a problem.

For further information refer to "Handbook of Noise and Vibration Control" published by Elsevier.

PAINT SPRAYING

There are three basic methods of spray painting: low pressure (the conventional method), high pressure and electrostatic painting. All three can be used for cold or hot spraying. They each have their own particular characteristics, and in order to decide which is the appropriate one for a particular application a number of factors have to be taken into account, such as economy in the use of paint, finish and capacity. Table 1 rates the various methods for suitability. Compressed air is used in each method, and particular care must be taken to ensure that the air is supplied dry and oil-free, if high quality work is required. Air may also be used for preparation of the surface prior to painting. Cleaning off the rust, old paint *etc.*, is done by blast cleaning using shot, steel sand or some other kind of abrasive particle.

Low pressure paint spraying

In this technique compressed air is used both for atomizing the paint and for carrying it to the spray gun. The air pressure used is about 6 bar. Paint is supplied to the gun by one of the three methods illustrated in Figure 1. The choice of method depends on the type and quantity of the paint to be used. For low viscosity paint, suction feed is adequate; gravity and pressure feed are better for higher viscosity paints. The pressure feed method is suitable for large capacity spraying. With suction or gravity feed, the capacity is up to 0.5 litre/min; pressure feed allows up to 2 litre/min. The air consumption varies from 2 litre/s to 10 litre/s according to capacity.

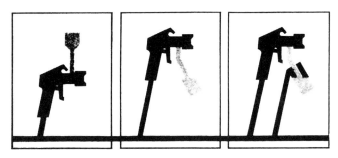

FIGURE 1– Different paint feed. Left: gravity feed. Middle: suction feed. Right: pressure feed.

TABLE 1

Features	Method of application							
	Low pressure spraying	High pressure spraying	Hot low pressure	Hot high pressure	Low pressure electrostatic	High pressure electrostatic	Hot low pressure electrostatic	Hot high pressure electrostatic
Controllability:								
Fan pattern	4	3	4	3	4	2	4	2
Paint quantity	4	3	4	3	4	2	4	2
Low capacity	4	2	4	3	4	2	4	3
High capacity	3	4	3	4	2	3	2	3
Penetration of paint particles	3	4	3	4	2	3	2	3
Uniformity of coat thickness	3	1	4	2	3	2	4	3
Atomization (finish)	4	1	4	3	3	2	4	3
Wrap-around effect	–	–	–	–	3	2	4	4

4 = very good. 3 = good. 2 = acceptable. 1 = poor.

TABLE 2 – Summary of spray systems

Type	Characteristics	Advantages	Disadvantages	Applications
Low pressure (up to 10 bar)	Compressed air used both for paint transport and atomization	Quick and easy adjustment of fan width and paint quantity. Low cost equipment	Heavy paint fog and high paint losses. Limited capacity (2 lit/min max with pressurised feed	Car bodies, office machines, refrigerators, furniture, high class work; spraying primers
High pressure (up to 360 bar)	Necessary pressure generated by a piston pump	High capacity; thicker coatings; minimal paint mist. Suitable for high viscosity paints	Higher cost equipment, less control of finish	Painting large objects, ships, buildings, etc; applying protective plastic coatings
Electrostatic	Can use either liquid or powder paints	Particularly suitable for auto-mated systems; wrap-around coating characteristics. Superior paint economy	High voltage equipment; objects must be earthed; more stringent safety regulations	Car bodies, steel tube items, cycle frames, fences, *etc*

TABLE 3 – Suitability for various types of paints and finishes

Type of paint or finish	System(s)	Remarks
Solvent type (drying by solvent evaporation)	Low pressure spray High pressure spray Electrostatic spray	Cold or hot spraying as applicable. (Dispersion-type paints are not sprayed) Chlorinated rubber paints may not be suitable for spraying
Air drying by oxidation	Brush painting preferred	
Evaporation and chemical reaction type	Low pressure spray High pressure spray Electrostatic spray	Electrostatic spraying limited to special paints
Stoving enamels	(i) Low pressure spray (ii) High pressure spray	(i) Hot or cold (ii) Hot Electrostatic spraying not generally suitable
Bituminous paints	Low pressure spray High pressure spray	Hot spraying preferred Hot spraying preferred
High zinc paints (organic)	Low pressure spray High pressure spray	Zinc-epozy paints may also be sprayed electrostatically Zinc-epoxy paints may also be sprayed electrostatically
High zinc paints (inorganic)	High pressure spray	

The flow rate and spray pattern is controlled by the needle valve in the spray gun. The paint is atomized by the air passing through the nozzle with the result that a mist of paint particles is present in the spray booth, which can be wasteful in the use of paint, particularly when painting open structures such as bicycle frames. The method is used for automobiles, kitchen machines and furniture.

High pressure paint spraying

This method relies on a high pressure pump to supply the paint to the gun. No air is employed to atomize the paint nor does any air issue from the gun, so the method is also known as "airless" spray painting. Air at 6 bar is supplied to the pump which generates a spray pressure up to 360 bar. The air consumption per litre of paint is lower than with low pressure painting.

The paint passes through a tungsten carbide nozzle with a small orifice and is atomized by the high pressure. Because no air is used at the nozzle, there is little or no mist created. Thick coats of high viscosity paint can be applied, so it is suitable for high capacity applications on large structures, ships and buildings. Up to 5 litres/min can be applied.

Electrostatic spraying

A drawback of both high and low pressure spraying is the wastage of paint when spraying open structures and small components. In electrostatic spraying, the method of atomizing and delivering the paint is the same as with either of the two methods described but, in addition, an electrostatic field of 50 to 100 kV is created between the paint and the sprayed object. The paint droplets follow the lines of the charge field, so it is possible to achieve "wrap-around", as shown in Figure 2, and obtain adequate coverage by spraying from one side only.

This method is useful for car bodies, bicycles and small objects; the coat thickness is even and of high quality, which is difficult to achieve on small objects with the other methods.

It is economical in the use of paint, and there are other savings such as lower cleaning costs.

The operator, spray equipment and workpiece are earthed and the paint is electrostatically charged in the gun. Safety is important with this method, so all the safety precautions must be observed; conductive footwear must be worn.

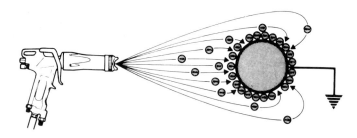

FIGURE 2– Principle of electrostatic painting.

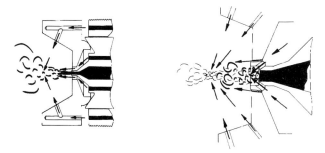

FIGURE 3 – Design of spray nozzle for low pressure spraying. On left, conventional nozzle. On right, improved nozzle showing low pressure mixing regime. *(Kompex Industrial Products)*

Cold or hot spray

Hot spraying allows a paint to be sprayed in a more concentrated form, flows better and produces a more homogeneous coating; it also results in better economy. Hot spraying is particularly suited to the application of thicker coats so is often preferred for priming coats. Low pressure hot spraying is the preferred choice for high quality,gloss finish coats.

Recent developments in the design of spray nozzles has, however, improved the quality of low pressure spraying so that the quality is claimed to be as good as with high pressure systems. The nozzles are designed to work with comparatively low pressure (up to 2.5 bar) and improve the atomization of the paint to produce not only a better finish but less wastage and consequently lower emission of fumes. The design of the new nozzle is shown in Figure 3.

The operating principle for hot spraying is illustrated in Figure 4. Paint is held in a heated container, using either direct or indirect heating. When spraying is not in progress, it circulates in a closed circuit through the pump, heater, and spray gun back to the suction

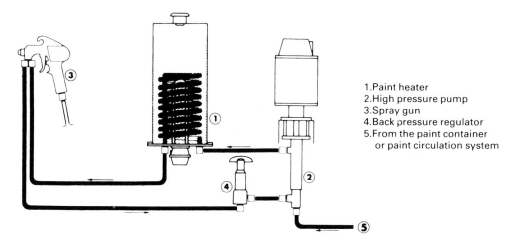

1. Paint heater
2. High pressure pump
3. Spray gun
4. Back pressure regulator
5. From the paint container or paint circulation system

FIGURE 4 – Operating principles for hot high pressure spraying.

side of the pump. A back pressure regulator is used to adjust the paint quality returned to the main circuit.

Paint transport and feed

The paint feed alternatives shown in Figure 1 can be considered as batch feed systems: the paint pot has to be refilled when empty, thus the process is discontinuous. A continuous paint feed system is illustrated in Figure 5. A pressure-operated circulation pump supplies paint to several guns through a ring main. There can be different spray systems on the same line, or the main can incorporate a heater.

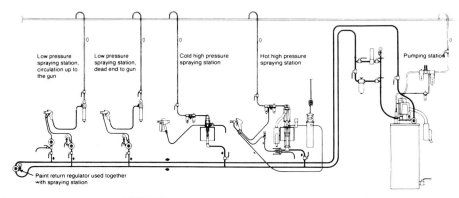

FIGURE 5 – Various circulation systems.

AIR SPRINGS

Pneumatic cylinders as air springs

The air in a closed chamber can be used in much the same way as a metallic spring, *i.e.* it can be used to resist deflection by the compression of the air. The air spring has its own characteristics which differ in some important respects from the metallic spring. It can be used as an alternative to mechanical springs for suspension and for vibration isolation. The simplest form of mechanical spring is the closed pneumatic cylinder, Figure 1.

In all applications concerning air compression, the behaviour depends on the nature of the compression cycle. For very slow speed movements, the compression can be considered to be isothermal, but for high frequency suspension systems the cycle is isentropic (no heat transfer to the surroundings). The following analysis is based on an assumed isentropic cycle, with γ, the index of compression equal to 1.4.

The spring of Figure 1 is assumed to contain an initial air pressure, P_1, which acts in the same way as a pre-load, F_1, in a mechanical spring. When the pre-load is taken up, the

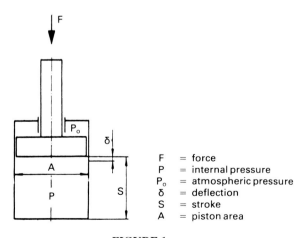

F	=	force
P	=	internal pressure
P_o	=	atmospheric pressure
δ	=	deflection
S	=	stroke
A	=	piston area

FIGURE 1

spring then has an elastic spring rate which, because of the nature of the compression, is non-linear.

$$F_1 = (P_1 - P_0) A$$

For F greater than F_1, the relationship between F and the deflection δ is given by:

$$F = PA \left(\frac{S}{S-\delta}\right)^\gamma - P_0A$$

and the equivalent spring rate is then given by

$$k = \frac{\gamma P A}{S-\delta} \left(\frac{S}{S-\delta}\right)^\gamma$$

When a spring is used for vibration isolation (a common application), the natural frequency is given by

$$f = 2\pi \sqrt{\left(\frac{k}{M}\right)}$$

where M is the supported mass

The above formulae are correct for consistent units. With F in newtons, p in bar, A, S and δ in metres and M in kg, the corresponding relations are:

$$F = 10^5 (p_1 - p_0) A \qquad N$$

$$F = 10^5 pA \left(\frac{S}{S-\delta}\right)^\gamma - p_0A \qquad N$$

$$k = 10^5 \gamma \frac{pA}{S-\delta} \left(\frac{S}{S-\delta}\right)^\gamma \qquad N/m$$

$$f = 2\pi \sqrt{\frac{k}{M}} \qquad Hz$$

In the above analysis, no account has been taken of seal friction, which can modify the relationships but is practically impossible to analyse. The main limitation to the use of pneumatic cylinders as vibration isolators is their reliance on elastomer seals, which are prone to wear and which cause friction. They are, however, useful for applications where a tension or a double acting spring is required.

The performance of a pneumatic spring can be modified by connecting the cylinders to a suitable external circuit as shown in Figures 2 and 3.

Gas springs are commercially available (Figure 4). These are primarily used for supporting horizontally hinged doors, but are also used for machine guards, hood supports and safety doors. They are filled with high pressure nitrogen rather than air and are heavily damped by hydraulic means. They are sealed units, factory set for the desired force characteristics, available for a maximum force up to 20 kN. For design characteristics refer to the manufacturer.

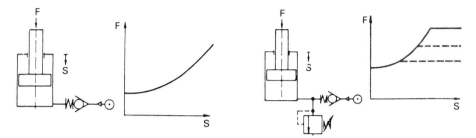

FIGURE 2 – Spring force diagrams – cylinder with non-return valve (left) and with relief valve limiting spring force (right).

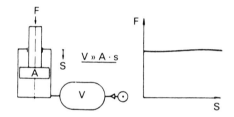

FIGURE 3 – Force diagram with air spring connected to a large reservoir.

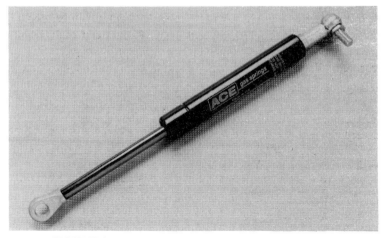

FIGURE 4 – Hydro pneumatic gas springs. *(Ace Controls)*

Flexible bellows

The advantages of the flexible bellows type of air spring, Figures 5 to 8, make them very versatile devices, applicable to both actuation and vibration isolation. The standard bellows has a two-ply construction of reinforced rubber with a maximum working pressure of 7 bar; four-ply construction is also available with a maximum pressure of 12 bar. A maximum force of 45 kN and stroke capability of 350 mm are possible. Special formulations of rubber allow low and high temperature operation (from -50°C to 110°C).

As actuators they can usefully replace conventional rams in short-stroke, high-force compression applications. Advantages include:

- no dynamic seal so no breakout friction,
- ability to stroke through an arc,
- ability to accommodate angular and side loads,
- low maintenance and generally lower cost,
- freedom from fatigue.

The force generated by any air spring depends on the pressure and the effective area. A cylinder has a constant area, but the effective area of a bellows changes as it extends. The force is greatest when the spring is collapsed and lessens as it extends; the variation can be as much as 50%.

As isolators, they have a substantially lower spring rate than a similarly sized rubber or steel coil item, allowing system natural frequencies as low as 1 Hz or lower, if an additional

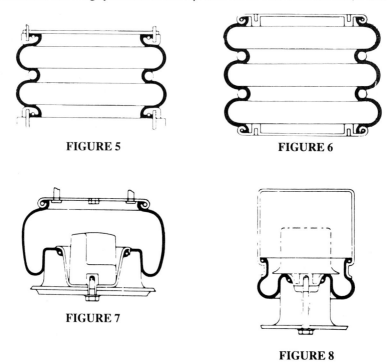

FIGURE 5   FIGURE 6

FIGURE 7

FIGURE 8

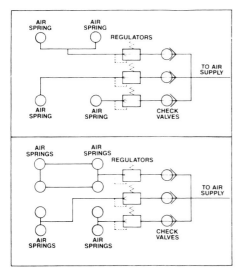

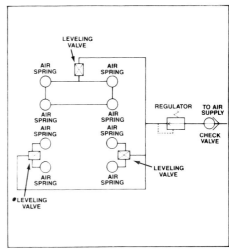

FIGURE 9 – Typical circuits for air supply to an air spring isolator.
Note that in each case there is only a three point control.

reservoir is provided as in Figure 3. They eliminate problems like fatigue and permanent set found in other isolators. The height of air spring isolators can be accurately controlled by regulating the air pressure. With a simple feedback device, such as a levelling valve, the height can be accurately controlled to ± 1 cm; with precision controllers, the height can be maintained to ± 0.25 mm. The air supply circuit needs careful consideration. If the air is sealed in the bellows with an inflation valve, there will be some leakage through permeation over a period of time (a loss of 2 bar a year is typical), so the pressure needs to be regularly checked. Otherwise a live pressure supply must be provided either to each spring individually or to groups of springs, as shown in Figure 9. The principle to follow is that there should be only three points of control on each structure.

Design Techniques when Using Bellows

Bellows are available in a wide range of sizes and configurations. Maximum diameters vary between 150 mm and 700 mm, in the styles shown in Figures 5 to 8. Up to three convolutions are available as a single unit which can be extended by bolting several units together in series with an intermediate steel plate. Suppliers of these units provide curves which enable the static volume/height and the force/pressure relationships to be determined for each style. When designing an actuator system, it is better to rely on these curves than to attempt to work from basic principles.

When used as an isolator, the static curves can be used as the primary information from which to calculate the dynamic spring rate. As in the case of the cylinder discussed above, isentropic compression can be assumed (*i.e.* the use of an index of compression = 1.4), although a value = 1.38 is recommended by one major manufacturer; the difference is likely to be small.

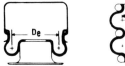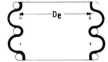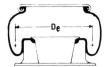

FIGURE 10

In a similar way to a cylinder, the forces can be calculated, but account has to be taken of the variation in the effective area, as shown in Figure 10.

$$F = (P_1 - P_0) A$$

$$F = PA \left(\frac{V_1}{V}\right)^\gamma - P_0 A$$

F_1 is the initial force;
V_1 is the initial volume at a pressure of P_1;
V is the volume and A the effective area at the pressure P.

The volumes and areas are obtained from the static curves.

In order to calculate the dynamic spring rate (where this value is not given in the published data), a first force at a height slightly above and a second force at a height slightly below the working height are calculated (a total height difference of 25 mm can be used); the spring rate is then calculated as the force difference divided by the height difference. Using this technique, the spring rate is given by a relationship of the form:

$$k = \left\{ P \left[A_2 \left(\frac{V}{V_2}\right)^\gamma - A_1 \left(\frac{V}{V_1}\right)^\gamma \right] - P_0 (A_2 - A_1) \right\} \frac{1}{L}$$

L is the height difference (25 mm, if that is the height difference chosen), and in this instance the suffixes 1 and 2 refer to values respectively above and below the working height.

The natural frequency can be calculated as before, using the calculated stiffness value.

The relationship between the forcing frequency and the natural frequency should be chosen to give the required degree of isolation, as determined from Figure 11.

General consideration in the use of air springs as isolators

It is recommended that, when using the air spring as a vibration isolator, the ratio of the forced frequency to the natural frequency should be about 3:1. This will give an isolation of about 90% (see Figure 11).

As the system is inherently soft, precautions must be taken to ensure stability. The ideal arrangement is to locate the springs in the same horizontal plane as the centre of gravity. Where this is not possible, the guideline to be used is: the narrowest distance between the mounting points should be at least twice the height of the centre of gravity above the plane of the springs.

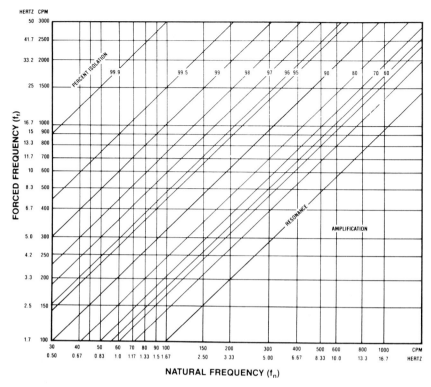

FIGURE 11 – Isolation chart. *(Firestone)*

The air spring should be used at the design height given in the published characteristics, because that height is the condition of maximum stability. Single and double convoluted springs can become unstable at a small distance away from the design height. The lateral stiffness (expressed in N/m) is of the order of 0.2 to 0.3 times the vertical stiffness. Rolling diaphragm types on the other hand can be designed to have a high lateral stiffness equal to or greater than the vertical stiffness.

The inherent damping in an air spring is about 0.03 times critical; this is so small that for most purposes it can be neglected. If damping is required, some external means has to be provided.

PNEUMATIC CONVEYING

Pneumatic conveyance in pipelines

Pneumatic conveying is the transport of bulk materials through a pipeline by air pressure or vacuum. Materials that can be handled range from asbestos with a bulk density of 100 kg/m^3 to crushed stone with a density of 1500 kg/m. The advantages of pneumatic conveying over mechanical conveying include safer working conditions (clean atmosphere and reduced fire hazards), greater flexibility, freedom from contamination and the ease in which a change of direction can be achieved.

It has to be admitted that the cost of air compression is likely to be higher than that of pure mechanical transport, particularly when the transported material requires a high degree of purity in the air. Fragile materials may not be suitable for pneumatic conveying, but most other materials and some manufactured components lend themselves to this form of transport.

The following types of pneumatic conveyors are available:

*   Vacuum system, similar in principle to a domestic vacuum cleaner.
*   Low pressure system, up to 1 bar.
*   Medium pressure system, from 1 bar to 3 bar.
*   High pressure system 3 bar to 8 bar.
*   Pulse phase system.
*   Combination vacuum/pressure systems.
*   Air activated gravity conveyor.

The choice of a suitable system will depend primarily on the material to be transported – its density, particle size, moisture content and abrasiveness. There is a great deal of skill needed in choosing a suitable conveying method, so anyone contemplating installing a system would be well advised to approach a company with a wide experience in the various techniques. One can do no more here than indicate some of the factors that should be considered.

Vacuum systems

This uses a high velocity (up to 40 m/s) airstream to suspend the material in the pipe, using a vacuum up to 400 mbar.

The materials suitable for vacuum transport are dry, pulverised and crushed granular with a small particle size and a low density. Under ideal conditions, it is claimed that the conveying distance can be as much as 500 m, but in practice the maximum distance is likely to be rather less. Optimum design would require the internal pipe diameter to increase in steps along its length so as to stabilize the velocity. As in all systems, the limitation on conveying length is the pressure loss that occurs through pipeline friction. With a vacuum system the pressure loss can be no more than the vacuum depression, but with a pressure system, the loss through friction can always be compensated by increasing the positive pressure. The energy consumption will be between 1.5 and 5 kW hr per tonne of material, depending on the density and conveying distance. This makes it the most expensive method in power consumption, but the simplicity of installation and low capital cost makes it appropriate for many situations.

Vacuum systems are ideal where several pick-up points are required in one line. Another advantage is the ease with which material can be introduced into the pipeline. The simplest way is from an open container such as a ship's hold where the material is admitted with the air. In this case the material may have to be lifted through a considerable height and the system is then known as a pneumatic elevator.

When the material is in a hopper, there are several forms of feed device. One such is a rotary feeder which has a star wheel rotating in a close-fitting housing. For dry materials which flow easily, a simple on/off valve can be used and the material falls into the pipe by gravity. A positive low pressure may also be used to fluidize the hopper.

The suction is usually generated by a turbo blower (centrifugal or axial). The same equipment can be used for both suction and low pressure conveying systems.

Low pressure systems

The distinction between low, medium and high pressure systems is related more to the means adopted for producing the pressure than to their application to different materials. Some materials can be conveyed at all pressure regimes, but the main application of low pressure conveying is for dry, low density materials. Turbo blowers or Roots-type blowers are commonly used for pressures up to 1 bar. For pressures up to 0.3 bar a simple fan may be used. The velocity in the pipe is limited to 20 m/s, so this method is more appropriate for fragile materials than the vacuum method. The power used is between 0.5 and 3.0 kW hr per tonne, depending on density and distance.

Because of the positive pressure in the line, special methods have to be used to introduce the material into the pipeline against the positive internal pressure. Close fitting feed mechanisms are required for two reasons: that air should not be wasted through leakage; and when dusty or unpleasant materials are being handled, they are not blown out of the hopper into the atmosphere. The system is suitable where there is only a single pick-up point with the option of multiple discharge points.

When using a pressure below about 0.2 bar, it is possible to use a venturi-type pick up, provided that the material is suitable and is carefully metered; a venturi feed cannot handle plugs of material or deal with an excess of capacity that would inhibit the venturi effect. A rotary feeder is the customary form of injecting the material into the pipe, but care must be taken that the feed mechanism does not damage the material, Figure 1.

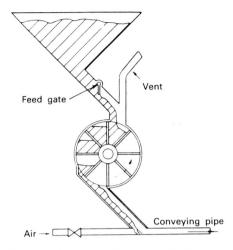

FIGURE 1 – Feed into air stream using a rotary lock to meter powder and
act as a pressure seal.

Medium pressure systems

In this system, the material has to be forced into the line through a feed pump. It is most successful when dealing with materials which can be fluidised and then behave like viscous liquids (known as fluid solids); dry and fine powders are most suitable. The most successful pump for this purpose is a rotating screw, of which several makes are available.

The Mono pump is one type in which a specially shaped rotor gyrates in a casing and so causes pockets of fluidized powder to be drawn in at the intake and pumped into the pipeline. Another type is the Fuller-Kinyon pump, which is a rotating screw conveyor incorporating a non-return valve at the delivery, Figure 2. Any pump used for this purpose has to be chosen to resist the abrasive or corrosive action of the powder. The pump chosen has to operate against the pressure in the line and be well sealed to prevent leakage of the air back into the intake.

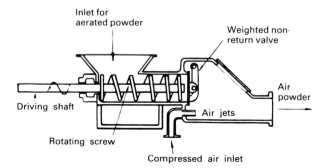

FIGURE 2 – Fuller-Kinyon pump.

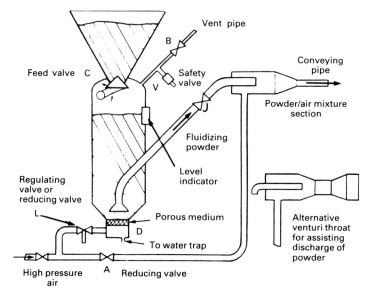

FIGURE 3 – Air pressure assisted gravity feed unit for powders needing
fluidisation, batch operations.

High pressure systems

When the air pressure is high, the material is transported in a dense phase compared with
medium and low pressure systems which use a dilute phase. Dense phase means that the
fluidized material moves as a compact slug along the pipe. It is suitable both for powders
that may be fluidized and for coarse and wet materials. It may also be used for a range of
materials such as manufactured components and slaughterhouse residue. The material/air
mass ratio is in excess of 50:1, so it is reasonably economical in the use of air. One method
of introducing powdered material into the air is through a blow-tank, Figure 3. The tank
has to be designed as a pressure vessel. The method is essentially a batch (non-continuous)
method, although it is possible to use twin tanks and switch between the two, approximating to a continuous feed.

Pulse phase systems

This is a type of medium pressure (between 1 and 2 bar) system where the material is
transported in discrete plugs. Material/air ratios in excess of 300:1 have been recorded; the
air consumption is low and so this can be a very economical method. In Figure 4, air is
injected into the vessel to fluidize the material; beyond the discharge valve at the base of
the hopper is an air knife which injects pulses of air into the conveying line, and as a result
the material is divided into plugs. When the full batch of material has been transported,
the vessel is returned to atmospheric conditions, the inlet valve opens and the cycle repeats
automatically.

Combination vacuum/pressure systems

This is a useful system when conveying from several pick-up points to several discharge

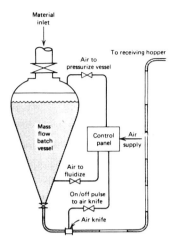

FIGURE 4 – Pulse phase concept. *(Sturtevant)*

points. The pick-up region is under vacuum and the delivery region is under pressure. The same blower can be used for both regions, but this places restrictions on the maximum positive pressure that can be generated, so it is more common to use both an exhauster and a blower and keep the two pressure regions separate.

Air-activated conveyor

This method (Figure 5) can be used when the material is to be transported over a distance with a vertical component. Air is used to fluidize the material which then moves along the incline under gravity. In normal conditions, a powder runs down an incline only when the angle of the incline is larger than the angle of repose. If the powder is fluidized, the natural angle of repose is reduced. As an example, a powder which has an angle of repose of 40° will require a chute set at an angle of 45°, but when fluidized the powder will flow down a chute of 2.5°. Even material which cannot be fluidized in the conventional sense may still benefit from a supply of air to the underside of a porous trough, which reduces the coefficient of friction.

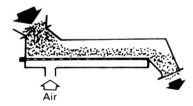

FIGURE 5

Separation methods

The material has to be separated from the air at the delivery point. Usually, a cyclone is needed for primary separation, with the outlet filtered. It is desirable to retain all the

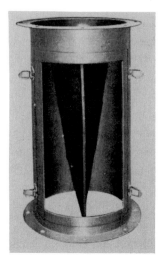

FIGURE 6 – Vacuum valve with access door removed. It consists of a steel funnel
and a rubber sleeve. The sleeve is held in a collapsed condition by the vacuum on the
conveying pipe. When the material in the hopper overcomes the vacuum,
the valve opens. *(Sturtevant)*

material in the line and avoid unpleasant dust, so an efficient filter system has to be
devised. For materials that do not cause problems if a small proportion escapes with the
exhausting air, a fabric bag is satisfactory. The cyclone is connected to a hopper from
which the material is taken by a rotary valve, this is similar to the method shown for
introducing material into the pipe as illustrated in Figure 1. Another useful device which
is suitable for vacuum lines is the vacuum valve, Figure 6.

Calculations on the power absorbed in pneumatic conveying

As indicated above, the design of conveying systems depends on practical experience. The
conveying speed for various materials and the most suitable pressure regime cannot
always be predicted without trials. However it may be helpful to indicate some of the
theoretical concepts that can be used to analyse the power required. The treatment below
is applicable to dilute phase systems (pressure or vacuum) only.

The pressure difference from one end of a pipeline to another is due to:

- Acceleration of the powder from rest
- Pipeline friction
- Changes of direction
- Gravitational forces.

The pressure difference required to accelerate the powder from rest is given by

$$\Delta P = \frac{F_1 V^2 \rho}{2}$$

where ΔP is the pressure difference, F_1 is the pick-up factor, V is the air velocity and ρ is the density of the powder/air mixture.

Usually, because the weight of the material being conveyed is so much greater than that of the air, the mixture density is accurately given by:

$$\rho = \frac{\text{Mass of material}}{\text{Volume of air at pipeline conditions}}$$

For some materials that are commonly conveyed, the value of the saturation is quoted; this is the reciprocal of the density. The pick-up factor varies with the feed design method, but it usually lies between 2 and 3; a value of 2.5 is customarily taken. The actual air velocity can be used, calculated from volumetric flow and pipe diameter. The ideal velocity for the common conveyed materials can often be obtained from published data.

Once the material has been picked up, it is conveyed along the pipe at constant velocity. The pressure drop caused by pipeline friction is calculated in a similar way to that of turbulent flow of a homogenous fluid.

$$\Delta P = \frac{F_2 L V^2 \rho}{2D}$$

F_2 is the conveying factor, L the pipeline length, D the internal diameter.

F_2 depends on a large number of factors – the size, shape and density of the particles, but primarily on the velocity of the flow. Figure 7 may be used if test results are unknown, with a generous factor of about 50% to account for uncertainty.

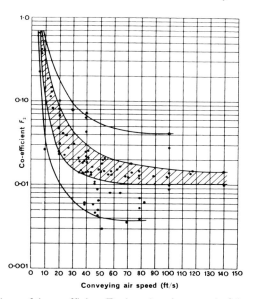

FIGURE 7 – Values of the coefficient F_2 plotted against speed of the conveying air for a large number of controlled tests. Note the abscissa is in ft/s.

Corrosion in the presence of salt-laden air is a problem which should always be borne in mind. Provided that the suppliers are made aware of the use to which their equipment is to be put and any special precautions, particularly in respect of lubrication and maintenance schedules, are strictly followed, there is no reason why most pneumatic equipment should not prove as satisfactory at sea as on land. Tools which are used close to the water line for work on descaling and removal of marine growth are likely to be frequently immersed in water and may require special cleaning after each shift. A useful technique consists in immersing the tool in paraffin or diesel oil to wash out the seawater and then thorough lubrication with oil; for short periods, the tool can be left immersed in the paraffin.

There are some compressed air applications which are of particular interest in offshore applications.

Prospecting with compressed air

Oceanographic research undertaken to improve our knowledge of the Earth's structure can make use of compressed air. Figure 2 shows how high pressure air (up to 350 bar) is fired in pulses from special guns at the area being surveyed. The air pulses create shock waves which echo from the survey area, either land mass or sub-strata. The echos are picked up by sensitive electronic devices to provide data on the structure and content of the survey area.

Drill string compensator system

This is a motion compensating system developed to nullify the effects of the heave of the drilling platform or vessel on the drill string or hook supported equipment. The compen-

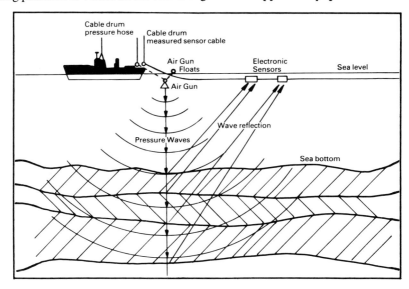

FIGURE 2 – Sketch showing principle used for firing compressed air in marine
seismic survey applications.

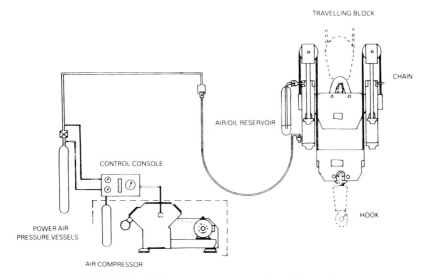

FIGURE 3 – Drill string compensator for offshore oil rigs.

sator shown in Figure 3 comprises two compression-loaded hydraulic-pneumatic cylinders. The cylinders are supplied with high pressure air which, in the application illustrated, is 163 bar. As the rig heaves upwards, the compensator cylinders retract and the hook moves downwards to maintain the selected drilling load. Thus while the drill platform and compensator move, the hook remains fixed relative to the Earth. Use of these compensators ensures that drilling can proceed in high heave conditions.

Use of pneumatic tools underwater

In modest depths of water, down to about 10 m, the standard air tool can be used without any difficulty apart from the worsened visibility caused by the exhausting air. The back pressure caused by the head of water acting on the exhaust port causes a progressive drop in performance; this is particularly true of percussive tools, which become completely ineffective at a depth of 30 m. From tests performed on a medium size percussive drill, operating at 6 bar, Table 1 has been prepared.

TABLE 1 – Performance of rock drill under back pressure

Depth (m)	Gauge pressure (bar)	Performance ratio (1)	(2)
0	0	1.0	1.0
7.5	0.8	0.59	0.75
15	1.5	0.22	0.44
30	3.1	0.0	0.20
60	6.2	–	0.0

Note: The performance ratio is the proportion of the output energy of the tool available at depth.

Condition (1) is for a constant 6 bar at the inlet; condition (2) is for a constant 6 bar pressure drop across the drill, achieved by increasing the inlet pressure.

It can be seen that it is possible to restore some of the lost performance by increasing the inlet pressure but only to a limited extent. It would be possible to design tools to work against a back pressure (a similar problem is faced when designing tools to work on a closed circuit system as discussed in an earlier chapter), but these are likely to remain theoretical concepts for the present.

In practice the best way of using tools at depth is to provide an exhaust boss which permits the connection of a hose to take the exhaust air back to the surface where it can be held bouyant at its open end by floats. Some manufacturers provide tools in this form, but often only to special order. Figure 4 shows one such modified tool. An exhaust hose gives the added advantage of improving visibility, but not all the air escapes through the exhaust port – there is plenty of leakage around the shank of the tool bit and from other unsealed joints, so ideally extra sealing is required for these regions. The exhaust hose must be generously sized, otherwise it would cause its own back pressure. Two further points are to be noted when using a return hose. Firstly, the inlet pressure must be higher than the local water pressure otherwise the water would leak into the tool. Secondly the return hose must be sufficiently robust so as not to collapse under external water pressure; conventional pneumatic hose has been found to be unsatisfactory in this application – either a wire wound hose (which is heavy and hard to manoeuvre) or a moulded PVC hose should be satisfactory.

Most modern tools as described in earlier chapters incorporate a noise muffler to meet onshore noise regulations, which is their main duty. To find a tool of the more traditional form with an exhaust port, which lends itself to the attachment of an exhaust hose, may

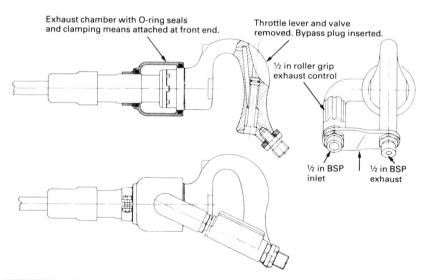

FIGURE 4 – Chipping hammer adapted for underwater use. Note the exhaust chamber seal and control on the exhaust. *(CompAir Holman)*

not be easy. A special tool may need to be designed and made. The general principles illustrated in Figure 4 should be followed.

An effective way of keeping a tool free from water, when not being used, is to transfer the on/off control from the inlet stem to the exhaust port; the tool is then kept continually under pressure and water is unable to get access to the working parts.

When drilling rock or concrete, high pressure water flushing has proved to be more efficient than air flushing. All tools require a hold-down force to keep them applied to the work; this may be difficult under water, so some means of reaction must be available to the operator, probably by anchoring him to the work surface.

AIR FILMS

The use of an air cushion to reduce friction is well known in various applications – air bearings, the floating of one machine surface over another and moving heavy loads. A similar analytical approach can be adopted for all these applications.

A thrust bearing can adopt one of the forms shown in Figure 1 – a central recess or an annular bearing.

The load that such a bearing can support is given by

$$F = C_L \, A \, P$$

F is the load carried, C_L is the load coefficient, A is the total pad area $(= \pi r_0^2)$ and P the supply pressure (gauge). The value of C_L can be estimated from Figure 2. It can be seen that for bearings with small pockets, a value of $C = 0.25$ is a reasonable approximation and this value is often taken as an initial estimate. The basic principle to be used for calculating the feed orifice diameter is that it should be smaller than the circumferential leakage area, *i.e.* the orifice area must be greater than $2\pi r_0 h$. This should ensure that the suspension system is stable, although if maximum stability is required, unpocketed orifices are preferred.

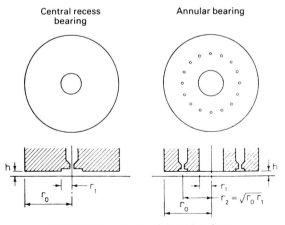

FIGURE 1 – Types of thrust bearings.

Air skates

These are suitable for moving machine tools and other heavy loads over a concrete or other level floor. The method of operation is shown in Figure 4, and a set of four bearings supplied ready for installation in Figure 5. Three or more of these skates are required for stability under load. The air pressure (usually shop air at 5 to 6 bar) inflates the urethane diaphragm creating a pressure zone which lifts the load until controlled air leakage forms a thin film between diaphragm and floor. Once floating on its air film, the load is virtually frictionless and may be moved in any direction with a minimum of effort. The ground pressure, which varies with the quality of the operating surfaces, is from 1 to 4 bar. The air consumption also depends on the quality of the surface as in Figure 6. It is really only practical to use these skates over a reasonably smooth surface. As can be seen from the

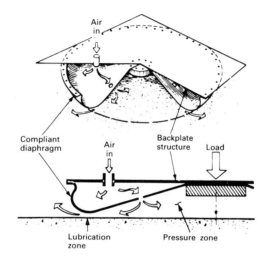

FIGURE 4 – Function of air skates.

TABLE 2 – Capacity of air skates

Load per skate (kg)		Dimensions (mm)	Bearing rise (mm)
Std duty	Heavy duty	Length x width x height	
500	1000	320 x 320 x 51	8 – 14
1000	2000	425 x 425 x 51	10 – 16
2000	4000	610 x 629 x 51	12 – 20
3000	6000	762 x 768 x 57	12 – 20
5000	10 000	914 x 914 x 57	11 – 19
7000	14 000	1067 x 1067 x 57	11 – 21
9000	18 000	1168 x 1168 x 57	15 – 27
15 000	30 000	1778 x 1232 x 57	12 – 27

Note that the standard duty rating operates at 1 bar ground pressure and the heavy duty at 2 bar pressure.

FIGURE 5 – Air bearing installation. Available from 255 mm diameter to 355 mm diameters and capacities from 250 kg to 600 kg. *(Hovair Systems)*

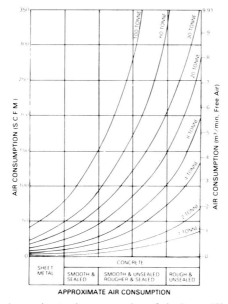

FIGURE 6 – Approximate air consumption of air skates. *(Hovair Systems)*

figure, the air consumption rises rapidly with increasing roughness; a temporary smooth surface such as a steel sheet may have to be laid where the surface is very rough.

To give some idea of the size and lift of a typical skate, Table 2 gives typical information.

FIGURE 7 – Engine supported on an air platform manoeuvred by one man
using an air trigger. *(Hovair Systems)*

Figure 7 shows how a 40 tonne engine, supported on an air film platform can be moved
by an air operated tugger.

Journal and thrust bearings

These are of two types: self-acting and externally pressurised.

Self-acting bearings are similar in principle to liquid hydrodynamic bearings. They are
used in the form of smooth surface cylindrical journals and flat thrust bearings. The main
drawback of air bearings in this form is that they lack the boundary lubrication property
of liquid lubricants so they have high friction at start and stop, limiting their use to about
0.5 bar over the projected bearing area. This is a subject where expert knowledge is
required for design.

Externally pressurised bearings are used in high speed (of the order of 25 000 rev/min),

FIGURE 8 – Air bearing for diamond machining incorporating an air operated chuck.
(Loadpoint)

low friction applications for such purposes as diamond machining and dicing (precision machining of semiconductor materials) as illustrated in Figure 8. Factors influencing the design are: pocketed or non-pocketed feed holes, bearing stiffness to achieve the desired resonant frequency, available pressure. For very high speed applications, water cooling is incorporated in the bearings. Charts are available for detailed design purposes, see for example Tribology Handbook published by Butterworth.

AIR BUBBLE TECHNIQUES

Air bubble techniques are based on the bubble barrier produced when compressed air is fed into a submerged, perforated hose, creating a series of bubble plumes rising from the holes. The rising bubbles cause a vertical current of air and water to flow to the surface, in turn generating a flow of water towards the barrier in the lower layer of water and away from the barrier in the upper layer. This has a mixing function, which can be useful in aerating a stagnant lake.

Table 1 describes and illustrates various techniques developed by a major manufacturing company.

TABLE 1 – Air bubble techniques *(Atlas Copco)*

Application	Diagrammatic	Description
Ice-prevention		The bubble barrier in this case transports warmer bottom water to the surface, creating an ice-free area along the barrier. It has been used successfully to prevent ice damage in yacht marinas.
Reduction of salt intrusion		Here the bubble barrier stops and reverses the intrusion of salt water into flowing fresh water, much of which is then carried back out to see by the fresh water stream. This scheme is used in several Dutch locks.
Underwater blasting		The bubble barrier can considerably reduce the effect of shock waves when underwater blasting by reducing the maximum amplitude of the shock wave, *ie* subduing peak pressures. It cannot, however, reduce the total energy of the shock wave.

TABLE 1 – Air bubble techniques (continued) *(Atlas Copco)*

Application	Diagrammatic	Description
Pneumatic breakwater		Provided the generated surface velocity produced by the bubble barrier is high enough – *ie* exceeds 25% of the propagation velocity of the waves – then steep waves will break or be reduced in height. High airflow rates are required. This system has been used in Japan.
Oil-protection barriers		A bubble barrier can prevent the spread of oil slick from an oil spill by the surface velocity generated by the barrier. This system has been shown to work well in practice, but is still subject to further development.
Lake restoration		Basically, in this case, an air bubble generator is used to treat polluted lakes by aeration; also to promote water circulation in deep lakes. Primarily it can offer a solution where the main problem is oxygen deficiency.

AIR GAUGING

One important application of compressed air is in air gauging. Although recent developments in electronic gauging have in some applications superseded the use of air systems,
there are situations in which it retains many advantages. It is a non- contact system, is self-
centring and self-cleaning. It is now possible, with the use of an air-to-electronic
transducer, to combine the benefits of the two systems.

Commercially available equipment is, in the main, confined to bore gauging and in this
application it is a very quick and simple technique, capable of identifying taper, ovality
and straightness of a bore.

Principles of operation

If the tube A which terminates in the jet J (Figure 1) is connected to a source of compressed
air held at constant pressure, air will flow through the jet to atmosphere at a constant rate.
If now the surface S is moved towards the jet, the escape of air will be impeded and the
flow will begin to decrease. Continued advancement of the surface will steadily reduce the
flow until finally with the surface in contact with the jet, the flow ceases. This simple
device is a displacement-flow transducer which permits detection of the movement of the
surface by observing the change in air flow.

In Figure 2, a regulator R maintains the incoming air at a constant pressure and a variable

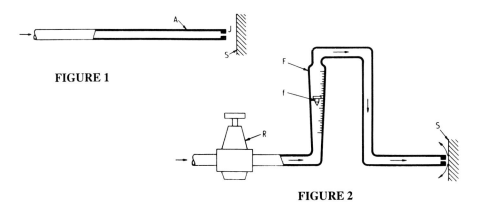

FIGURE 1

FIGURE 2

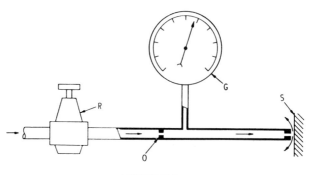

FIGURE 3

area flow meter F measures the flow of air through the jet. The flow responds to the change in flow, rising as the flow increases and falling as it decreases. The flowmeter can be graduated in units of length to give a scale which can be used to measure the displacement of the surface S.

It is not, however, essential to measure directly the changes in air flow; these changes can be converted into changes in air pressure by the method shown in Figure 3. A restriction O, called the control orifice, is introduced between the regulator and the jet, and the air pressure between this restriction and the jet is measured by means of a suitable pressure indicator G. This indicator, shown in Figure 3 in simple diagrammatic form as a pressure gauge, will register a lower pressure when the air flow through the jet increases and a higher pressure when it decreases. As before the scale of the instrument can be graduated in units of length and used for gauging. This arrangement is a displacement-pressure transducer. Most commercially available systems are of this type.

These are the basic ideas of air gauging and are simple in concept. What is important is that they can be used to build robust measuring instruments of extremely high accuracy and stability for use in precision engineering. Their magnification, *i.e.* the ratio of the movement of the indicator (the float in Figure 2 or the pointer in Figure 3) to the movement of the surface which produces it, varies from relatively low (1000) to very high (100 000). Magnifications of 10 000 or 20 000 are common and permit accurate inspection of close tolerance components. The true size of the components are obtained and errors of form can be investigated. The air gauging system thus offers advantages over inspection by limit gauging and can, when required, be used for selective assembly of mating components such as glandless spool valves. The gauging jet can be separated from the indicator (flowmeter or pressure indicator) by a suitable hose connection, so that remote reading can be arranged.

The open jet never makes contact with the work being gauged and may be described as a non-contact gauging element; it is suitable for many applications of air gauging. For others, *eg* when measuring components which have a rough surface finish or porous surfaces, it is better to use a contact gauging element, comprising a stylus, obturator and jet. This may take several forms, but the principle is shown in Figure 4 where it will be seen how the movement of the stylus, which is in contact with the work being gauged, changes

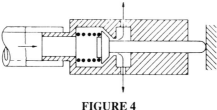

FIGURE 4

the flow of air through the jet. As in the case of the open jet, the changes of flow can be measured directly by a flowmeter or pressure indicator.

An air gauging system comprises an air gauge unit and a measuring head. The air gauge unit contains both the means to display the measured sizes (or to generate signals based on them), certain other items depending on type; when the variable measured is pressure it will contain the control orifice. The measuring head may contain a single gauging element, either contact or non-contact or two or more such elements. The form of the head and the number of gauging elements used will depend on the type of measurement to be made: length or thickness, internal diameter, external diameter, straightness, squareness *etc.*

The air supply for the gauging system will usually be drawn from the factory air line, but a local compressor may be employed.

The air gauge must be operated at constant pressure, which requires a pressure regulator, usually incorporated in the air gauge unit. It is essential that the supply pressure is significantly larger than the operating pressure. Systems using flow measurement normally operate at 0.7 bar, whereas systems using pressure measurement are normally from 3 to 7 bar. The air supplied to the unit must be clean and dry and free from oil vapour. The air consumption of a single unit is small (typically about 1 m³/hour).

The air gauging principle shown in Figures 2 and 3 provides a means for comparing the sizes of like objects, *ie* it acts as a comparator. As with all comparators it requires standardisation before it can give a true size. Suitable setting standards are therefore required for use with the instrument and are employed in conjunction with simple controls for datum setting and fixing the magnification. In some cases, it is convenient to fix the magnification by using master jets which are substituted for the measuring head.

However precise the instrument used in measuring, it cannot give the correct answer if a poor inspection technique is employed. Change of temperature alters the size of the components being measured; difference of form between the standard and the component or errors in form in the component, may lead to false results.

Typical examples of air gauging

Some of the more direct applications of air gauging are illustrated in Figures 5 and 6, in which the gauging element is represented by an arrow. Figure 5 illustrates a single jet, and Figure 6 multiple jet systems.

One of the features of this system is its flexibility and in consequence its wide field of operation. It is particularly valuable for the simultaneous inspection of several dimen-

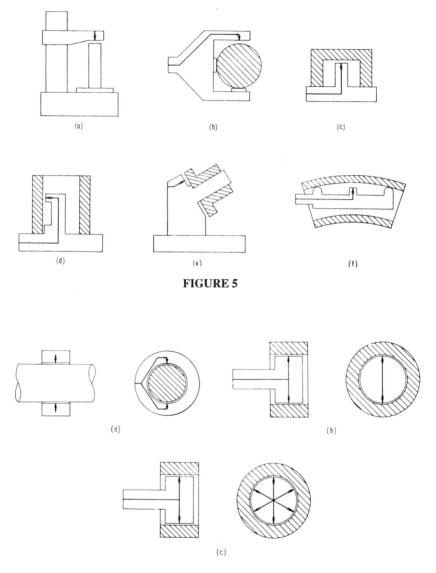

FIGURE 5

FIGURE 6

sions, and it can conveniently be employed on a machine tool to give measurement during production. Signal retainers and signal inverters can be incorporated; measurements can be converted into electrical signals for automatic control.

Circuit layouts are shown in Figures 7 and 8. Note that the system of Figure 8 incorporates means for regulating the pressure and setting the zero.

A wide range of air plugs are commercially available to measure tapers, ovality, multi-diameters, straightness and squareness.

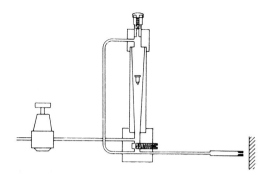

FIGURE 7 – System using flow indicator.

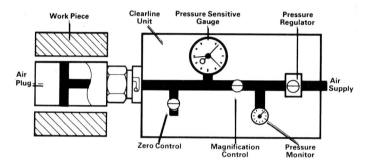

FIGURE 8 – System using a pressure gauge. *(Mercer Brown & Sharpe)*

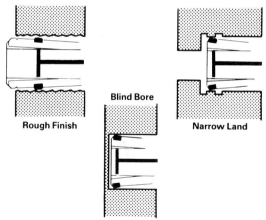

FIGURE 9 – Mechanical contact air gauging. The air plug has tungsten carbide pads mounted on reed springs. *(Mercer Brown & Sharpe)*

For mechanical contact air gauging, the air plug of Figure 9 can be used. This has pads which contact the measured surface and allows the air to pass through the gap made by the displaced pad.

Air/electronic gauging

The advantages of air gauging can be combined with an electronic display or for further control purposes. An air/electronic transducer can be incorporated in the system as shown in Figures 10 and 11. The transducer consists of a diaphragm which is sensitive to variations of pressure, whose movement is measured by an electronic probe. The signal can then read by an appropriate display unit, Figure 12.

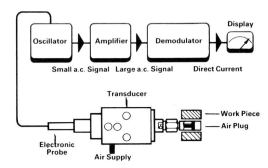

FIGURE 10 – Air/electronic gauging system. *(Mercer Brown & Sharpe)*

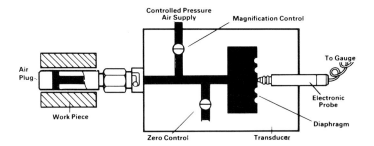

FIGURE 11 – Air/electronic transducer. *(Mercer Brown & Sharpe)*

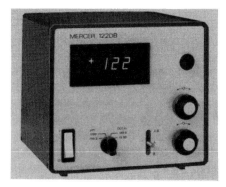

FIGURE 12 – Digital display unit.

SECTION 6

Valves and Sensors

CONTROL VALVES

AIR FLOW MEASUREMENT

PROXIMITY SENSORS

VALVE CONSTRUCTION

CONTROL VALVES

Types of valves

Pneumatic valves can be divided into three main categories defined by the function they perform: directional control, flow control and pressure control. They can be further divided into those valves which control the airflow to carry out a power function and those which perform a control function. The latter includes a number of specialised valve types which only perform a logic function in control circuits.

They can be also be classified by the method of construction – seated valves, sliding spool valves and variable orifice valves.

Directional control valves

Directional control valves are categorised by the number of port openings or ways; a two-port (two-way) valve either opens or closes a flow path in a single line; a three-port (three-way) valve either opens or closes alternative flow paths between one or other of two ports or a third port. Other basic configurations are four-port or five-port.

Such valves can also be described by the number of positions provided and also whether the outlet is open or closed in the non-operated position, *ie* normally open or normally closed respectively. Description may be simplified by giving number of ports and positions as figures separated by a stroke *eg* a 4/2 valve is a four-port, two-position valve. Figure 1, for example, shows a three-way, two-position directional control valve in a simple diagrammatic form.

The three-way two-position valve is thus a logical choice for a single acting cylinder

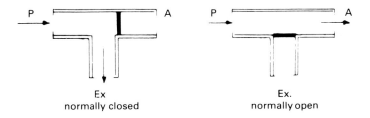

Ex
normally closed

Ex.
normally open

FIGURE 1

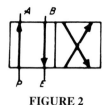

FIGURE 2

control circuit control or for any other single circuit where the down stream air has to be exhausted. It is not necessarily a complete answer to such circuits and there are cases where the use of two two-way, two-position valves may be preferred.

Increasing the number of ports, and if necessary the number of positions, extends the switching capability of the valve. The four-way, two-position valve of Figure 2 has two through-connections. In one position P is connected to A, and B can exhaust downstream air through E. In the second position, P is connected to B, and A can exhaust downstream through E. A four-way, two-position valve can therefore be used to operate a double acting cylinder or any other device requiring alternate pressure and exhaust in two connecting lines.

While this is a logical control valve choice for a double-acting cylinder there are instances where the use of two three-way two- position valves may be preferred for close coupling or the mounting of valves directly on the cylinder ports.

The five-way two-position valve shown in Figure 3 is similar to the four-way valve except for the provision of an extra exhaust port. Thus when either A or B is switched to exhaust, it operates through a separate exhaust port; this can have advantages in particular applications.

The other directional valve is the four-way, three-position valve which can have three possible modes (Figure 4). In the first (left) the function is similar to a four-way two-position valve except that an additional position is available with all ports blocked, *ie* no

FIGURE 3

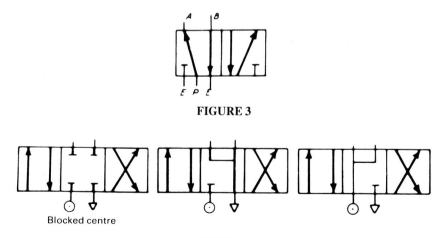

Blocked centre

FIGURE 4

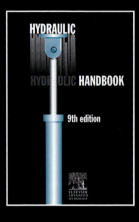

flow is possible through the valve in either direction; this is the normal position. A typical application is the control of a double acting cylinder in which a hold facility is available where air can be trapped on both sides of the piston.

In the second (centre), conventional four-way switching is available from the two extreme positions, but the mid position shuts off the supply and connects both downstream lines to exhaust through the valve. This type is used either with double acting cylinders where it is required to free the piston by exhausting air from both sides of the cylinder, or with motors which are required to free wheel in the off position.

In the third mode (right), normal switching is available from the two extreme positions, but the centre position provides pressurised air to both downstream lines, with the exhaust closed. This provides an alternative hold facility for double acting cylinders with both sides of the cylinder continuously pressurised. This will not give true hold on a single-rod cylinder unless the pressure is dropped in the line to the blank side of the piston to compensate for the loss of effective piston area on the rod side.

Spool valves

Spool valves are the usual choice for directional control, because they can be manufactured to give complicated connections in a simple sliding mechanism. They are straightforward to manufacture, although for good sealing a fine surface is required on both the spool and the barrel with close tolerances to ensure minimum clearances. Glandless spool valves normally require match grinding or a lapped fit between spool and body. Spools may be either stainless steel or aluminium; stainless for the glandless type and aluminium for the sealed type.

Spool valves have two elements: a cylindrical barrel in which slides a plunger. Port blocking is provided by lands or full diameter sections on the spool, separated by waisted sections which provide port interconnections through the barrel. This provides multi-way and multi-position switching. Spool valves have another advantage: the forces on them are balanced when in one of their selected positions; the forces are not necessarily in balance when in movement from one position to another

Spool valves with seals are of simple construction and design. Seals, which may be O-rings or square section, are positioned between the valve spaces, so that there is a seal between each port and on the outside of the two outer ports (see Figure 5).

The seals may be located in the valve barrel, in which case they are inserted into grooves

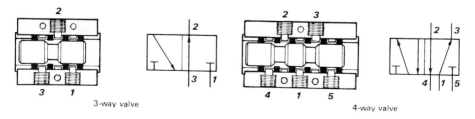

FIGURE 5

machined in the inner surface, or in plastic or metal cages mounted on the inner face of the barrel. Alternatively the seals may be inserted in grooves in the spool. The disadvantage of valves with elastomer seals is that they have to move over the port openings in the case of seals on the spool or by the edge of the spool in the case of seals in the barrel. In either case they may be subject to cutting action or distortion. Some valves have ports drilled with a number of fine holes designed to overcome the possibility of seal cutting.

O-rings are manufactured to wide tolerances which can cause problems: either the seal is loose, which allows leakage, or is tight which increases the static friction (stiction). It is preferable, in valves which are required to have a long operating life, to use square section or other special section seals; these are often backed by PTFE rings.

Elastomer bonded spools have also been developed to overcome the difficulties of using O-rings. These comprise a metal core to which a thin coating of elastomer is vulcanised. The coating is chemically hardened and ground to a fine tolerance, providing a seal material with good elasticity and compression properties. They are claimed to have a life of 20 million cycles compared with more than 100 million cycles for a match ground spool.

Rotary valves

Rotary valves have the advantage that they are adaptable to multi-outlet working. There are two types of rotary valve. One is a form of selector valve with the moving element rotating over a fixed plate to open or block ports drilled in the plate. Channels cut in the rotary plate can provide transverse flow. Sliding plate valves are restricted to two-, three- or four-way operation, see Figure 6.

An example of a versatile form of rotary valve is shown in Figure 7. This consists of a square or hexagonal body with four or six peripheral ports, communicating with a central chamber. An operating spindle passes through this chamber, operating spring-loaded poppets in individual ports by cam action. By choosing different combinations of spindle and port spools a variety of different switching combinations are possible. A number of valves of this type may be ganged together to form a multi-bank valve.

FIGURE 6a – Semi-rotary valve. *(Vickers)*

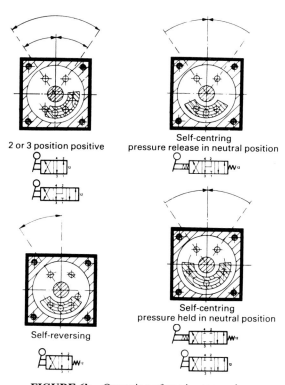

FIGURE 6b – Operation of semi-rotary valve.

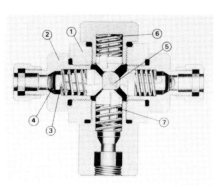

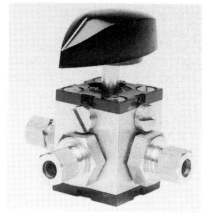

1 Main body; 2 Removable port incorporating connector;
3 Spring loaded stem; 4 O-ring seal; 5 Operating spindle;
6 Blanking plug making top port inoperative;
7 Truncated stem for fully open port

FIGURE 7 – Interlock valve. *(Drallim)*

Seated valves

These can be either poppet, ball or disc valves with metal to metal seats or metal to elastomer seats. The latter are preferred for pneumatic use where good sealing is preferred for conserving air.

Poppet valves require a minimum movement to achieve full opening; they have a low flow resistance when open and give a good seal when closed. They require a large operating force because of the inherent pressure unbalance on the poppet, which in practice limits their use to solenoid or pilot operation.

A simple poppet valve designed for a pressure of 10 bar may require an operative force as large as 100 N because of its unbalance. Full opening is achieved with only a small travel giving a rapid flow response to a signal, which can be of advantage in high speed applications in pilot circuits.

A limitation of a poppet valve is its restricted switching ability. The basic poppet is capable of only an on–off switching function, so a combination of poppets is needed to perform a three- or four-way function; spool valves are normally used for such duties.

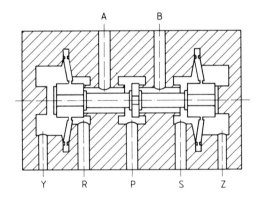

FIGURE 8 – Pneumatic 5/2-way valve. *(Freudenberg Simrit)*

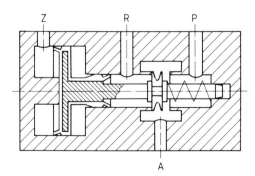

FIGURE 9 – Pneumatic 3/2-way valve. *(Freudenberg Simrit)*

Figure 8 shows a pilot operated 5/2 poppet valve. The diaphragms at the two ends are crucial for operation of this valve. Pilot pressure in ports Y or Z causes the triggering of the valve, which admits air from P to either A or B. For solenoid operation of this valve, either a solenoid replaces the pilot (if the operating force is small) or the pilot pressure itself can be controlled by a solenoid. An alternative method of operation is shown in Figure 9 where pilot pressure selects triggering in one direction, and a spring return in the other.

Ball valves

These are used as shut-off valves or to direct flow to, or isolate flow from, a circuit. The construction consists of a spherical ball, located by two resilient sealing rings backed by O-rings, with a hole through one axis which connects inlet to outlet (Figure 10). Full bore flow is possible when the hole is aligned with one axis of the valve. A quarter turn is sufficient to close the valve. A three-way version is also available to port the flow to one of two lines; this requires a half turn. Bore sizes are available up to 50 mm. With low friction seals, the torque required to operate them is small: a 50 mm valve of the design shown in Figure 10, for example, requires only 12 Nm. The self-compensating design ensures leak free operation with a long life.

For pneumatic circuits it is often desirable to make sure that when the supply is cut off by means of such a valve, the downstream leg is vented. A venting ball valve has a venting passage machined into the ball which does this.

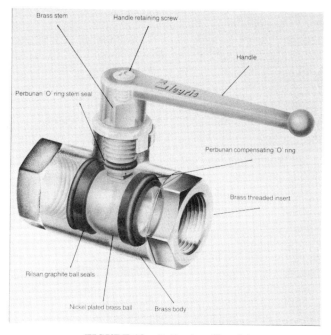

FIGURE 10 – Ball valve. *(Legris)*

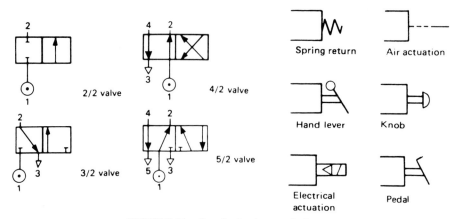

FIGURE 11 – Standard valve symbols.

Standard valve symbols

Symbols used for designating valves on pneumatic circuits are conventionally used in the form of adjacent squares, each square representing one position of the valve, and this system is adopted whatever the detailed construction of the valve. A two-position valve would have two adjoining squares, and a three position valve three adjoining squares. Interior connections between points in each position are then indicated by an arrow or arrows within the appropriate square, together with some indication of the unconnected points. Standard valve symbols are shown in Figure 11. Additional symbols may also be appended showing the method of operation. A complete set of International Symbols are give in the chapter on Graphical Symbols.

In addition main ports may also be identified by numbers:

    1 = Normal inlet, *ie* main supply port.

    2 = Normal outlet port on 3/2 and 3/3 valves.

    2 and 4 = Normal outlet ports on 5/2 and 5/3 valves.

    3 = Normal exhaust port on 3/2 and 3/3 valves.

    3 and 5 = Normal exhaust ports on 5/2 and 5/3 valves.

Port 1 is the main supply port, other odd numbers denote exhaust ports and even numbers the outlet ports. Port 3 is always internally connected to port 2 when the valve is in one position. In the reverse position on a five-port valve, port 5 is always connected to port 4.

Although this numbering is standardised, it is not always used and other numbering systems may be used for particular applications.

With increasing automation the majority of valves used in factory installations are sub-base mounted, *ie* with valve bodies in block form for mounting on a common baseplate or manifold; all lines are connected to the base. This arrangement is very flexible in use, because it allows for easy replacement of valves and servicing. For mobile equipment and for heavy duty applications such as offshore, most valves are of in-line configuration, with direct connection of inlet, outlet and exhaust lines.

With sub-base or manifold mounting the valve body has no tapped connections. All internal ports are brought to the base of the valve and connection is completed by mounting on a matching sub-base or manifold, carrying corresponding ports. Joints are sealed with gaskets or O-rings.

Shuttle valves

The shuttle valve is an automatic type of control valve with three-way two-position characteristics. This is based on a simple three-port body with a moving element in the form of a free spool, a shuttle or a ball (Figure 12).

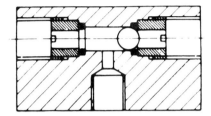

FIGURE 12a – Shuttle valve with ball element.

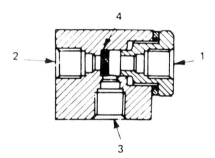

FIGURE 12b – Shuttle valve (4) allowing ports 1 or 2 to be connected to port 3.

A shuttle valve gives an outlet from either one or two pressurised inlets (whichever has the greater pressure): the higher pressure inlet moves the shuttle to a position where it blocks the other inlet. It is thus more of a special valve than a true directional control valve and its main use is for logic OR switching.

Non-return valves

These allow the flow of air in one direction only, the other direction through the valve being blocked at all times to the airflow. The valves are usually designed so that the check is pressurised by the downstream pressure and a spring loaded ball or cone.

The simplest type of non-return valve is the check valve, which completely blocks airflow in one direction and allows flow in the opposite direction, with minimum pressure drop across the valve.

As soon as the inlet pressure in the free flow direction overcomes the internal spring force, the check which can be either a ball or poppet is lifted clear of the valve seat. Alternatively the check may be lifted off the valve seat by some means, as in the case of a quick connect coupling.

Check valves are installed where different components need to be isolated, or restriction of flow through a component in one direction only is required for safety considerations.

Quick exhaust valves

An exhaust valve (quick relief valve), as shown in Figure 13, is designed to increase the speed of movement of a cylinder by allowing the exhaust air to vent directly to atmosphere. Ideally a quick exhaust valve should be screwed directly into the cylinder ports to obtain maximum effect.

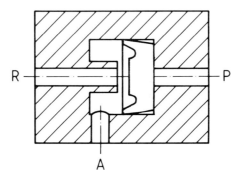

FIGURE 13 – Pneumatic quick relief valve. *(Freudenberg Simrit)*

Air pressure entering the inlet from a selector valve forces a cup seal against the seat of the exhaust port and flows into the cylinder port. The exhaust air flowing from the cylinder port presses against the lips of the seal, forcing it against the inlet port to close it. The exhaust air then flows from the cylinder directly to atmosphere without having to go through the selector valve.

Flow control

Air flow is controlled by using throttling devices. The geometry can range from that of a fixed orifice to an adjustable needle valve or shaped grooves, capable of supplying progressive throttling. The basic function may be combined with a second function such as non-return in the opposite direction.

Speed control valves

These valves for pneumatic applications are usually restrictor check valves. The throttling function is provided by a flow control orifice and the incorporation of a check function also makes them non-return valves when the orifice is closed. Usually the throttle is adjustable to permit regulation of the air flow through the valve, with throttling in one direction of the flow only. In the other direction, free flow is provided through the check valve.

Figure 14 shows a needle exhaust port flow regulator. This type of valve is for use with a five port fully balanced spool valve for cylinder speed control. The setting of the needle regulates the rate of flow of air to exhaust through the control valve, regulating the speed

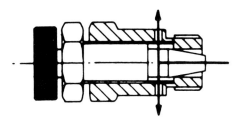

FIGURE 14

of movement of the cylinder. A flow regulator is required in each of the normal exhaust ports of a five-port valve to control the speed of the cylinder in both directions.

A secondary effect of exhausting air directly to atmosphere from the valve is that it tends to be noisy. This can be overcome by the fitting of exhaust silencers. Exhaust flow regulators can be supplied as complete regulator/silencer assemblies comprising the regulator and silencer element screwed to the regulator (Figure 15).

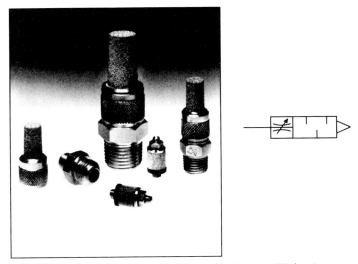

FIGURE 15 – Exhaust port regulators with silencer. *(Vickers)*

The regulator unit can adjust flow from zero to maximum within six or seven turns, depending upon design.

Speed control

The speed of operation of an air cylinder can be controlled by throttling either the inlet or outlet or both. The degree of speed control available from such a simple method is suitable for a wide range of applications, although the actual response time will be variable, because of the varying compressibility of air against different loads.

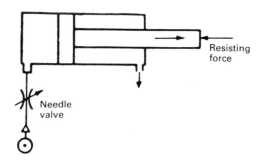

FIGURE 16

Actual control of speed, however, is often less important than reducing the speed of operation to a satisfactory level. Excessive speed can result in dangerous inertial loading, increase cylinder wear and waste air.

Speed control accomplished by metering the inlet air will only produce a constant speed if the resistive load is constant (Figure 16). Pressure will first build up in the cylinder until the force available overcomes the static friction and the resistive load, when the piston will start to move. It will then continue to move at constant speed, so long as the resistive load is constant, with speed governed by the rate of air admission. Any change in the resistive load will cause a corresponding change in the speed of movement.

Speed control by metered exhaust (Figure 17) will tend to give an initial rapid movement followed by a slowing down as air is compressed on the exhaust side of the

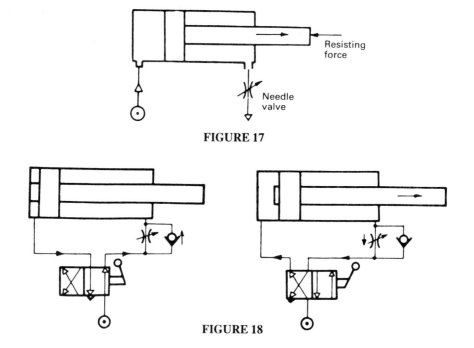

FIGURE 17

FIGURE 18

piston; the movement is likely to be jerky. Much better control is achieved if the modified circuit of Figure 18 is used where the piston is held in the retracted position under pressure.

When the control valve is operated, the piston will start to move forward through pressure applied to the head end before the air has had time to leak away from the rod end. The result is freedom from initial jerk and smooth movement throughout, whilst any tendency to jerk forward, should the resisting force be removed, is minimised.

The general recommendation is that speed control should be applied by throttling the exhaust. A combination of methods may be needed in some circumstances. Often a degree of trial and error is required.

Provided that the load is reasonably constant and unidirectional, good accuracy can be obtained by simple methods. With very low piston velocities, friction may become the most important parameter. To achieve consistent results with speed control it may be necessary to use cylinders with low friction seals and well-finished bores.

For more precise speed control, and to take care of fluctuations in load and line pressure, surge damping may be applied to the system. This can be done hydraulically with an external dashpot or the use of an oversize system with a surge tank in the circuit. More sophisticated methods may sometimes be required; these include velocity feed back with servo-valves.

Slow start flow control valves

A slow start (or soft start) valve allows pressure to increase gradually after the system has been vented for some reason. It ensures that cylinders or other protected devices return slowly to the end of their stroke after the pressure is restored. Depending on the desired behaviour of the circuit on restoration, a slow start valve may be mounted next to a single direction control valve or further back in the circuit where it controls several cylinders. In the latter case, the cylinders will return to their end-of-stroke position, one after the other

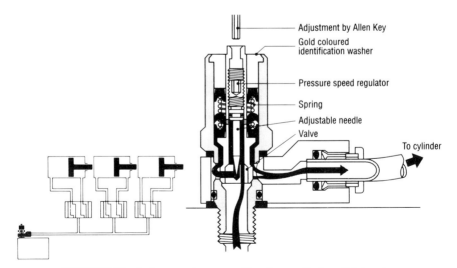

FIGURE 19 – Soft start valve controlling three cylinders. *(Vickers)*

depending on their relative resistance to motion. These valves incorporate a needle valve which regulates the speed of pressurisation. Once the soft start system has opened fully, the system works normally, with no restriction to flow. See Figure 19.

This type of valve may also be used to control pressure decay when the system is shut down.

Time delay valves

This type of valve puts a time delay into the system. It may be used when it is necessary to ensure that a particular sequence of movements is completed before another sequence is commenced. The second sequence is protected by a time delay valve. It consists of a restrictor valve which charges a reservoir in a controllable way until the pressure in the reservoir has built up to a level where it can trigger a pilot valve (Figure 20); a non-return valve allows the reservoir to be emptied rapidly for the next timing cycle. The time delay depends on two factors: the airflow through the restrictor and the volume of the reservoir. It may be considered to be the pneumatic equivalent of a resistor/capacitor electronic circuit. These valves normally incorporate a small reservoir in the body of the valve and rely only on the size of the restrictor to control the time delay, which in a typical commercial valve is adjustable up to 60 seconds . For longer time delays of several minutes, the size of the reservoir may have to be increased externally.

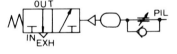

FIGURE 20 – Time delay valve. *(SMC)*

AIR FLOW MEASUREMENT

Choice of methods of flow measurement

There is a variety of different methods of flow measurement, each of which has its own advantages and characteristics. The choice in any particular circumstances will depend on a number of factors, *eg* the pressure, temperature and desired accuracy. For determination of compressor performance, the method of measurement is prescribed, as in BS 1571 Parts 1 and 2 (also ISO 1217). Refer to the chapter on Compressor Acceptance Tests.

Use of nozzles in measurement

For the most accurate method of measurement of compressor delivery, the method of BS 1571 Part 1 is recommended. Part 2 offers a simplified method. The principle adopted is that where the pressure drop through an orifice is small, the flow through it is adiabatic, sub-critical and so is proportional to (pressure drop)$^{0.5}$. The size and shape of the orifice are chosen so that these conditions are met.

The nozzle can be placed either on the intake or the delivery side of the compressor. When on the delivery side, the discharge can be to the atmosphere or into a pipeline under pressure. In order that the flow through the nozzle shall be smooth and free from pulsations, a smoothing receiver and a perforated plate must be used (Figure 1). Similar

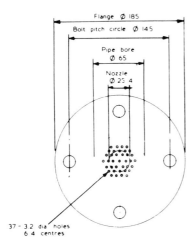

FIGURE 1 – BS perforated plate for flow smoothing.

TABLE 1 – Dimensions of nozzles (BS 1571 : 1975)

| Nominal size | | | A | B | C | E | F | G | H | J | K |
Nozzle diameter d in	Pipe bore D	d									
7/32 5.56	25	5.56±0.01	3.18±0.06	—	—	—	19.05	44.45	—	0.40	—
3/8 9.5	25	9.53±0.01	5.77±0.06	2.87	3.15	1.91	19.05	44.45	—	0.40	—
5/8 15.9	40	15.88±0.01	9.60±0.06	4.78	5.23	3.18	26.19	60.33	—	0.40	—
1 25.4	65	25.4±0.03	15.37±0.08	7.62	8.38	5.08	38.89	85.73	11.11	0.80	29.77
1½ 38.1	90	38.1±0.03	23.06±0.10	10.29	12.57	7.62	51.59	111.13	11.11	1.60	44.45
2½ 63.5	150	63.5±0.05	38.43±0.13	19.05	20.96	12.70	84.93	177.80	11.11	2.40	74.22
4 101.6	270	101.6±0.05	61.47±0.13	30.48	33.53	20.32	141.29	292.10	12.70	3.20	117.67
6 152.4	375	152.4±0.10	92.20±0.18	45.72	50.29	30.48	206.38	425.45	15.88	4.80	177.80
10 254.0	600	254.0±0.18	153.67±0.25	76.20	83.82	50.80	323.85	660.40	19.05	8.00	293.10
15 381.0	900	381.0±0.25	230.51±0.38	114.30	125.73	76.20	497.43	971.55	19.05	12.20	419.10

All dimensions are in millimetres

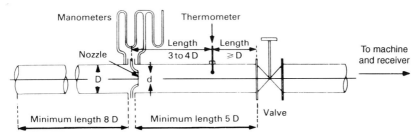

FIGURE 2 – Testing on the intake side.

methods can be used for accurate measurements of the flow rate in other applications than in compressor acceptance tests, but in these cases, simpler methods are preferred. The arrangement of the test method is given in Figure 2.

A receiver of adequate size must be fitted between the compressor and control valve to dampen out any pulsations and thus ensure accurate flow measurement.

Where the pressure drop across the control valve is greater than 30% of the upstream pressure, an adequate capacity receiver is one which would contain, at the receiver pressure, the output of the compressor or exhauster during 50 complete pulsations.

Where the pressure drop across the control valve is less than 30% of the upstream pressure or where no control valve is used between the receiver and the nozzle, a larger receiver may be necessary.

Other damping arrangements may be used provided that it can be shown that no pulsations exceeding ±2% of the differential pressure occur at the nozzle inlet.

The pipe connecting the machine and receiver should be as short as possible to minimise the risk of false readings due to a resonance effect. Pipe flanges are preferably faced so that a metal-to-metal joint is made with the nozzle. If jointing material is used between the flanges and the nozzle, its thickness must not exceed 1.6 mm and the inner diameter of the joint ring must be not less than the appropriate value of the dimension G given in Table 1.

Testing with nozzles on the delivery side

The control valve may be either the same nominal size as the nozzle approach pipe or smaller. The latter may be preferred because of ease of control and it then should be connected to the nozzle approach pipe by a cone piece. In either case a perforated plate must be fitted at the inlet flange of the approach pipe to smooth the flow, which leaves the control valve in a violently turbulent condition. The perforated plate must have an hexagonal pattern of holes drilled in the central portion, the centres of the holes being uniformly spaced and located at the corners of equilateral triangles. The number, size and spacing of the holes, and thickness of the plate are specified in Table 2.

The pressure drop across this perforated plate is about four times the nozzle head, hence if it is desired to test up to the maximum nozzle head of 1000 mm H_2O the minimum pressure would need to be about 0.5 bar gauge when discharging to atmosphere. If tests are to be made at lower supply pressures, the nozzle size can be chosen to work at a lower head.

TABLE 2 – Details of perforated plate (BS 1571 : 1975)

Nominal nozzle diameter		Nominal pipe bore	Details of perforated plate			Thickness of plate
			Number of holes	Diameter of holes	Spacing of holes	
d in	mm	D mm		mm	mm	mm
7/32	5.56	25	19	1.6	3.2	3.2
3/8	9.53	25	19	1.6	3.2	3.2
5/8	15.88	40	61	1.6	3.2	3.2
1	25.4	65	37	3.2	6.4	3.2
1½	38.1	90	61	3.2	6.4	3.2
2½	63.5	150	217	3.2	6.4	3.2
4	101.6	250	169	6.4	12.7	6.4
6	152.4	375	331	6.4	12.7	6.4
10	254.0	600	217	12.7	25.4	12.7
15	381.0	900	469	12.7	25.4	12.7

Generally the air pressure on the outlet side of the nozzle is only slightly in excess of atmospheric, because the exit pipe discharges freely into atmosphere or through an exhaust silencing system to atmosphere.

Downstream from the perforated plate the approach pipe to the nozzle must be straight with a length not less than 8 pipe diameters. The pipe bore must not be greater than 2.5 nor less than 2.25 nozzle diameters, except for the two smallest nozzle sizes in Table 1 where it must be 25.4 ±3.2 mm.

The exit pipeline from the nozzle to atmosphere must be straight for a distance not less than 4 pipe diameters and of the same diameter as the approach pipe.

The roughness of the internal surface of the pipes need not be reduced below that of ordinary commercial finish, but there must be no obstructions in the pipelines such as thick welding ridges.

Testing with nozzles on the intake

When the nozzle is on the intake, the approach pipe to the nozzle must have minimum straight length of 8 pipe diameters and the exit pipe a minimum straight length of 5 pipe diameters. If aspiring from atmosphere, the inlet end may be plain, flanged or flared but must not be restricted. As the control valve is on the downstream side of the nozzle, a perforated plate is not necessary.

If drawing from another pipe system, special precautions should be taken to avoid swirl or other flow disturbances eg by using an appropriate perforated plate at the entry to the straight pipe of 8 pipe diameters length.

Nozzle selection

The following empirical rule can be used to ascertain a size of nozzle suitable for a particular test, when the pressure at the nozzle is approximately atmospheric. Calculate 0.25 times the maximum expected volume flow in l/s and select a nozzle with a nozzle

TABLE 3 – Rounded nozzle:
Nozzle constants and volume flow (BS 1571 : 1975)

Nominal nozzle diameter		Approx. pipe bore	Constant	Approximate volume flow for nozzle outlet pressure near atmospheric			
d		D	k	h=10mm H_2O	h=100mm H_2O	h=400mm H_2O	h=1000mm H_2O
in	mm	mm		litres/s	litres/s	litres/s	litres/s
7/32	5.56	25	0.1488	—	1	2	3
3/8	9.53	25	0.4432	—	3	6	8
5/8	15.88	40	1.254	—	8	16	24
1	25.4	65	3.248	—	20	40	60
$1\frac{1}{2}$	38.1	90	7.377	14	43	85	140
$2\frac{1}{2}$	63.5	150	20.516	38	125	250	380
4	101.6	250	52.429	100	310	620	1000
6	152.4	375	118.41	220	700	1400	2200
10	254	600	328.00	560	1900	3900	5600
15	381	900	737.68	1450	4400	8700	14500

constant, k, next larger than that number. Nozzle constants for BS rounded nozzles are given in Table 3. For pressures other than atmospheric, the approximate value of k may be found from the value of Q (see later) using a value of h within the range given in Table 3.

Observations and measurements

The following observations are necessary to conform to British Standard practice.

(i)   Atmospheric pressure
Measured in mm Hg by means of a barometer. If there is no barometer on site, readings can be obtained from a local recognised weather bureau and corrections applied to allow for the difference in altitude.

(ii) At the agreed inlet point
The difference in pressure between the agreed inlet point and the local barometer reading. For the purpose of calculation, the pressure at the inlet point is the barometer pressure less the depression at the agreed inlet point.
The temperature of the air at the agreed point in °C. The thermometer shall be so placed that it is not affected by radiant heat from the surroundings.

(iii) At the receiver
The receiver pressure in bar measured by means of a pressure gauge. This is taken as a record of performance but is not used in the calculation of the flow except where necessary to allow for any small change occurring during test. Such changes shall not exceed 2% of the absolute pressure at the receiver.

(iv) At the nozzle
The head across the nozzle, measured as the difference in pressure given by the reading of a double-leg manometer.

The difference in pressure between the downstream side of the nozzle and the atmosphere. This shall be measured by means of a second manometer connected as shown in Figure 2.

The temperature in the pipeline, measured by means of a thermometer as shown in Figure 2. The thermometer shall be placed between three and four pipe diameters from the mouth of the nozzle and not closer than one pipe diameter to an open end.

Instruments used for these and other measurements are summarised in Table 4.

TABLE 4 – Measuring equipment and methods
(as specified in BS 1571 : Part 2)

Quantity	Instrument(s)	Exclusions	Precautions
Temperature	Mercury in glass thermometers with etched stems (or other methods as given in BS 1041)	Conventional or industrial metal-cased thermometers	Air velocity not to exceed 30 m/s at point of measurement
Pressure (above 2 bar)	Calibrated Bourdon gauge, dead-weight gauge, mercury barometer	Diaphragm gauges	Pressure readings to be between $1/4$ and $3/4$ full scale using Bourdon gauges
Pressure (below 2 bar)	Manometer or columns (or vacuum gauge)	Closed mercury columns for vacuum measurement	Eliminate pulsating flow (eg with receiver)
Air flow	BS nozzle		See text
Shaft speed	Tachometer or speed counter		Instrument should be checked regularly for accuracy
Power	Electric motor (BS 5000 : Part 99) or dynamometer		Calculation of output based on average of increasing and decreasing loads

Method of calculation (BS 1571 Part 2)

The capacity Q in l/s is calculated from the formula

$$Q = k \left(\frac{T_1}{P_1} \right) \sqrt{h \frac{P_2}{T}}$$

where:
Q is the inlet volume in l/s
h is the pressure drop across the nozzle in mm H_2O
k is the constant for the nozzle

T_1 is the absolute temperature at the agreed inlet point
T is the absolute temperature downstream from the nozzle
P_1 is the absolute pressure at the agreed inlet point in mm Hg
P_2 is the absolute pressure at the downstream side of the nozzle in mm Hg

The absolute pressures are obtained by adding to or subtracting from the atmospheric pressure, as given by the barometer, the water gauge reading obtained from the appropriate manometer. The water gauge readings are converted to mm Hg by dividing by 13.6.

Alternative tests

The discharge may have to be made into a pipeline into pressure instead of into the atmosphere, or it may be desirable to measure the flow at the delivery pressure in order to reduce the dimensions of the pipe line required. The test layout should be the same as that shown in Figure 3, with the addition of a second perforated plate across the downstream pipe at least five pipe diameters away from the nozzle and before the flow reaches any bends, valves or obstructions. The pressure at the downstream side of the nozzle is measured by a pressure gauge substituted for the second manometer and the reading converted to mm Hg.

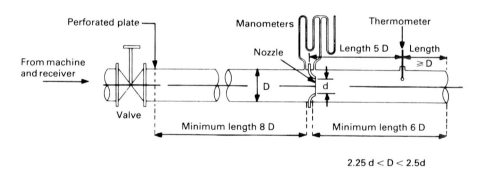

2.25 d < D < 2.5d

FIGURE 3 – Arrangement for testing with a nozzle on the delivery side.

Bourdon gauges are mounted vertically and must be readable to 0.5% of the full scale reading. They must be used only between one-quarter and three-quarters of their full scale reading. They must be calibrated against a dead-weight gauge before and after the test.

Care must be taken to ensure that the high pressure pipeline into which the air is discharged is free from pulsations.

The calculations are made in the same way as for discharge to atmosphere.

A round-edged orifice is preferable for highest accuracy. With nozzles manufactured to the dimensions given in BS 1571, the nozzle coefficient is very close to the theoretical value of unity (in fact, the value is 0.985). A rounded orifice is expensive to manufacture, so as a cheaper alternative, a sharp-edged orifice plate can be used, using the appropriate value of the flow coefficient. The pressure loss through an orifice plate can be considerable.

Selection of the primary device to be used

When determining the flow by measuring pressure difference across a nozzle or orifice, a number of factors influence the choice of the primary device. Apart from the rounded nozzle discussed, it is possible to use a simple orifice plate, a classical venturi tube or a venturi nozzle. A venturi has pressure tappings upstream and at the throat, rather than upstream and downstream as in the case of the nozzles . All the different methods are described in BS 1042 Section 1.1, where the flow coefficients and the accuracy of each method is given.

For each primary device there exists a limiting value of pipe diameter, diameter ratio, β, and Reynolds number. If the chosen value of differential pressure and flow rate for an orifice plate exceeds the permissible limit for β, a nozzle may be used since it allows the use of a lower β value.

The pressure loss through an orifice plate or nozzle is 4 to 6 times the loss in a venturi nozzle or classical venturi tube.

A classical venturi tube requires shorter straight lengths of pipe than the other devices, but the mounting distance of venturi nozzles and tubes is significantly larger than for nozzles or orifice plates.

Orifice plates are cheaper and simpler than any of the other devices.

If the flow is pulsating, the accuracy of flow measurement is suspect.

High pressure nozzles

For slightly less accurate determination of the flow rate of compressors, high pressure nozzles can be used. The recommended technique is the use of a circular arc venturi nozzle under critical flow conditions. This method is described in BS 1571 Part 1 and uses a nozzle of the shape given in Figure 4, to the dimensions of Table 5. The accuracy is quoted

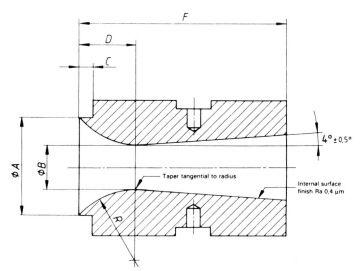

FIGURE 4 – Circular arc venturi nozzle.

TABLE 5 – Nozzle dimensions

Flow rate	A	B	C	D	R	F
l/s	mm	mm	mm	mm	mm	mm
12 to 40	16.00	6.350	2.40	9.96	12.70	60.5
24 to 90	24.00	9.525	3.60	14.95	19.05	91.0
50 to 160	32.00	12.700	4.60	19.93	25.40	121.5
100 to 360	48.00	19.050	7.10	29.89	38.10	182.0
180 to 650	64.00	25.400	9.60	39.85	50.80	243.0
280 to 1000	80.00	31.750	12.00	49.82	63.50	303.5
400 to 1500	95.00	38.100	14.20	59.38	76.20	364.0

as ± 2.5%. It is recommended as being satisfactory for determining the flow rate of compressors when subject to EC noise testing, where the only purpose of the flow measurement is to determine the noise emission category into which the compressor is placed; but it can be used in other applications where the lower accuracy is adequate. As with other nozzle methods, a flow straightener is placed upstream of the venturi and both temperature and pressure are to be taken. Provided that the nozzle size is chosen so that there is sonic velocity at the throat, only the upstream pressure and temperature need be taken. The advantage of this method is its simplicity. The expression for the volumetric flow is given by the expression:

$$q = 9 \times 10^{-3} \frac{B^2 P T_0}{P_0 \sqrt{T}}$$

q is the flow rate in l/s at the reference conditions (usually 1.013 bar and 15°C),
B the diameter of the nozzle in mm,
P the upstream pressure in bar,
P_0 the reference pressure in bar,
T the upstream temperature in kelvin,
T_0 the reference temperature in kelvin.

For the usual reference conditions, this reduces to

$$q = 2.63 \frac{B^2 P}{\sqrt{T}}$$

Strictly, these formulas apply where the temperature is between 20 °C and 70 °C, and the pressure between 2 and 8 bar; outside this range, a different multiplying constant can be found from BS 1571.

Because the flow is in the critical regime, the flow rate is directly proportional to pressure, rather than to $(pressure)^{0.5}$ as in the case of the other nozzle methods.

Flow measurements using alternatives to nozzles

In many instances it is not practical to adopt the complexities of measurement by nozzles. A major drawback in their use is the limited range over which a given nozzle can be used.

If the flow varies over a wide range, it is not practical to use a nozzle. Also a direct measurement method is more appropriate for applications such as assessing the flow rate in various parts of a main circuit, for determining the air consumption of tools and in process applications. In studying the velocity pattern across large pipe diameters or at the exit of a compressor impeller, more localised techniques are needed. The direct reading methods are more suitable when included as part of an overall monitoring facility. Some of the modern flow measuring methods use electronic instrumentation and so make it easier to provide a remote readout of the flow. A manometer is not easy to adapt to remote readout.

Some of the alternative metering methods suitable for air flow determination are:

- Pitot tube
- Hot wire anemometer
- Variable area flowmeters (spring loaded or dead weight)
- Multi-blade turbine meters
- Vortex shedding
- Coriolis effect mass meters

Pitot tubes

This measures the dynamic and static head of the flowing gas, Figure 5. The measurements can be taken by a manometer in the same way as in the previous methods. These are used for establishing flow patterns in large pipes and in unconfined passages such as wind tunnels. As with nozzles and venturi meters, the flow velocity is proportional to (pressure drop)$^{0.5}$.

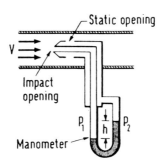

FIGURE 5 – Pitot tube.

Hot wire anemometer

This determines the flow velocity by measuring the cooling effect of a flowing air stream. A hot wire is inserted into the stream and the heat convection from the wire produces an imbalance in a bridge circuit. The relation between the flow velocity and the bridge potential is not readily determined, so they have to be calibrated against another measurement method.

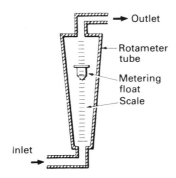

FIGURE 6 – Variable area flowmeter.

Variable area flowmeter

There are two common types: the cone and float meter, and the variable orifice spring-loaded meter.

The cone and float meter, Figure 6, can have either a glass or metal tube. Gas flows vertically through a conical tube in which a free float or bobbin is located. The float has a constant weight, so increasing flow requires it to move up the tube to increase the area of the aperture. They are suitable for measuring over a large flow range, from zero up to the maximum allowed through the flow area. The use of a glass tube allows the float to be viewed directly, but in applications where safety is of prime concern, stainless steel or non-ferrous tubes may be used and the float position determined magnetically. As supplied, these meters are calibrated at a specified temperature and pressure, typically 7 bar and 20°C. A table of correction factors is usually supplied with the meter, but if these are not available, the factors can be calculated:

$$f = \sqrt{\left(\frac{P_2}{P_1} \frac{T_1}{T_2} \right)}$$

Actual air flow = (indicated air flow) x f
P_2 = operating pressure
P_1 = calibrated pressure
T_2 = operating pressure
T_1 = calibrated pressure

The same flowmeter can be used for gases other than air, in which case a further factor using the specific gravity is introduced. The pressure loss through this device is the weight of the float divided by its cross-section area; this is likely to be small. The accuracy depends on the accuracy of manufacture of the tapered tubes; typically it is of the order of ± 1%.

When using this equipment for the measurement of the air consumption of tools, it is customary to build a portable test facility (commonly known as a "pig"), consisting of a

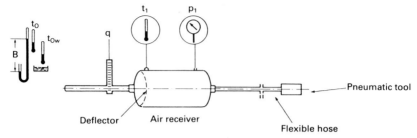

FIGURE 7 – Measuring points for performance testing of percussive hand tools.
For rotary tools the pressure of the compressed air is measured at the inlet of the tool.

flowmeter, an air receiver, a pressure gauge and a thermometer as in Figure 7. With a freely floating bobbin, the damping is practically non-existent, and without a receiver to suppress the pulsations generated by the varying tool demand, the meter scale would be unreadable.

The size of the receiver is calculated from

$$A > 7 \times 10^{-5} \frac{q}{P}$$

where q is the maximum free air flow (l/s),
P is the pressure in bar and
A is the cross-section area of the receiver.

It may also be necessary to incorporate a pressure reduction valve to reduce the mains pressure to the desired test pressure. This valve can sometimes produce its own pulsations, or there may be pressure variations from other components in the main circuit, in which case a second air receiver upstream of the flowmeter is required.

Spring loaded variable-area meters work on a similar principle, except that the spring allows it to be used at any angle, see Figure 8. The pressure loss through this type of meter is several times that of the float type described above. The accuracy is of the order ±5%. Glass (or more often plastic) tubes allow visual sighting; metal tubes with magnetic sensors are available, as in the float type.

The spring-loaded meter has greater damping and is less accurate than the float type, so it is usually employed in less critical measurement situations. Although it may not be

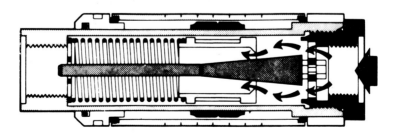

FIGURE 8 – Typical view showing flowmeter operation.

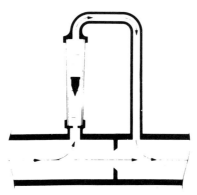

FIGURE 9 – Orifice plate VA flowmeter function.

worthwhile incorporating a flow straightener, it makes sense to take reasonable precautions to situate such a meter in a place where pulsating flow has minimal effect.

A combination of a variable area and an orifice plate can also be used; this arrangement is shown in Figure 9. It is suitable for direct reading, with the flow through the float meter scaled in flow units through the orifice plate. It is useful for insertion in large pipes.

Multi-blade meters

This type of meter is an indirect device in that the flow rate is obtained from the rotation of a turbine. In Figure 10, the unit consists of a turbine, mounted on low friction bearings and inserted directly into the pipe. The flow rate is proportional to the rotational frequency of the turbine. The accuracy is of the order of ±1%. For accuracy of measurement, a flow straightener fitted into a straight length of pipe is essential as shown in Figure 11. A turbine, because of its rotational inertia, has an averaging effect in the presence of a

FIGURE 10 – Turbine flowmeter suitable for flow rates up to 80 m³/mm. *(Quadrina)*

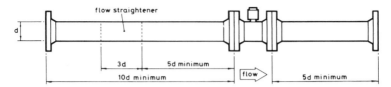

To achieve the maximum performance from a turbine flowmeter it is essential that straight pipe sections are fitted immediately upstream and downstream of the flowmeter position. For liquid flows the dimensions indicated should be considered a minimum. For gas flow they should be multiplied by 4.

FIGURE 11 *(Quadrina)*

pulsating flow, so it may not be apparent that the meter is affected by the pulsations. If pulsations are known to exist or are suspected, the same kind of air receiver should be employed as for the variable area type of meter.

The rotational speed of the turbine is measured by an external pickup, of which there are two types – magnetic and electronic. The magnetic pickup is simple and inexpensive; however due to magnetic attraction, drag is imposed on the rotor and at low flow rate its performance is unreliable. It should not normally be used for airflow measurement, except where an approximate reading is adequate. The alternative electronic pickup uses the principle of variable reluctance so there is negligible drag on the rotor; a further advantage is that it produces a high output signal of constant amplitude over the whole of the operating range of the meter.

The signal from either type of pickup is fed into a counter or a direct reading frequency panel meter. The flow is proportional to frequency so it is easy, when using a panel meter for direct read-out, to calibrate the scale in flow units.

A turbine meter can also be used to measure the velocity profile across a large diameter

FIGURE 12 – Pelton wheel type flowmeter. *(Quadrina)*

pipeline. An insertion flowmeter is mounted on the end of an insertion tube, which allows movement of the measurement head to a point across the diameter of the pipe.

A meter similar to a turbine meter is a pelton wheel meter, in which the flow impinges on a rotating wheel tangentially. This is suitable for accurate measurement of small flows. The meter of Figure 12, for example, with a $1/4$ in fitting has a measuring capacity of 0.12 l/s.

Vortex shedding meters

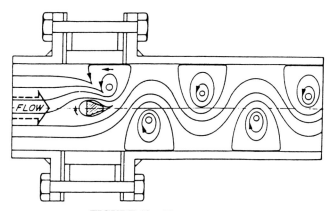

FIGURE 13 – Vortex meter.

Figure 13 shows a section through a vortex flowmeter. The gas flows past a fixed flow element (a pillar with a delta-shaped cross-section) and generates a succession of vortices which are formed alternately on opposite flanks of the flow element. The frequency of vortex shedding is a linear function of velocity. Within the range covered by a given flowmeter, it is independent of density, viscosity and temperature and is given by:

$$n = \frac{S V}{D}$$

n is the frequency, V is the velocity, is the width of the flow element and S is the Strouhal number or shape factor. From a knowledge of the velocity and the cross-sectional area, the total flow can be determined. For the flow to generate vortices, it has to be in the turbulent regime, which condition is dependent on the Reynolds number.

The velocity and pressure of the gas in contact with the flow element oscillate in time with the vortex frequency. A variety of different methods of measuring the frequency are possible according to the manufacturer:

- Differential pressure fluctuations at the shedding element measured by a pressure sensor.
- Measurement of the twisting of the shedder with a piezo-electric sensor.
- Pressure fluctuations at the shedder causing a disc to oscillate, which is measured by a magnetic pickup.

FIGURE 14 – Vortex shedding flow meter. *(Schlumberger)*

- Temperature sensors measuring the change in velocity at the shedder.
- Ultrasonic measurement of the pressure fluctuations downstream of the shedder.

As with all velocity metering devices, an upstream flow straightener is essential, which can be a straight length of pipe at least 15 times the internal pipe diameter. The accuracy is ±1%, the measuring range is 10:1 and the pressure drop is 0.01 bar. Determination of the flow, as in all indirect methods, relies on the accuracy of the manufacturer's calibration.

Figure 14 shows a vortex shedding flowmeter in which the vortices are measured by pressure fluctuations at the shedder.

Coriolis effect mass meters

This is one of the few techniques which directly measures mass flow rather than velocity.

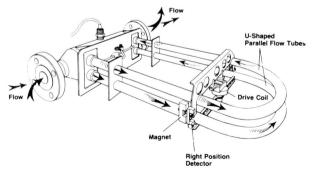

FIGURE 15 – Coriolis effect mass meter. *(Schlumberger)*

It is useful where the flow rate is required in conditions of varying temperature and pressure. A further advantage is the absence of any requirement for upstream flow straightening. The major applications of this technique are for difficult liquids and slurries, but it is equally suitable for gas measurements.

Figure 15 shows the principle of operation. One or two U-shaped tubes are vibrated at their natural frequency by a drive coil. The flowing fluid is forced to take on the vertical movement of the vibration, increasing as it flows towards the bend and decreasing as it flows away from the bend. The forces induced by the fluid on the tubes are the Coriolis forces and although they are very small, they are sufficient to cause the tubes to twist. When the tube moves upward, during half the vibration cycle, the tube twists in one direction and when the tube moves downward during the second half of the cycle, the tube twists in the opposite direction. The twist is proportional to mass flow and is sensed by the pair of magnetic position detectors. These detectors feed the information into an electronic processor for subsequent display.

PROXIMITY SENSORS

The most common form of pneumatic sensor is the interruptible jet, the principle of which is shown in Figure 1. Supply and receiver pipes are aligned axially, separated by a gap. The intrusion of any solid object into the gap, interrupting the jet, causes the pressure in the receiver to fall to atmospheric. This change in pressure is used to operate a switching element controlling an appropriate circuit, for example a counting circuit. The switching element is normally a transducer giving an electric signal output.

This type of sensor has the advantage that it is not critical on gap dimensions, can change its state rapidly for counting purposes and is not sensitive to shape or texture. Its main disadvantage is that atmospheric air is entrained in the receiver and if contaminated, can interfere with the performance.

The limitation can be overcome by lightly pressurising the receiver in the opposite direction (Figure 2). The main jet then impinges on the secondary jet, applying back pressure which increases the receiver pressure. An object interfering with the main jet cuts off this back pressure and the receiver pressure falls back to its normal level. Under both conditions there is always outflow from the receiver which cannot therefore entrain ambient air.

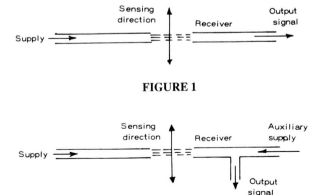

FIGURE 1

FIGURE 2

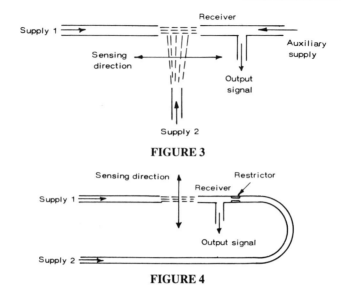

FIGURE 3

FIGURE 4

Both of these types require small nozzle sizes in the supply to achieve laminar flow. The effective gap length is the distance over which the laminar flow can be maintained.

An alternative shown in Figure 3 uses a third jet placed at right angles to the main supply jet. The receiver may or may not be pressurised. This works on the principle of a turbulence amplifier, with the main jet flow, normally laminar, being rendered turbulent by the impingement of the side jet, reverting to laminar flow when the side jet is interrupted. The practical measuring gap, which is between the side jet and the main jet can be made much larger for the same pressure difference in the receiver.

A more practical form of interrupted sensor with back flow from the receiver is shown in Figure 4, back flow being obtained from the same supply but with pressure reduced by a restrictor. Sensors of this type can be expected to have a maximum gap of about 20 mm working off a supply pressure of about 0.1 bar.

The system shown in Figure 5 is capable of working with larger gaps at the same pressure levels. Here a collector is incorporated to supply a second external gap (the sensing gap), with output signal derived from the same position as before. This device is known as an airstream detector.

Back pressure sensors are effective when the object to be sensed can pass close to the

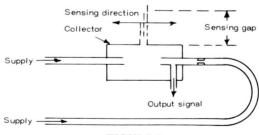

FIGURE 5

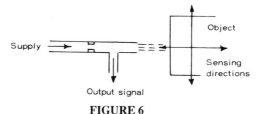

FIGURE 6

jet. A single jet is used in this case, the presence of an object modulating the flow and causing a pressure change at the output (Figure 6). It can be used to measure objects moving towards or away from it as well as across it, with suitable signal amplification.

The conical jet sensor is a further type where the jet emerges from an annular nozzle and is of divergent conical form. Air inside the cone is entrained, resulting in a region of reduced pressure, sensed by the output; the depression and therefore the output signal strength, is modified by the asymmetric distortion of the conical jet resulting from its impingement on a downstream object. It is rather slower in response than other types of interruptible jet sensors and requires a higher air flow.

The reverse configuration or cone jet, Figure 7, employs an annular nozzle to produce a convergent conical jet. When this impinges on an object, pressure is increased within the cone, creating a back pressure through the outlet.

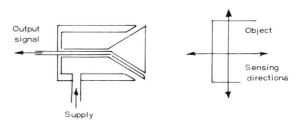

FIGURE 7

All single jet devices are free from contamination in conditions of dirty ambient air, however they do work with small orifice sizes, so the quality of the supply air has to be good at least with a filtration level of 5 micron.

Bleed sensors

Where contact with the sensed object is possible there are some other forms of pneumatic sensors as shown in Figure 8: a simple incorporated bleed sensor, a ball roller bleed sensor

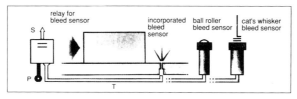

FIGURE 8 *(Telemecanique)*

Dimensions [mm]

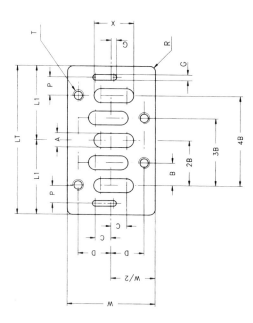

Size	A	B	C	D	G[2]	L₁ min.	L₁ max.	P	R max.	T[3]	W	X	Y[4] min.	csa slots mm²
ISO I	4,5	9	9	14	3	32,5	65	8,5	2,5	M5x0,8	38	16,5	43	70
ISO II	7	12	10	19	3	40,5	81	10	3	M6x1,0	50	22	56	143
ISO III	10	16	11,5	24	4	53,0	106	13	4	M8x1,25	64	29	71	269

2) "G" is the minimum width of the slots.
3) The thread depth is at least twice the nominal thread diameter.
4) Dimension "Y" is the distance between the centrelines of adjacent blocks.

FIGURE 1 – ISO sub-base dimensions.

TABLE 1

ISO number	Recommended port sizes (in)	Orifice area mm²
1	¹/₈, ¹/₄	79
2	¹/₄, ³/₈	143
3	³/₈, ¹/₂	269
4	¹/₂, ³/₄	438
5	³/₄, 1	652
6	1, 1¹/₄, 1¹/₂	924

so as to allow 50% more flow than the corresponding port size, so that there is no resistance to the flow in the sub-base. See Table 1. The ISO standard does not cover electrical entry.

Sub-bases allow for side or back connection. It is possible to have stacking sub-bases and combine two or more ISO bases with a transfer module, Figure 2.

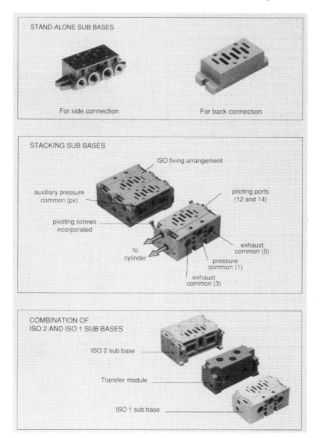

FIGURE 2 – Sub-base combinations. *(Telemechanique)*

For miniature valves smaller than ISO 5991 size 1, sub-bases are made either to the manufacturer's own design or to the German standard VDMA 24563. The latter standard, where it overlaps the smallest ISO size (ISO 1) has a smaller body. So, for example, ISO 1 has a body width of 43 mm, the VDMA standard has body widths of 18 and 26 mm.

Other sub-base patterns may be found, for example DIN, CNOMO and ANSI as well as manufacturer's own designs.

Manifold mounting

In-line valves can be mounted together on a manifold for compact installations. The valve body has to be designed to allow for connection in this form. The inlet and exhaust connections are in the manifold and the cylinder ports are in the valve body. The number of valves that can be connected in this way can be large; depending on the manufacturer there can be 20 or more. One typical type is shown in Figure 3.

FIGURE 3 – Multivalve sub-base. *(SMC)*

Sub-base valves can either have a single sub-base block or more usually be manifolded together on a manifold block with all the connections in the sub-base.

Manifold valves are usually pilot or solenoid operated (or both in the same block). It is not usually possible or practical to manifold together manually or mechanically operated valves.

Valve operation

Movement of a valve element (spool poppet or slide) from one position to another can be accomplished by any of the following methods:

• Manually by a push button, lever or pedal
• Mechanically by spring, plunger or roller

- Electrically by a solenoid or electric motor
- By fluid pressure (air or hydraulic operation)
- By self actuation

Standard methods of designating these in a circuit diagram are shown in Figure 4. See also the chapter on Graphical Symbols.

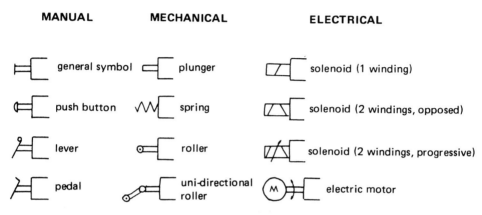

MANUAL   **MECHANICAL**   **ELECTRICAL**

general symbol   plunger   solenoid (1 winding)

push button   spring   solenoid (2 windings, opposed)

lever   roller   solenoid (2 windings, progressive)

pedal   uni-directional   electric motor
                  roller

FIGURE 4 – Standard symbols for valve control methods.

Manual and mechanical operation are basically similar in that a mechanical force is applied directly to the moving element. With a lever, the mechanical advantage can be adjusted to allow for the available force or displacement. It may be necessary to incorporate a spring detent to hold the moving element in a particular position (either of the selected positions or neutral). It may also be necessary to incorporate a spring return so that the valve is held in the selected position only so long as the actuating force is applied.

Manual and mechanical operation

These valves may have direct or indirect operation of the spool or poppet. In the case of direct operation, the spool is actuated by a mechanism or by human actuation (hand or pedal).

Indirectly operated valves are where the spool is moved by the application or removal of an air supply on the end of a valve spool or on a pilot piston. The thrust exerted by the air pressure acting over the surface area of the pilot piston overcomes the friction of the valve and any return spring that may be fitted and causes the spool to move. Valves can be either dead movement or spring return.

Pedal valves are foot operated, suitable for such purposes as operating machinery or safety guards. They can be operated by a cam mechanism that prevents over-travel of the spool. A compact unit with a small movement gives less operator fatigue during repeated operations.

Two handed control units are intended to operate a pressure operated spring return valve which controls an air cylinder which may be used directly to move a press or other machine. The control unit protects the person operating the buttons. If there is a danger of persons being injured, guards or interlocks are necessary.

Figure 5 shows a two handed control unit which has push buttons built into the end covers; the positioning of the push buttons is arranged so that they cannot be operated by one hand nor readily bridged by a tool. The unit should be positioned so that it cannot be operated by one hand and another part of the body. In addition the push buttons are shrouded to guard against accidental operation, while it is still possible to wear gloves. The control unit will supply pilot pressure only if the two buttons are pressed within one second of each other. No signal will be given if one button is continually depressed or under any circumstances other than the two buttons being pressed together.

Manual- and mechanically-operated valves can be supplied with a variety of operating

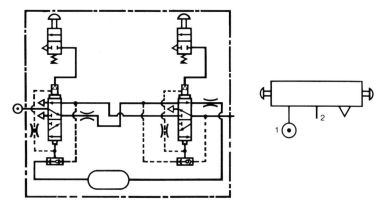

FIGURE 5 – Two-handed control unit, simplified and detailed circuit symbols.

FIGURE 6 – Pilot valves with a variety of mechanical triggering. *(Vickers)*

mechanisms, examples of which are shown in Figure 6. Mechanical operation includes operation by palm, plunger, roller, trip, foot, push button, switch, key, rotary selector.

Air operation

In the case of air operation, this may be direct, indirect or pilot operated. In a direct air operated valve, signal pressure is applied directly to a piston formed by an end cap or full bore section on the ends of the valve spindle. Air operated movement can be in one direction (with spring return) or in both directions; either method is suitable for two-position valves.

For three-position valves, a central position can be given by applying signal pressure to both ends simultaneously, releasing pressure on one side or the other to achieve movement. Alternatively, the spool can be spring biased towards the central position and pressure applied to one side or the other with the opposite side vented.

The signal pressure is often substantially less than the line pressure carried by the valve, so that lower signalling pressures can be used. However if the available signal pressure is too low, the valve may be operated indirectly by a low pressure signal through a suitable actuator such as a diaphragm, Figure 7. One-direction or two-direction actuation is possible by feeding an air signal to one or both sides of the diaphragm simultaneously. Pilot-operation provides the main valve with a pneumatic actuator capable of moving the valve when a low pressure signal is applied.

Air-actuated valves can be designed to provide a delay in shifting the valve after admission of pilot air (timed-in) or cessation of the pilot supply (timed-out). For timed-in operation, air is admitted slowly so that a few seconds are required for enough air to accumulate in the head to shift the valve. For timed-out operation the process is reversed, causing a delay in the exhausting of the pilot air from the head. The valve therefore remains actuated for a few seconds after the pilot supply air has been turned off. The delay mechanism can be adjusted to provide variable time delays. This kind of valve introduces the possibility of having an automatic sequence of operations without the need for operator

FIGURE 7 – Diaphragm operated pilot valve responds to a pressure of 0.24 bar. *(Vickers)*

intervention; pilot air can be supplied by some other operation and the subsequent steps can be delayed for a preset time.

Solenoid valves

These valves are becoming increasingly important with the extension of computer control and other electrical control systems for automation and robotics.

When control is to be initiated from an electrical signal, some form of solenoid operation is required in order to generate the force necessary to operate the valve. A simple solenoid valve comprises an integrally mounted solenoid with an armature directly linked to the valve movement. Alternatively (and commonly) in order to minimise the electric power required, the solenoid may only act as a pilot control, the valve itself then being operated by the fluid pressure available.

Solenoid poppet valves

By arranging the solenoid armature to work in a sealed tube with the solenoid coil enveloping it, the sealing glands can be dispensed with, simplifying the construction and eliminating one source of leakage. The principle is illustrated in Figure 8.

The valve is T-shaped with two ports opposite each other, while the third is at right angles to them. The plunger of a corrosion-resistant ferrous material is spring-biased so that when un-energised it closes the lower orifice while leaving the other open.

When energised, the plunger is pulled up so that the lower orifice is opened and the upper closed. The spring can be arranged to bias the plunger in the opposite direction. The plunger is provided with plastic valve discs, nylon or synthetic rubber. Because the plunger is unbalanced, the force due to the pressure must be limited. The size of the orifice and therefore the flow and pressure drop is related to the power. Sealing is normally bubble tight, but this is dependent on the cleanliness of the air.

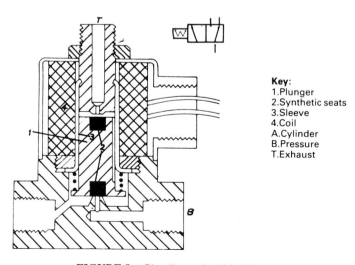

Key:
1.Plunger
2.Synthetic seats
3.Sleeve
4.Coil
A.Cylinder
B.Pressure
T.Exhaust

FIGURE 8 – Glandless solenoid valve.

Glandless valves may be installed in any position and will withstand appreciable shock loads. Response time to energise or de-energise a miniature valve of this kind is of the order of 5 ms on 24 V d.c.

The valve as illustrated is an in-line type but sub-base or manifold versions are also available.

Solenoid pilot valves

An example of this type is shown in Figure 9. The solenoid controls a pilot supply for the indirect operation of the main spool. When the solenoid is de-energised, the pilot air is blocked off and the pilot chamber is vented to exhaust by way of axial slots in the armature of the solenoid. Energising the solenoid applies a pilot air to the pilot piston on the spool. Thus a small pilot solenoid operates a spool that would otherwise require a large solenoid with its armature directly connected to the spool. The single solenoid-operated spool valve is normally supplied with a spring return, although any other mechanism could be fitted to reset. A second solenoid could be used which would give a latching function from two independent pulses. By fitting low power valves to a larger pilot operated valve, low power, high flow valves can be produced.

Single- and double-solenoid valves can be mounted on an ISO 5599 standard sub-base or on a manifold as described above.

The pilot air can come through a separate port which allows extra flexibility of control in that both pilot air and an electrical signal have to be present before spool movement is possible. If that is not required, the pilot air can come from an internal passage in the valve.

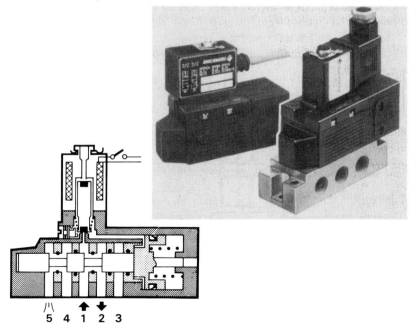

FIGURE 9 – Solenoid pilot valve. *(Asco Joucomatic)*

Functional diagram

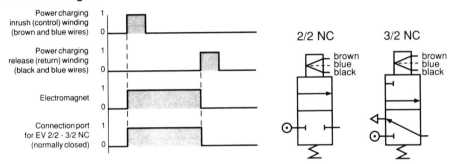

FIGURE 10 – Impulse solenoid valve. *(Asco Joucomatic)*

Impulse solenoid valves

These are a type of solenoid valve which move on the application of an impulse power signal and therefore consume no power in the steady state. The advantages of these zero-watt coils are no continuous power consumption, negligible heating effects and safety, which is realised because the valve is bi-stable, remaining in its last position in the event of a power failure. This is of particular value in battery operated equipment.

A typical valve is shown in Figure 10. The actuator consists of an epoxy-encapsulated coil containing windings for pull and throw function. A permanent magnet incorporated is completely separated from the fluid medium.

The impulse coil for plunger-armature systems works in such a way that switching is achieved by means of a short power impulse on the electro-magnet. The permanent magnets enable the valve to retain this position without the need for continuous application of electrical energy. Only when a second impulse is applied does the valve switch back. No power is required to maintain the operated position. Consequently the heating effect of the coil is negligible and the seal materials of the valve are not subjected to any thermal loading. The advantages of this impulse coil can be appreciated in applications involving the control of multi-way pneumatic valves.

Valves of traditional design require an electrical control unit for switching pneumatic cylinders, which have to provide a continuous electrical or pneumatic signal during the entire period of solenoid valve operation. The impulse coil simplifies matters considerably when used in conjunction with position sensors for the control of pneumatic cylinders.

SECTION 7

Actuators

ACTUATORS

CONSTRUCTION OF PNEUMATIC CYLINDERS

SELECTION AND PERFORMANCE OF CYLINDERS

ACTUATORS

Types of actuator

Compressed air can be used to move, clamp or vibrate some sort of output device. Pneumatic cylinders are the most commonly found actuators, the construction of which is considered in the next chapter. In this chapter some of the other mechanisms are discussed.

Diaphragm cylinders or thrusters

A diaphragm cylinder is a large diameter short stroke cylinder fitted with a diaphragm rather than a piston; it is usually catalogued as a thruster. The construction takes the form of a pair of shallow convex housings, with a diaphragm sandwiched between them and a piston rod attached to the diaphragm. They are usually of a single-rod type, either single acting with spring return or double acting.

They are capable of generating high forces with short strokes, which is what is required for component clamping. The thrust force is equal to the product of the diaphragm area and the pressure. The stroke is limited to about one-third of the diameter, within the flexibility of the diaphragm element.

The thruster body can be made from aluminium castings or steel pressings, clamped or

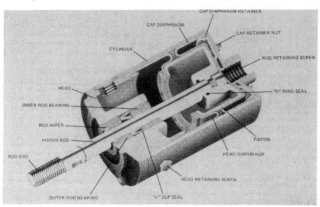

FIGURE 1 – Function of doubling acting membrane cylinder. *(Bellofram)*

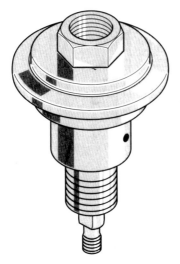

FIGURE 2 – Short stroke, spring return membrane cylinder.
Force at 6 bar = 500 N.
(Kuhnke)

bolted together as in Figures 1 and 2. The diaphragm is made of neoprene or fabric-reinforced nitrile rubber.

The cross-section of the these diaphragm thrusters may be round or assume any other convenient shape for special purpose clamping as illustrated in Figure 3.

Example of application:

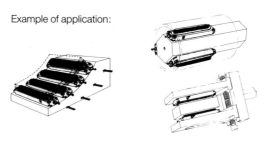

FIGURE 3 – Selection of diaphragm thrusters. *(Festo)*

Rodless cylinders

For long strokes, the conventional rod-type cylinder has severe space limitations: it has an overall length when extended of twice the stroke plus the length of the end caps; the cylinder rod when extended can flex, resulting in excess wear on the seals and bearings.

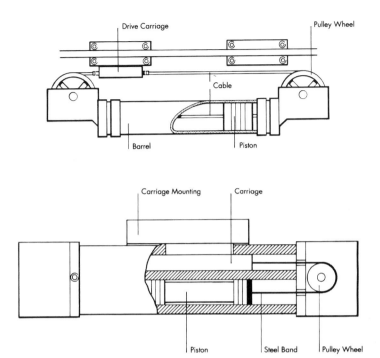

FIGURE 4 – Cable and band cylinders.

The first attempt to overcome these problems was a cable cylinder. This consists of a piston sliding in the cylinder barrel attached to a steel cable instead of a piston rod. Pulley wheels are mounted on the cylinder ends and the drive carriage is attached to the cable. This results in a cylinder with an overall length little more than the stroke (Figure 4). It is possible with some ingenuity to imagine applications where the cable can go round corners and where a rotary drive can be taken from one of the pulleys. The drawbacks are the stretch and wear of the cable which necessitates continual lubrication, adjustment and maintenance; the carriage requires its own supports which can make for a bulky construction.

A band cylinder is similar to a cable cylinder, except that the cable is replaced by a stainless steel band. This is more compact but still suffers the same problems of stretch and need for maintenance.

A magnetic cylinder is intended to overcome some of these difficulties (Figures 5 and 6). In this design the piston contains a series of permanent magnets operating in a non-magnetic, stainless steel tube; a carriage is mounted on the outside of the tube and houses a second set of magnets creating a magnetic coupling between the rod and carriage. This is satisfactory for light, slow moving loads, but the carriage tends to separate under high inertial loads, and the positional register is lost. For slow moving loads, this turns into an advantage because when magnetic break-away occurs, addition mechanical protection is given to the machinery; coupling is only restored when piston and carriage come back into

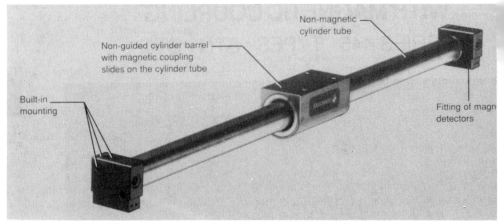

FIGURE 5 – Magnetic coupled cylinder – non-guided. *(Asco Joucomatic)*

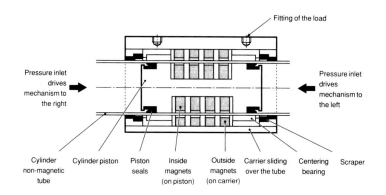

FIGURE 6 – Principle of magnetic cylinder. *(Asco Joucomatic)*

alignment. Additional guidance may also be required, particularly if there is a tendency for the load to rotate. These cylinders are used for special purposes, where a compact design is important and for such applications as door opening and material handling.

Modern rodless cylinders are based on the band principle, but instead of a continuous band running over pulleys, there is only one pair of bands running from one end cap to the other and connected to the carriage through a slot in the wall of the cylinder. The slot is sealed by the bands made of stainless steel or elastomer. Figure 7 illustrates the construction. The stroke length is little less than the overall length of the cylinder.

The cylinder is made of aluminium extrusion. The slot in cylinder reduces its torsional rigidity, which has to be compensated by additional material in the cross section which can be of a complex form. The piston is also made of aluminium, rigidly coupled to the external carriage.

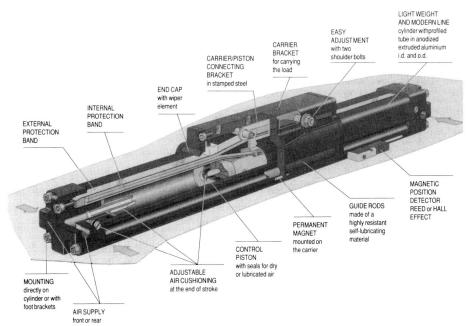

FIGURE 7 – Rodless cylinder. *(Asco Joucomatic)*

Despite the apparent complexity, these rodless cylinders work very well and are of use where space is at a premium. The absence of a piston rod means that the whole of the piston area is available for pressure in each direction. The carriage can be supported on guide rods, as shown in Figure 7, or more usually in the body of the cylinder, and requires no additional support

A further ingenious development of the rodless cylinder uses a compound cylinder with two bores each with their own carriage. The two carriages move in opposite directions along the length of the cylinder. If one carriage is now taken as a fixed point, it can be seen that the other carriage will move almost twice the length of the cylinder (Figure 8). This design incorporates a toothed band to obtain the extended motion.

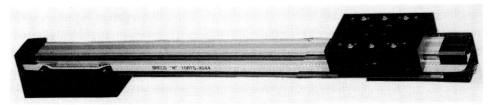

FIGURE 8 – Double extension rodless cylinder. *(Norgren)*

Rotary actuators

To convert pneumatic pressure into a torque, various devices are available. Continuous rotary motion, as in a pneumatic rotation motor, is dealt with in other chapters. The name rotary actuator implies a limited, rather than a continuous, motion with a torque which is calculated on the basis of pressure and area and is maintained as long as the pressure is applied. It is a quasi-static rather than a dynamic device.

Of the several kinds of rotary actuators, the following are found:

- rotation by a linkage attached to the rod end of a cylinder
- rotation by a pair of cylinders acting in opposition
- a vane actuator, similar to a vane motor but of limited angular movement
- rack and pinion, where the rack is operated by a cylinder
- linear movement of a piston acting on an angled surface, producing limited rotation of a shaft
- rotary indexing tables with a pawl and ratchet mechanism
- linear/rotary devices with a cam track on the piston of the cylinder.

Vane actuators

A vane actuator can have one or two vanes as illustrated diagrammatically in Figure 9, which also shows the range of sizes. The single vane type consists of a vaned shaft with

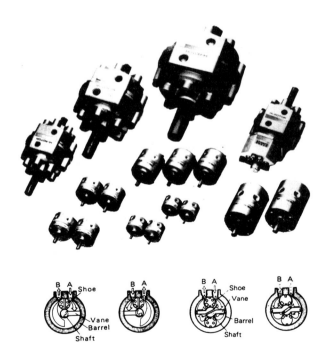

FIGURE 9 – Vane type rotary actuators showing the operation of single and double vanes.
(CompAir Maxam)

seal, a shoe (stop) and two barrels. If air enters through port A, it pushes the vane, turns the shaft and generates a clockwise torque. The air in the opposite chamber is discharged through port B, until the vane strikes the shoe and stops. If air enters through port B an anti-clockwise torque is generated. The possible total angular movement for a single vane is about 270°. The total angular movement is adjusted by means of the stops.

A double vane type has two vanes, two shoes and two barrels. It is capable of generating approximately twice the torque of the single vane, but the angle of rotation is limited to about 90°. This particular model is available in a range of sizes which at 10 bar pressure can produce a torque of up to 200 Nm for a single vane and 400 Nm for a double vane.

Rack and pinion type actuators

There are several designs available, operated by a single acting cylinder with spring return, a double acting cylinder or twin cylinders. The basic principal is the same in each: the piston rod of a pneumatic cylinder becomes a rack which rotates the pinion shaft. In a twin cylinder arrangement it is possible to have a three- or four-position actuator according to which ports are pressurised, as shown in Figure 10. The angular rotation is limited in

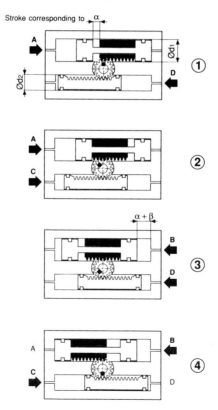

FIGURE 10 – Twin rack and pinion actuators showing how four-positions are obtained according to which ports are pressurised. *(Asco Joucomatic)*

principal only by the stroke length of the cylinder, but in practice, standard units rarely exceed about 360°. Up to 400 Nm is possible with this type of design.

Inclined plane devices

In these the linear movement of two pistons produces via angled surfaces a rotation of the drive shaft (Figure 11). The maximum rotation is about 90°. These are miniature devices with a torque capacity of about 2 Nm.

FIGURE 11 – Inclined plane actuator with spring return. *(Kuhnke)*

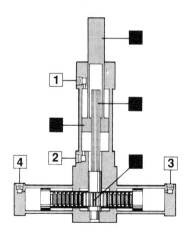

WORKING PRINCIPLE
The main components of the Actuator consist of a cylinder and rack-and-pinion type rotary actuator. Linear motion of the rod (D) is produced when port 1 or 2 is pressurised. Rotary motion of the rod (D) is produced when port 3 or 4 is pressurised causing pinion gear (A) and spline bar (B), which are coupled together, to rotate coupled piston (C).

FIGURE 12 – Combined rotary and linear actuator. *(PHD inc)*

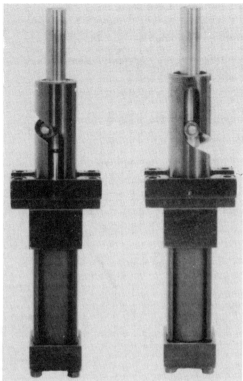

FIGURE 13 – Cam-type actuator. *(E and E Engineering)*

Combined linear/rotary devices

These are of two kinds: in one the linear and rotary motion are independently controlled; in the other the rotation and linear motion are geometrically related through a cam mechanism.

In the former, the rotation unit and cylinder are built into one body with separate air supply to the two units, as illustrated and described in Figure 12. This range is available to produce a torque of 360 Nm combined with a direct extension force of 4.7 kN.

In the latter, the shape of the cam track has to be chosen to produce exactly the right kind of motion. Two types are shown in Figure 13: linear followed by rotary and rotary followed by linear. Simultaneous or more complex motions may be created for special purposes. These are mainly used for clamping purposes.

Air chucks and grippers

These also are intended for clamping and gripping, often used on the end of a robot arm for manipulating components for machining or processing. A variety of different grippers are available: parallel, angled, radial and multi-point, Figure 14. They may rely solely on the force produced by the air or may incorporate a toggle mechanism to hold the

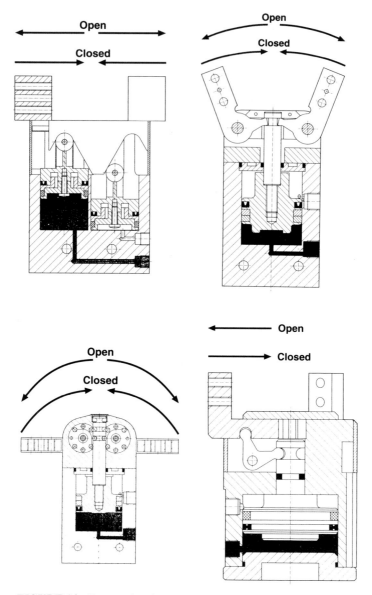

FIGURE 14 – Pneumatic grippers showing the operation of parallel,
angled, radial and three-point types. *(Festo)*

component even if the pressure drops, Figure 15. The short-stroke pneumatic cylinders,
which are at the heart of these grippers, may, as with such cylinders, be either double
acting, spring return (spring close or spring open) or spring assist (spring assist open or
close). The features of these are shown in Figure 16.

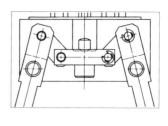

FIGURE 15 – Toggle-type gripper. *(SMC)*

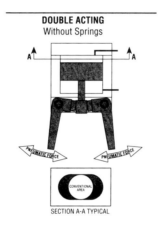

DOUBLE ACTING
Without Springs

SECTION A-A TYPICAL

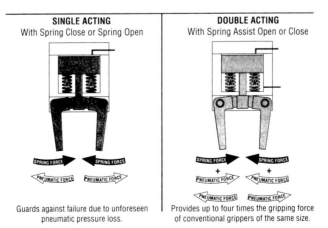

SINGLE ACTING
With Spring Close or Spring Open

DOUBLE ACTING
With Spring Assist Open or Close

Guards against failure due to unforeseen pneumatic pressure loss.

Provides up to four times the gripping force of conventional grippers of the same size.

FIGURE 16 – Forms of return mechanism of grippers. *(PHD inc)*

Air bellows

Air bellows are used for short stroke, high thrust applications, available as single, double or triple convolutions, Figure 17. They have no metal reciprocating parts and so can provide virtually frictionless thrust compared with a conventional cylinder. They are single acting only; the return stroke is provide in part by the natural spring action of the bellows, but more usually by the load itself. The simplicity of construction provides an extremely long, maintenance free service life under arduous conditions.

Due to the flexible construction, the mounting of bellows is less critical than with conventional cylinders. They will operate with axial misalignment up to 15° and a maximum axial misalignment of 10 mm. Since the effective diameter reduces as the bellows extend, the force is correspondingly reduced, and this should be taken into account by reference to manufacturers' literature. The end plates are made of plastic (polyamide) in smaller sizes and steel or aluminium for larger sizes.

In operation, the bellows should not be allowed to bottom out or achieve its maximum height. Sizes are available (in imperial sizes) from $2^3/_4$" up to 26" and can generate a force up to 3400 daN per 1 bar pressure. The maximum working pressure can be up to 15 bar.

For further information refer to the chapter on Air Springs.

Single convolution

Double convolution

Triple convolution

FIGURE 17 – Air bellows actuators. *(Parker)*

CONSTRUCTION OF PNEUMATIC CYLINDERS

Pneumatic cylinders can be classified in various ways such as single-acting, double-acting, through-rod, tandem and duplex, cushioned, rotating and miniature. A wide choice of mountings is available: foot mounting, pin end, neck and flange mounting, hinge brackets, spherical rod ends, clevises *etc*. Pneumatic cylinders of these types are manufactured on similar lines to hydraulic cylinders, except for a difference in the materials of construction and the relative rod sizes due to the lower working pressures of compressed air. A wide range is available commercially to meet the needs of most users and are readily made from standard sections, end caps and pistons.

If possible one should adopt a standard design to allow for interchangeability.

The following standards should be referred to for consistent dimensions of cylinders.

- BS ISO 6431 for 32 mm to 320 mm bore (32, 40, 50, 63, 80, 100, 125,160, 200, 250 and 320 mm)
- BS ISO 6432 for 8 mm to 25 mm bore (8, 10, 12, 16, 20, 25 mm)
- BS 5755 (ISO 4393) for recommended stroke lengths
- BS 6331 (ISO 6099) for mounting styles
- BS 5755 (ISO 4395) for piston rod threads
- ISO 6430 for mountings integrated in the main body of the cylinder.

Cylinders made to these standards are usually listed by manufacturers as ISO cylinders. The former British standard BS 4862 has now been withdrawn.

Other standards such as VDMA 24562, DIN 24335, CNOMO/AFNOR NFE are also to be found.

When considering interchangeability, the level of standardisation has to be considered. When replacing ISO cylinders, it is not always possible to change just the cylinder: it may be necessary to change both cylinder and mountings. With VDMA and CNOMO/AFNOR, the bare cylinder is interchangeable.

Imperial (inch sizes) are still manufactured, mainly in the U.S.A. but to no particular standard dimensions.

Various cylinder designs are illustrated in Figures 1 to 5.

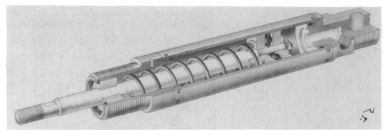

FIGURE 1 – Single-acting (spring return) cylinder. *(Asco Joucomatic)*

FIGURE 2 – Double-acting cylinder. *(Asco Joucomatic)*

FIGURE 3 – Heavy duty cylinder with tie rods to ISO standards. Pressures up to 16 bar. *(Norgren)*

FIGURE 4 – Light duty cylinders to ISO 6431. Cylinder body in aluminium alloy, piston rod in stainless steel, die cast aluminium end caps. End caps attached by steel screws. *(Norgren)*

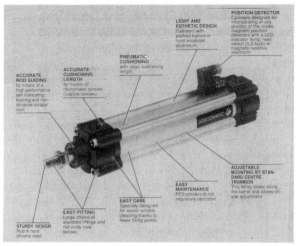

FIGURE 5– Features of a double-acting cylinder. *(Asco Joucomatic)*

Tube materials

Material selection must ensure adequate resistance to wear, corrosion and accidental damage. Particular attention must be given to the choice of materials in special environments such as explosive atmospheres where aluminium alloys are not permitted or where corrosive atmospheres are present. Refer to Tables 1 to 5 for a summary of available materials. Plastics were introduced for light weight applications but have now practically disappeared, although for special applications glass- or carbon-fibre reinforced resin is used.

TABLE 1 – Cylinder tube materials

Light-duty	Medium-duty	Heavy-duty
Plastic Hard drawn aluminium tube Hard drawn brass tube	Hard drawn brass tube Aluminium castings	Hard drawn brass tube Hard drawn steel tube Brass, bronze, iron or steel castings Welded steel tubes

TABLE 2 – End cover materials

Light-duty	Medium-duty	Heavy-duty
Aluminium stock (fabricated) Brass stock (fabricated) Aluminium castings	Aluminium stock (fabricated) Brass stock (fabricated) Bronze stock (fabricated) Aluminium, brass, iron, or steel castings	High tensile castings

TABLE 3 – Piston materials

Light-duty	Medium-duty	Heavy-duty
Aluminium castings	Aluminium castings Brass (fabricated) Bronze (fabricated) Brass, bronze, iron or steel castings	Aluminium forgings Aluminium castings Brass (fabricated) Bronze (fabricated) Brass, bronze, iron or steel castings

TABLE 4 – Piston rod materials

Material	Finish	Remarks
Mild steel	Ground and polished Hardened, ground and polished Chrome-plated	
Stainless steel	Ground and polished	Generally preferred Less scratch-resistant than chrome-plated rods

TABLE 5 – Mount materials

Light-duty	Medium-duty	Heavy-duty
Aluminium castings Light alloy fabrications	Aluminium, brass and steel castings	High tensile steel castings High tensile steel fabrications

Cylinders can be made in magnetic or non-magnetic versions.

Cylinder tubes are cheaply made from drawn metallic tubing. For light duty, aluminium is commonly used; for medium duty, although less commonly, brass tubing is used. For the most robust applications, both stainless and alloy steel tubing is used.

If non-stainless steel is used, attention must be given preventing corrosion. Large size cylinders may be cast in aluminium or steel. Welded steel tubes are also used.

To ensure low friction and to minimise seal wear, a fine finish in the bore is essential to 600 μm or less when elastomeric seals are employed. This may require honing even with hard drawn seamless tubes. Hard chrome plating, honed and polished is used on alloy cylinders, and hard anodising on aluminium tubes.

End covers

These may be cast or fabricated from stock, normally in the same material as the tube to avoid the possibility of electrolytic corrosion. This is an important consideration, particularly since the air may be saturated with moisture.

The method of attachment of the end covers varies. Covers may be threaded, welded or mechanically secured. A common construction for medium and heavy duty cylinders is

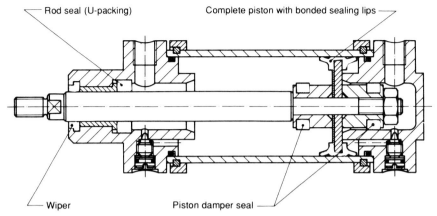

Rod seal (U-packing) Complete piston with bonded sealing lips

Wiper Piston damper seal

FIGURE 6 – Double-acting cylinder showing adjustable cushioning and seals.
(Freudenberg Simrit)

the use of tie rods (usually four) made of high tensile steel. Square section tubes allow for attachment of the end caps by screws threaded into the body of the section.

The end covers incorporate the inlet and outlet ports and the cushion chambers where fitted. Pneumatic cushioning is required in order to absorb the dynamic forces at each end of the stroke. The cushioning can be fixed or adjustable by means of a needle valve, as in Figure 6. Where the additional feature of pneumatic cushioning is not required, a simple elastomeric washer built into the end cap may suffice.

One or both of the end covers carry the rod bearings and rod seals according to whether the design is of single rod or through-rod designs. A through-rod cylinder is symmetrical in end cap design.

Tie rod cylinders

The main requirement of a tie rod construction is that the end covers cannot move when the cylinder is under pressure. This implies tensioning the rods to a load equivalent to the maximum pressure in the cylinder. For long cylinders there may be a central tie rod support ring. Any stripping down of the cylinder must be followed by the correct tightening of the tie rods to the manufacturer's specification.

Pistons and piston rods

Pistons are usually aluminium alloy stock, castings or forgings. Other materials are steel, brass or bronze.

They may be of one-piece or multi-piece construction. Cup seals require multi-piece construction. O-ring or square section seals can be fitted into a one-piece construction.

An additional nylon wear ring (or two rings) may be fitted to provide support for the assembly. One of the design problems of a long cylinder is its stability under compression, so the providing of this extra support to the piston is worth incorporating. This is less a problem with through cylinders where the opposite rod gives the extra stability in compression.

For operation at high temperature, beyond the range of elastomeric seals, automobile

pressure in the cylinder without any twisting or bending loads caused by the method of mounting. End fixings are preferred to central trunnion mounting.

Rigid mountings are used where the cylinder has to have good support. They may be single or double, the latter being suitable for long stroke cylinders. Foot and centre-line mounts are normally incorporated in the construction of the cylinder end covers. Single foot mounts are only used in light duty applications. Pedestal and flange mounts are attached to the end covers and supplied by the manufacturer.

Rigid mounts can absorb the end thrust on the cylinder centre-line or in a plane parallel to it but removed from the centre-line. The former is to be preferred where the forces are heavy, otherwise bending is induced in the mount. The mounting bolts are then only stressed in tension, provided that there is no misalignment.

Examples of mounts which absorb the forces on the cylinder centre-line are centre-line lugs, mounting by extension of the side rods, flange mounting and screwed nose mounting. These mounts may tolerate some misalignment when the piston rod is extended but little or none when contracted.

End-mounted cylinders may require extra support at the free end of the cylinder. Double mounts are chosen when a long stroke is overhung. Long cylinders may have an additional extra support.

Foot mounts which do not absorb forces on the centre-line are unsatisfactory where the forces are heavy or where shock loads are involved. The mounting bolts are subject to compound stresses which may be indeterminate, the cylinder itself is subject to bending or buckling and the platform on which the mounts are attached may distort. A bent cylinder may require a greater air pressure to operate or the friction induced may be such as to prevent operation altogether.

Rod ends

These are finished in a variety of ways, some of which are shown in Figure 8. These are standard forms supplied by the manufacturer. A clevis or tongue end or a universal joint is to be preferred where there is any chance of misalignment. Bending forces on the piston rod are to be avoided at all cost.

Non-rotating rod cylinders

In the majority of applications for cylinders it is permissible to allow the rod and piston to rotate freely, any angular orientation being taken care of by the mechanism to which the cylinder is attached. Where this is not possible, a non-rotating cylinder can be employed. Such a cylinder maintains the angular orientation of the rod and its end fitting without any further guide mechanism. A variety of ways are available to achieve this, Figures 9 to 14:

- external guides,
- twin piston rods,
- guide splines in the piston,
- oval pistons,
- non-circular (oval, square or triangular) section piston rods,
- linear ball supporting splines in the piston rod.

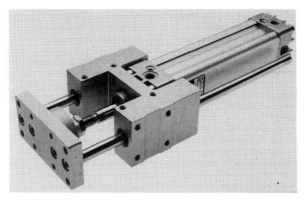

FIGURE 9 – Cylinder with external guide rods. *(Norgren)*

FIGURE 10 –
Cylinder with twin
piston rods.
(Asco Joucomatic)

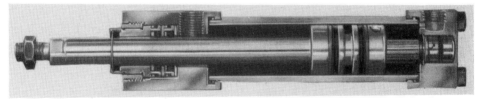

FIGURE 11 – Cylinder with twin piston rods. *(PHD)*

FIGURE 12 – Double-acting cylinders of oval cross section. *(Festo)*

FIGURE 13 –
Non-rotating
cylinder with
triangular piston
rod.

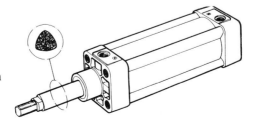

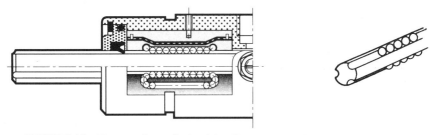

FIGURE 14 – Non-rotating cylinder rising linear ball bearings. *(Asco Joucomatic)*

External guides are the most usual because they can be designed to take substantial torque and bending loads, Figure 9. Guide units can be attached to the end caps of standard cylinders, which is an advantage in standardisation. In addition to the non-rotating function they are also used where there is an overhung load on the piston roads which requires extra support.

Mechanisms which are built into the cylinder itself can tolerate only a limited torque and no extra bending load. The one exception to this is the twin-rod type which can accept transverse loading in the plane of the axis through the rods. Suppliers of special cylinders can supply design curves giving the maximum permissible transverse loading and torque at the various stroke lengths.

Compact (short stroke) cylinders

Where space is limited and the required stroke is short, special compact cylinders, Figure

FIGURE 15 – Compact cylinders, approximately one-third of the length of the corresponding ISO cylinders, rotating and non-rotating types. *(Norgren)*

15, can be used; their most common use is for clamping purposes. They can show considerable space savings over the ISO standard cylinder (a reduction in overall length of up to 80%). They can be machined out of solid, with the end caps integral with the body, or using drawn tubing the ends can be retained by circlips. Porting is machined into the body of the cylinder. They can be mounted by through bolts in the body. Non-rotating versions of compact cylinders are also available.

Composite and special cylinders

By combining cylinders together end-to-end it is possible achieve some special effects.
With two cylinders back to back (Figures 16 and 17) a multi-position cylinder is

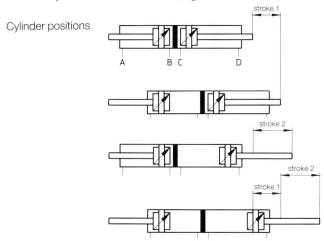

FIGURE 16 – Diagram showing how a 3-position duplex cylinder functions. *(Festo)*

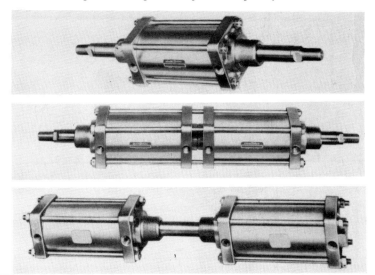

FIGURE 17 – Special cylinders through-rod, duplex, duplex with common piston rod. *(Parker)*

obtained. Usually the cylinders are of equal diameter. If they are of the same stroke length, a three-position device is produced, if of unequal length a four-position device. A similar effect is achievable with a common piston rod

A through rod cylinder with a common piston is completely balanced, giving equal thrust in both directions. This can be used where one end of the rod is inside the machine, leaving the other end to trip a valve.

Position sensors for cylinders

In automation applications, it is often required for a signal to be generated when the cylinder piston reaches a particular position along its stroke, so that the control system can initiate the next phase of the operation. This position may be at either end of the stroke or some point intermediate between the ends. There are several ways in which this can be done (Figure 18):

- The piston rod trips a micro-switch or pneumatic valve.
- Pressure threshold sensors respond to a drop in exhaust pressure when the piston stops moving.
- Magnetic sensors mounted directly on the cylinder barrel sense the magnetic field created by a permanent magnet incorporated in the piston and trigger a reed switch.

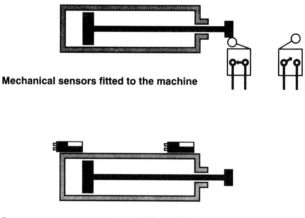

Mechanical sensors fitted to the machine

Pressure sensors mounted on the cylinder

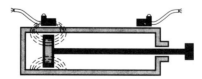

Magnetic sensors mounted on the cylinder

FIGURE 18 – Types of cylinder position sensors. *(Telemechanique)*

- Hall effect sensors (solid state) triggered by a magnetic piston.
- Pneumatic reed valves, triggered by a magnetic piston.
- Miscellaneous such as photoelectric, inductive and capacitive detectors.

Mechanical trip switches are the traditional method for position sensing and are still widely used and reliable. Pressure threshold sensors respond less to actual position than to piston velocity, so they are only suitable for detecting the end of the piston travel; they may work unreliably with slow moving pistons.

Both of these are gradually being superseded by the two types of magnetic sensors.

Most manufacturers can supply pistons incorporating permanent magnets, which are suitable for the two main types of magnetic sensor. The cylinder has to be made of a non-magnetic material such as stainless steel, brass, aluminium or plastic. In the reed switch, the contacts are made mechanically by the presence of the magnet. Some switches incorporate an LED or warning lamp in the body of the switch as in Figure 19. Hall effect sensors have no moving part, but the presence of a magnetic field causes a current to flow in the switch. Reed switches can equally well carry a.c. or d.c., Hall effect switches only d.c. The main advantage of a Hall effect switch is its long life and so it is more suitable for rapid switching rates; there is also no possibility of switch bounce.

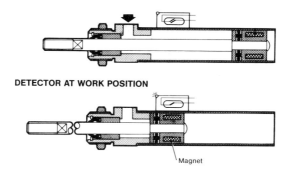

DETECTOR AT WORK POSITION

Magnet

FIGURE 19 – Reed switch mounted on a cylinder. *(Asco Joucomatic)*

These magnetic switches can be clamped to the outside of the cylinder barrel or fitted into a fastening track (Figure 20). With some types of attachment it is possible to have two switches located at the same point of the cylinder giving the security of redundancy.

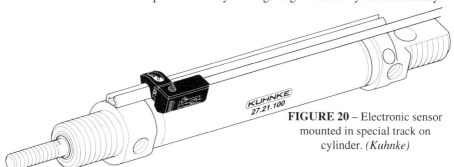

FIGURE 20 – Electronic sensor mounted in special track on cylinder. *(Kuhnke)*

Commercially available cylinders are designed with a generous safety factor. Cylinders of conventional design do not have to meet the criteria for pressure vessels as discussed in the chapter on Air Receivers

Rod strength

It is more likely, when designing a cylinder, that the critical element is the piston rod.
When stressed in tension, the material stress is given by:

$$f = F/A$$

where F is the tensile load and A is the cross-sectional area of the rod.
If the cylinder is used in dynamic applications, a generous factor of safety should be applied. A factor of 5 or 6 on ultimate tensile strength should be used.

The attachment of the rod to the piston and the sizing of the pin or shackle should also be checked in tension or shear.

When the cylinder is used as a strut at maximum extension, particularly with a long stroke length, the piston rod should be checked in buckling. The relationship is:

$$F = \frac{\pi^2 EI}{S^2 f}$$

where: F is the buckling load,
            E the modulus of elasticity,
            I the second moment of area,
            S the buckling length and f the safety factor. This is correct for consistent units.

The buckling length S is related to the stroke length of the cylinder. S is the stroke length x stroke factor obtained from Table 1. Figure 1 can be used to determine the buckling strength.

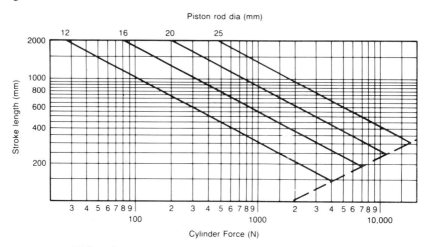

FIGURE 1 – Buckling chart based on a stroke factor of 2.0.

TABLE 1

Rod End Connection	Application		Stroke * Factor
Fixed and rigidly guided	I		.50
Pivoted and rigidly guided	II		.70
Supported but not rigidly guided	III		2.00
Pivoted and rigidly guided	IV		1.00
Pivoted and rigidly guided	V		1.50
Pivoted and rigidly guided	VI		2.00

*Stroke factor should be modified according to the application

If the rod is solid, the second moment of area is given by:

$$I = \frac{\pi D^4}{64}$$

where D is the outside diameter of the rod.

As before the safety factor should have a minimum value of 5.

If the cylinder rod carries a bending load caused by side forces on the rod end, this must also be checked. Manufacturers can supply strength data for this purpose.

Cylinder cushioning

Air cylinders are inherently fast acting in the absence of throttling or inertia loading. If loads are light, the piston will continue to accelerate rapidly until brought to a stop at the end of the stroke. The travel can be arrested either by reaching the end of the cylinder or by a mechanical stop built into the operated mechanism.

When the cylinder itself forms the stop, the kinetic energy of the piston and rod is dissipated in the form of a shock load on the cylinder end cover, which may over-stress the end attachments or tie rods. This can be avoided by incorporating some means of decelerating the piston in a uniform manner as it approaches the end of its stroke; this is known as cushioning. A simple form of cushioning is an elastomer (customarily polyurethane) washer inserted between end cap and piston. Integral cushioning can be done pneumatically via a controlled bleed, thus providing an adjustable damping or braking effect over the last few millimetres of travel. The piston is brought to rest gradually and the kinetic energy is dissipated in heat. Cylinders without cushioning should only be used if the stroke limitation is external.

Cushions can be fitted at one or both ends of the cylinder, and most manufacturers can supply them as an option. The cushion chamber may be incorporated in the end covers or in the form of a cushion collar. The cushion boss or nose of the piston is usually tapered to provide smooth entry, but produces a plunger-type seal once it has fully entered the cushion chamber, trapping the air volume in the chamber which is then compressed by further motion. The rate of bleed and thus the cushioning effect is adjustable by means of a needle valve (see the chapter on Construction of Pneumatic Cylinders).

Detail design varies with manufacturer. Originally cushions were of a simple plunger type with both the cushion chamber bore and the piston nose of metal and this arrangement is widely used. Metal–plastic combinations may be preferred as being less affected by corrosion or dirt. Elastomeric seals may also be used in place of plunger seals.

Another type is based on a spring-loaded seal or an arrangement of face seals which forms its own cushion chamber once the seal is brought into contact with the seating surface at the end of the cylinder. This eliminates the need for a separate cushion chamber but requires a needle valve in both the blind end of the cylinder and the piston rod to bleed off the trapped air. See Figure 2.

The weakest point of all cushioning systems is usually the needle valve or bleed control, which can be affected by dirt or corrosion with a resulting change to the cushioning action.

A cushioned cylinder also requires some modification to the porting. On a double-acting cylinder provision must be made to supply air directly to the piston area and not

CUSHION BLOCK STYLE

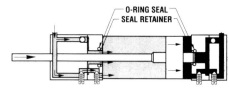

POPPET STYLE

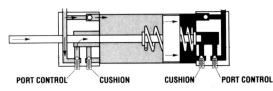

FIGURE 2 – Types of pneumatic cushioning showing combined port control and cushioning controlled by needle valves. *(PHD inc)*

through the cushion chamber, otherwise the initial thrust available would be too low; this is a matter of suitable geometry.

With seal-type cushions, the flexibility of the seal can be such that it may collapse when air pressure is applied from the reverse side, allowing flow to the full piston area. Flow to the full area may also be encouraged by distribution channels in the end cover or piston face.

Cushioned cylinders are most suitable for low to moderate speeds, At speeds above 500 mm/s, they are relatively ineffective, the initial shock as the piston reaches the cushion area being too high unless a long cushion is used with progressive entry. Many automation applications such as machining operations and pick-and-place units require faster movements up to 1 m/s or greater. In these cases it may be necessary to arrange for some form of external cushioning, either by use of a buffer cylinder or some form of speed control which provides retardation over a suitable proportion of the stroke. Hydraulic cushions are often used for this purpose.

Two further limitations of integral cushioning are – (a) they are effective only if the full stroke of the piston is used and (b) they are relatively ineffective on pre-exhausted cylinders or those with quick exhaust valves.

In the former, the piston never enters the cushion area, being brought to rest by external stops. The use of cylinders with standard stroke lengths makes it unlikely that the exact length will be suitable for any particular operation.

In the latter, there is only a small amount of air trapped in the cushion and so the cushioning effect is correspondingly small.

Hydraulic cushioning

One solution to end of stroke cushioning is the use of an hydraulic shock absorber, which can be mounted directly on the cylinder or on the load. One example is shown in Figure 3, mounted on a rodless cylinder. These units are rated by the energy to be absorbed (obtained from the final velocity and the weight being moved) and the cycles per minute. Since shock

FIGURE 3 – Hydraulic shock absorber fitted to a rodless cylinder. *(Asco Joucomatic)*

Fast approach facility

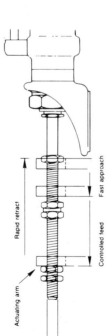

Basic outward checking unit

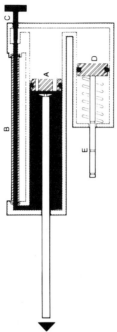

A. Piston with plate valve to permit rapid flow for fast retract.

B. Oil transfer tube.

C. Manually set speed regulator needle valve.

D. Spring loaded piston compensates for varying internal piston rod volume. Cylinder provides oil reservoir.

E. Indicator rod for oil content.

FIGURE 4 – Hydraulic cylinder, to be attached to pneumatic cylinder. *(Parker)*

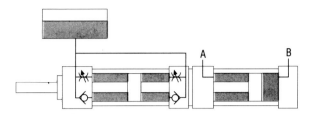

FIGURE 5 – Pneumatic and hydraulic cylinders coupled together in tandem.
Oil is metered from one side of the piston to the other through a transfer tube. *(PHD inc)*

absorbers convert the kinetic energy into heat, it is important not to exceed the maximum permissible cycles per minute for the given energy absorption. Up to 60 cycles per minute is possible for the smaller sizes and up to 5 m/s maximum velocity. Two types are available: automatic adjustment to give optimum absorption performance and user adjustable.

To control the velocity throughout the full stroke of the pneumatic cylinder, it is necessary to have an hydraulic cylinder working in parallel with it. This can be done by mounting the hydraulic cylinder on the back of the pneumatic one (Figure 4) or coupled to it by a common piston rod (Figure 5). In Figure 5, the reservoir is optional and should be fitted in rapid cycling applications. The whole assembly can occupy a considerable amount of space and should only be resorted to if other options fail. With this type of speed control, velocities from about 100 mm/s virtually down to 0 mm/s can be controlled.

Installation

The installation of standard pneumatic cylinders follows a similar procedure to that of hydraulic cylinders and depends on the mounting style chosen. This can be decided while the layout is being arranged.

With special cylinders such as rodless, the machine can be designed to take advantage of the special features of the cylinders and the manufacturer's instructions should be followed.

Before fitting and operating the equipment, it is advisable to arrange a flexible air line complete with a blow gun for test purposes. By this means it is possible to see that the installation is correct and the cylinders can move freely.

It is important that where cylinders are operating guided slides, linkages and levers the axis of normal cylinder movement follows that of associated guided components in all planes. The cylinder axis can be lined up by the use of dial gauges located on the outside of the cylinder.

When the cylinder has been lined up it should be possible manually to move the load between the stroke extremities to make sure that the slides move freely. If the mechanical arrangement is such that manual operation is not possible due to the loading, the blow gun can be used at appropriate ports to actuate movement. Only when the movement is smooth and steady will the alignment be satisfactory. This test is not conclusive of free movement under full load, because it takes no account of deflections caused by reactive forces on the cylinder supports, but it should be done as a first stage, to be followed by a final load check.

For trunnion mountings, it is necessary to ensure that the cylinders are free to oscillate on their mountings and the cylinder pipe connections must be clear to flex naturally under normal oscillation.

Finally all trunnions and bearing slides must be lubricated following the manufacturer's instructions and a test of the complete system carried out.

SECTION 8

Seals

PNEUMATIC SEALS

PNEUMATIC SEALS

Seals for valves and cylinders

Seals for cylinders and valves are required to withstand lower pressures than hydraulic seals, normally not more than 10 bar. At the same time operating forces are lower and speeds are higher, which calls for low friction seals. For valves, this implies the use of O-rings (which may be used in conjunction with harder plastic back-up rings such as PTFE), T-section or similar simple seals. For cylinder pistons and rods, where diameters are larger, pressure energised U-ring seals are preferred.

Additionally, in a dirty or abrasive environment, wiper seals can be fitted to piston rods and may be combined with gaiters. See Figure 1 for the classification of pneumatic seals.

High conformity is required from a seal, because air is more difficult to seal than fluids. Pneumatic seals are often used at higher temperatures than seals for liquids, so the permeability of the seal material becomes increasingly important; this may affect the choice of seal material. At normal temperatures, the permeability of most elastomers is low and similar to each other in magnitude, with the exception of silicone rubber. There are however marked differences at elevated temperatures see Table 1. All elastomers have operating temperature limits above which they become hard and brittle.

Refer to BS 7714 for general guidance on storage, identification and care when fitting sealing materials.

Seal lubrication

Pneumatic seals are increasingly required to operate in dry (*ie* with air which is both dry and oil-free) conditions. Such seals are expected to survive on a one-and-only initial lubrication. Most elastomers, with the exception of silicone are appreciably less resistant to 'dry' heat than 'wet' heat. Unless circumstances preclude the use of lubricated air, it is preferable to install a lubricator in the system, filled with a lubricant chosen to be compatible with the seal materials. The higher the temperature and the faster the speeds, the more important it is to lubricate the air. The absence of oil-mist lubrication can make it difficult to keep a dynamic seal sufficiently well lubricated. There is also the deteriorating effect of the oxygen in the air. Manufacturers recommend that once equipment has been used with extra lubrication, this has to be continued.

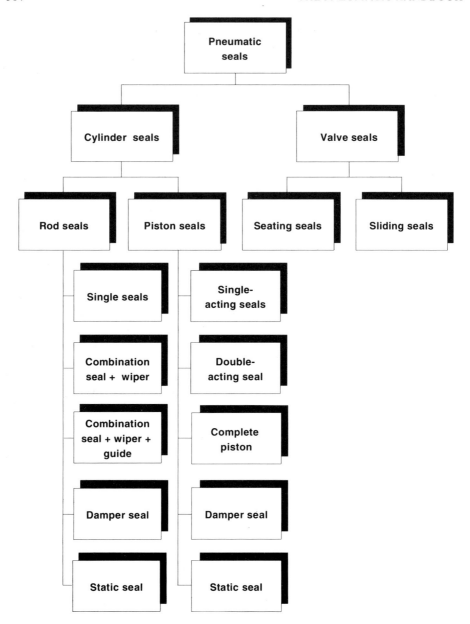

FIGURE 1 – Classification of pneumatic seals. *(Freudenberg Simrit)*

In the case of a seal operating dry, the apparent permanent set of an elastomer may appear much higher (typically twice as high) than the same seal used on oil. This may call for a greater initial interference when assembling the seal and subsequently higher friction

TABLE 1 – Air permeability of elastomers
(permeability coefficient Q x $10^8 cm^2/s/bar$)

Elastomer	40	60	80	Temperature °C 100	120	140	160	180
Acrylic	5	13	30	38	45	60	75	93
Butyl	0.5	1.8	4.5	10	19	28	40	60
Neoprene (Cr)	2	4.5	10	17	26	38	52	71
Nitrile	2	4.5	8	15	22	31	46	63
SBR	3.5	9	18	30	45	70	100	115
Natural rubber	9.5	20	4.3	56	70	100	110	120
Silicone	290	380	475	580	660	755	900	1000
Fluorocarbon	1	3	9	20	36	60	90	150
Polyurethane	2	4.5	10	19	31	40	52	70

during the running-in period when the permanent set is taken up. Selection of a suitable interference is more difficult and errors may either cause the seal to continue to bind and develop high friction and wear, or to run-in to a loose fit with consequent leakage.

A permanently binding seal is more likely than a loose seal, particularly if the seal is dimensioned as a dry type and subsequently lubricated by oil-mist. Some oil will be absorbed by the seal and cause it to swell. Consequently a common fault with seals which operate under oil mist lubrication is a failure to provide for the amount of swell the seal is likely to undergo during its conditioning period, which may be of several weeks duration.

Not all seals can be used in the absence of lubrication, so care must be taken when replacing a seal with another of a different specification that the new seal can be used in the prevailing conditions.

Seal materials

The most common materials for pneumatic sealing purposes are artificial rubbers (predominantly nitrile) in a variety of formulations, which may be used with PTFE or other hard plastics. There are also composite seals using more than one material, such as fabric reinforcement, PTFE/rubber combinations and seals with metal inserts. Natural rubber is rarely used because of its temperature limitation (a maximum of about 80°C). Leather was formerly much used but, despite a variety of tanning processes available, is obsolescent for pneumatic applications. It does however have good wear and extrusion resistance and is self-lubricating, so for special applications may still be found.

In choosing which of the seal materials are suitable for a particular application, consideration must be given to the temperature regime and the rubbing velocity, see Table 2.

TABLE 2 – Seal and guide materials

Material formulation	Hardness (A) Shore A	Temperature °C	Speed m/s	Resistant to:	Characteristics
Acrylonitrile-butadiene NBR	70–90	–30 to 100	≤ 1	mineral oils water to 60°C	general purpose for all seals
Polyurethane AU,EU	90–94	–30 to 100	≤ 1	mineral oils water to 50°C	mainly dry applications
Fluoro-rubber FKM	90	–20 to 200	≤ 1	dry heat mineral oils	high temperature high cost
Acrylate rubber ACM	–	–25 to 150	≤ 1	mineral oils not water	for moderately high temperatures
Silicone rubber VMQ, PVMQ	–	–60 to 200	≤ 1	water to 100°C	high temperature, high cost seldom used for pneumatics
Polyamide PA	–	–30 to 100	≤ 1	mineral oils	excellent sliding properties used for guide rings
Polytetrafluoroethylene PTFE	–	–100 to 200	≤ 2	nearly all liquids	low friction, used in combination with rubber

Notes
1. Temperature and speed limitations are for continuous usage. For short periods the range may be extended.
2. Within all grades, special formulations may give particular characteristics for special applications.

General principles in design of seals

The modern trend in design of sealing systems is to use simple designs and assemblies. U-rings are favoured for larger diameter dynamic applications; and O-rings and their derivatives are used for static and small diameter dynamic applications; the commonest composite seal is rubber with PTFE. For the large majority of pneumatic sealing duties in cylinders and valves one of these types will be satisfactory, and should be the first choice before trying some other type of seal.

O-rings (toroidal sealing rings)

These are ring-shaped seals of circular cross-section, primarily used for static sealing but also for dynamic seals, both axial and rotating. For low pressure (<100 bar) pneumatic seals, the hardness should be about 70 Shore A. For higher pressures, about 90 Shore A hardness should be chosen. If the gap to be sealed is greater than about 0.3 mm, consideration should be given to the use of a backup ring made of PTFE, see Figure 2. The back-up ring acts solely as an anti-extrusion device and is usually split at an angle. BS 5106 gives the sizes of spiral rings and their housings.

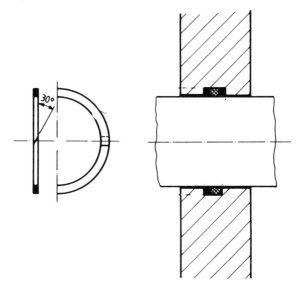

FIGURE 2 – Combined O-ring with PTFE anti-extrusion rings. *(Freudenberg Simrit)*

The sealing action of an O-ring under static pressure is illustrated in Figure 3. The initial pressure on the ring caused by compression of the rubber is increased by the pressure in the fluid; as the fluid pressure increases, the effectiveness of the seal increases so that very high pressures can be sealed. It will be seen from Figure 3 that the locating groove must be large enough both to allow for pressure to act over a large portion of the ring surface and to cater for any volume increase as the seal absorbs water or oil; at least 25% extra volume should be allowed. British Standard BS 4518 (BS 1806 for Imperial sizes) gives

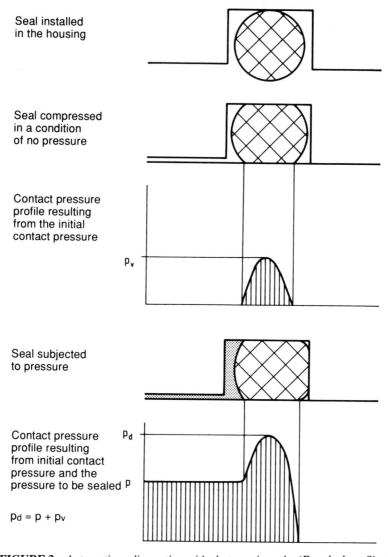

Seal installed
in the housing

Seal compressed
in a condition
of no pressure

Contact pressure
profile resulting
from the initial
contact pressure

p_v

Seal subjected
to pressure

Contact pressure
profile resulting
from initial contact
pressure and the
pressure to be sealed p p_d

$p_d = p + p_v$

FIGURE 3 – Automatic sealing action with elastomeric seals. *(Freudenberg Simrit)*

the recommended groove sizes for both hydraulic and pneumatic installations. These or data supplied by the ring manufacturers should be followed as for example in Table 3. The radial compression in this table varies from 10% in the larger sections to 25% in the smaller.

BS 4518 also contains recommendations for the choice of O-ring size according to the diameter to be sealed.

TABLE 3 – O-ring groove sizes for static sealing with radial compression *(Freudenberg Simrit)*

	Groove size	
d_2	T	B
1.50	1.10	1.9
1.60	1.20	2.1
1.78	1.30	2.3
2.00	1.50	2.6
2.40	1.80	3.1
2.50	1.90	3.2
2.62	2.00	3.4
3.00	2.30	3.9
3.50	2.70	4.5
3.53	2.75	4.5
4.00	3.15	5.2
4.50	3.60	5.8
5.30	4.30	6.9
5.50	4.50	7.1
4.00	3.15	5.2
4.50	3.60	5.8
5.00	4.00	6.5
5.33	4.30	6.9
5.50	4.50	7.1
5.70	4.65	7.4
6.00	4.95	7.8
6.50	5.40	8.4
6.99	5.85	9.1
7.00	5.85	9./1
7.50	6.30	9.7
8.00	6.75	10.4
8.40	7.15	10.9
8.50	7.25	11.0
9.00	7.70	11.7
9.50	8.20	12.3
10.00	8.65	13.0
10.50	9.15	13.6
11.00	9.65	14.3
11.50	10.10	15.0
12.00	10.60	15.6
12.50	11.05	16.2
13.00	11.55	16.9
13.50	12.05	17.5
14.00	12.55	18.2
14.50	13.00	18.8
15.00	13.50	19.5

O-rings used in dynamic conditions

O-ring seals are typically used for sealing spool valves and for pistons in pneumatic cylinders and similar reciprocating applications. Two regimes are possible:

- With radial compression of the O-ring. The ring is compressed radially in its groove, with a compression of the cross-section of from 2% to 6%, according to thickness. This ensures good sealing with comparatively long life even with inadequate lubrication. See Table 4 for dimensions.
- Without radial compression of the O-ring, Figure 4. It has a floating fit in the groove and no contact with the base of the groove. There may be some leakage at low

TABLE 4 – O-ring groove sizes for dynamic installation with radial compression *(Freudenberg Simrit)*

d_2	T	B	Z
1.50	1.35	1.9	1.0
1.78	1.55	2.3	1.1
2.00	1.80	2.4	1.2
2.40	2.15	2.9	1.4
2.50	2.25	3.0	1.4
2.62	2.35	3.1	1.5
3.00	2.75	3.6	1.6
3.50	3.25	4.2	1.8
3.53	3.25	4.2	1.8
4.00	3.70	4.8	2.0
4.50	4.20	5.4	2.3
5.00	4.65	6.0	2.5
5.33	4.95	6.4	2.7
5.50	5.15	6.6	2.8
5.70	5.35	6.9	3.0
6.00	5.65	7.2	3.1
6.50	6.10	7.8	3.3
6.99	6.60	8.4	3.6
7.00	6.60	8.4	3.6
7.50	7.10	9.0	3.8
8.00	7.60	9.6	4.0
8.50	8.00	10.2	4.2
9.00	8.50	10.8	4.3
9.50	9.00	11.4	4.4
10.00	9.50	12.0	4.5

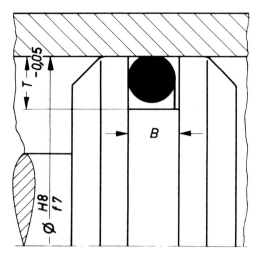

FIGURE 4 – Installation of O-ring without radial compression. *(Freudenberg Simrit)*

pressure, but the friction and wear are low. The external diameter of the ring should be 2% to 5% greater than that of the cylinder bore. See Table 5.

If the O-ring is properly assembled and adequately lubricated, and if the groove is correctly dimensioned, it should not twist under reciprocating motion, because the contact area in the groove is far greater than that with the sliding surface. Also, the friction in the groove is static friction which is lower than rubbing friction. However, it has to be admitted that ideal conditions are often not met. If there is any circumferential variation in the friction, some parts of the ring surface may slide while other parts roll, with the result that the ring may eventually suffer a characteristic spiral failure. This is the most common type of ring

TABLE 5 – O-ring groove for dynamic installation without radial compression
(see Figure 4) *(Freudenberg Simrit)*

	Groove size	
d_2	T	B
1.78	1.90	2.0±0.05
2.40	2.55	2.7±0.05
2.62	2.75	2.9±0.05
3.00	3.15	3.4±0.05
3.53	3.70	4.0±0.10
5.33	5.50	6.0±0.10
5.70	5.90	6.4±0.10
6.99	7.20	7.9±0.10

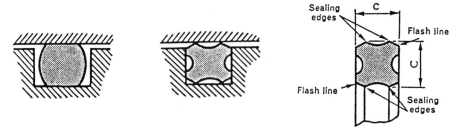

FIGURE 5 – Non-circular section O-rings. *(Pioneer Weston)*

failure and is easily recognised. It may also show as a variation in shape and size of the section.

One difficulty that sometimes occurs with poor quality rings is inadequate removal of the flash produced during the moulding process. Even with high grade rings, the flash removal disturbs the surface on which sealing depends.

The advantages of dynamic O-rings are

- simplicity of housing design
- availability in small sizes, which makes it a first choice for small bores
- bi-directional sealing (in contrast to lip seals which have to be used in pairs)
- lower friction than U-rings.

Non-toroidal O-rings

In order to overcome some of the drawbacks in the use of circular section rings for dynamic sealing whilst retaining the simplicity of the groove design, non-circular-section rings

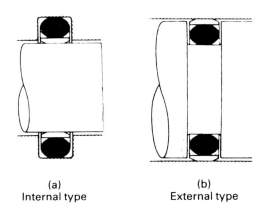

(a)   (b)
Internal type   External type

FIGURE 6 – Rubber/PTFE combination seal.

have been developed. They are intended to be direct replacements for standard O-rings and are designed to be fitted into the same size grooves.

Square section rings are not subject to spiral twist, but they tend to have a high friction caused by the large surface area in contact with the moving component.

This has led to a special form of O-ring which is of a basic square section with relief on the faces, see Figure 5. This design ensures that the contact area is small, reducing the friction whilst ensuring that spiral twist is impossible. Although the rings can be fitted into standard O-ring grooves, it is preferable to make the groove widths rather narrower. This is possible because radial compression is accommodated by distortion of the cross-section rather than lateral displacement. When designing from scratch, the groove dimensions recommended by the manufacturer should always be followed.

Although these rings can be used as static seals, there is no real advantage in fitting them. They can be used for rotary applications where space limitations rule out a rotary shaft seal.

O-ring combinations

A further variation on the basic O-ring is its use with PTFE slip rings; here the sealing pressure comes from the ring and the PTFE provides a low friction rubbing surface. These seals can be of an external or internal type as in Figure 6. Depending on the manufacturer, they can be used as direct O-ring replacements fitting into the same standard groove. When comparing a simple O-ring with a PTFE combination, the frictional force when lubricated is very similar, but when run dry the PTFE is much superior.

U-rings

The U-ring suitably proportioned and in a relatively soft rubber, is capable of providing good sealing in with low friction at low pressures. It is mainly used in dynamic applications for piston seals. Desirable features are thin lips for response to pressure energisation, generous length for good flexibility, with the lips designed to have low lip loads for low friction and at the same time enough contact area and pre-load for good sealing at low pressure. A square heel section is preferable to minimise heel wear.

There is a difference in design philosophy between U-rings intended for hydraulic and pneumatic installations. The friction and wear between two bodies sliding over one another is reduced if the surfaces are separated by a layer of lubricant. the formation of this film depends on the relative velocity, the viscosity of the lubricant and the pressure profile between the surfaces. The velocity and viscosity are predetermined so the formation of the film is influenced only by the pressure profile. A comparison between the two regimes is shown in Figure 7. Most pneumatic seals get a one and only lubrication during fitting, so it is important that the lubricating film should be retained during the lifetime of the seal. This is in contrast to hydraulic seals which are being constantly lubricated and the seal is required to wipe off the lubricant during each stroke. It can be seen from Figure 7 that the geometry of the two kinds of seal is different; when the pneumatic seal is assembled into its groove, the angle between the seal and surface is symmetrical, ensuring that the lubricant film is maintained.

U-rings are designed to seal from one side only so in double-acting cylinders, two

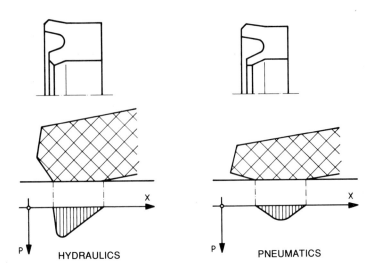

FIGURE 7– Differing lip geometry and contact pressure profile with
hydraulics and pneumatics. *(Freudenberg Simrit)*

U-rings have to be fitted into the piston. The seals are often separated by a bearing ring
which has little sealing function in itself but helps to take side loads. A piston containing
these elements can be long, which is undesirable in compact designs, so there has been
developed a variety of different piston seals combining several elements in one. Very
compact designs consist of an integrated piston and seal, with the seal material bonded to
the piston and replaceable as a unit.

Wiper seals

For sealing on the piston rod, a single acting U-packing can be used, similar to a piston seal.
An additional feature is required on rods: in dirty and dusty environments, protection has
to be given to the rod to ensure that no contamination is drawn into the system. Special
wiper seals are used for this purpose; they do not require pressure behind them (indeed if
they did, it would indicate that the rod seal was not functioning properly) and rely solely
on rubber flexibility to maintain contact with the rod.

Combined piston and wiper seals are also available

See Figure 8 for a variety of pneumatic seal combinations.

Housing design

Care should be taken to design the seal housing in accordance with the instructions of the
seal supplier. For O-rings it may be an advantage to make the sides of the groove slightly
angled (about 5°) for ease of manufacture. The surface finish of the dynamic sealing
surfaces should be 0.4 µm R_a and on static sealing surfaces 0.8 µm R_a, surface texture
in accordance with BS 1134. A lower quality of finish is acceptable in the housings,
1.6 µm R_a.

U-packing

Piston seal with asymmetric profile. Excellent sealing capacity and low friction due to special pneumatic sealing edges.

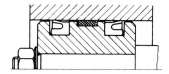

Compact seal

Double acting compact seal with grooves on the front face for pressure activation. The compact design allows for narrow piston designs. The rounded seal profile in combination with the flexible body help achieve good sealing action at low friction, while maintaining an effective lubrication film.

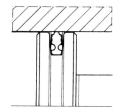

Complete piston

Ready to fit, double acting complete piston with compact dimensions, light alloy body, snap-in seal and integrated guide. Simple fitting to the piston rod without any additional seal.

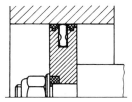

Wiper

NBR rod wiper without metal case for less demanding applications in standard pneumatic cylinders. Easy snap fitting.

Combination seal

Non metal reinforced rod seal with integrated wiper for miniature pneumatics. Special pneumatic sealing edges provide for excellent sealing capabilities at low friction, while maintaining an effective lubrication film.

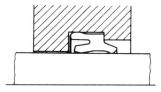

FIGURE 8– Piston and rod seal combinations. *(Freudenberg Simrit)*

Consideration should be given at the design stage to how the seal is to be assembled into its housing. Entry tapers should be provided on the cylinder bores and piston rods. Special mandrels which cover any threads or sharp edges may be of assistance. During manufacture, all sharp edges on housings adjacent to the seals must be removed.

FLOW ANALYSIS FOR CYLINDERS AND VALVES

Circuit analysis

In choosing the correct sizes for circuit components, it is necessary to establish the required forces in each cylinder, the speeds of operation and the air flow rate through the system. The analysis is sometimes made more complicated than it need be by the absence of appropriate data from manufacturers and the non-standard ways of expressing flow characteristics. The situation has been improved in recent years by the availability on the market of valves and cylinders to ISO standards.

The methods outlined here are offered as a practical way of circuit analysis and, as far as possible, are methods that are adopted by the majority of manufacturers.

Flow data for pneumatic valves

There may be a certain amount of confusion caused when reading manufacturers' catalogues and trying to compare the data as presented.

In the past, the flow characteristics of pneumatic valves were determined by measuring the volume of water passing through them under pressure. This method produces a parameter known as the flow coefficient, K_v. The definition of K_v for a valve or other component is the flow in litres of water which passes through it with a pressure drop of 1 bar.

The corresponding value in American units is the flow in gallons per minute with a pressure drop of 1 lbf/in^2 ; it is usually given the symbol C_v. K_v is related to C_v by the relationship $K_v = 14.3\,C_v$. British data are sometimes quoted in terms of Imperial gallons, so it is important to check the basis of measurement, when a value for C_v is quoted.

Under this system, the relationship between flow and pressure for pneumatic flow is given by:

$$q = 0.47 K_v \sqrt{(P_1 - P_2)}$$

The factor 0.47 has been introduced to allow for the change in density between water and air. q is the flow in litres and P_1 and P_2 are the upstream and downstream pressures in bar.

Since this relationship has been developed for an incompressible fluid (water), it is only accurate for circumstances where the flow approximates to incompressible, *ie* where the density change through the valve is small, which implies a small pressure drop. For flow typical of a pneumatic valve supplying a cylinder, this is not usually the case, although it is applicable for valves on a main flow circuit, where a large pressure loss would be unacceptable. A valve feeding a cylinder will behave more often as an orifice, for which the orifice formulae will apply. Of course, for a small pressure drop, the orifice formula and the incompressible (square root) formula give the same result.

The basis of presenting valve flow data using the generalised form of the orifice formula has been discussed in the chapter on Pipe Flow Calculations, and manufacturers now usually adopt this method when giving flow data for valves. These data will have been determined experimentally in accordance with the method described in BS 5793, and so can be used with confidence.

Under this method, the equation for flow through a valve is given by the pair of equations:

$$q = CP_1$$

for critical flow where $P_2/P_1 \leq b$

and

$$q = CP_1 \left(1 - \left[\frac{P_2/P_1 - b}{1 - b}\right]^2\right)^{0.5}$$

for sub-critical flow where $P_2/P_1 > b$

P_1 and P_2 are the upstream and downstream pressures respectively in bar (absolute) and q is the flow rate in litre/s. The value C is known as the conductance with units of litre/(s bar). b is the critical pressure ratio. Note that C is not the same as C_v.

In order to take account of the practical difference between a valve and a simple orifice, the values of C and b are found experimentally and so, in general, they are different from the theoretical values that would be obtained from treating the valve as an orifice with an area equal to that of the minimum passage area through the valve. Failing any better information, the theoretical values for a sharp-edged orifice may be used. For air at 15°C they are:

$$C = 0.2 A$$

$$b = 0.53$$

A is the effective cross-sectional area in mm , which may not be the same as the area of the inlet connection.

Where C_v or K_v are quoted but not C, the value may be obtained from:

$$C = 4.0 C_v$$

$$C = 0.28 K_v$$

The value for b can be taken as 0.53 in most cases.

It should be emphasised that the conversions for C are semi-empirical and should only be used when the correct values are not otherwise available. The value of C, however obtained, can only be used in the above equations to determine q and only if the correct units are used.

It is possible, knowing the value of C and b, to calculate the flow for any values of upstream and downstream pressure. For some purposes, *eg* when incorporated in a computer program, this may be the preferred option. In their catalogues, manufacturers often include sets of curves which can be used directly to find the flow rate. A sample of these curves is given in Figure 1. While the general shape is typical of all valves, the actual flow rates, even for the same nominal size, will vary from manufacturer to manufacturer.

Valve characteristics in this form apply not only to selector valves but also to shut-off,

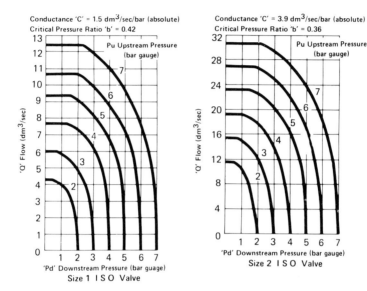

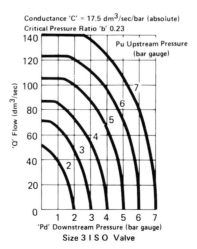

FIGURE 1 – Typical flow cures for pneumatic valves made to ISO standards.
Note dm³/s is the same as litre/s.

shuttle, check and restrictor valves. If used at a temperature significantly different from 20°C, a correction factor can be applied, although this is usually small.

Cylinder performance

The thrust available from an air cylinder is given by:

$$F = 0.1 \, f \, P \, A$$

F is in newtons, P is the gauge pressure in bar and A is the effective piston area in mm². f is the cylinder efficiency, taking into account seal friction.

A is the area of the piston when extending, or the area of the piston minus the area of the piston rod when contracting. f is usually taken as 0.75 (75% efficiency). One manufacturer quotes 0.8, except for cylinders larger in diameter than 100 mm, in which case, 0.65 is suggested.

The air consumption of a cylinder is:

$$q = 10^{-6} \, P \, L \, A \, N$$

q is in litre/s (free air), P is the absolute pressure in bar, L is the stroke in mm, A is the cross-section area in mm² and N is the frequency of operation in cycle/s. For single-acting cylinders (no return air), A is the piston area; for double-acting cylinders A is the sum of the piston area and the return area. This formula takes no account of the volume of air in the pipework or the clearance volume in the cylinder. It should be borne in mind that a cylinder frequently strokes through a shorter distance than the actual geometric stroke available. If that is the case, then a wasted volume has to be filled both on the outward and inner strokes, and it should be taken into account in the calculation.

Manufacturers' catalogues usually give tables of air consumption and thrust. These values are readily calculated and so are not reproduced here, but Figure 2 gives a ready method of determining the air consumed for a stroke of 10 mm. Consumption for other strokes can be calculated by proportion.

Piston velocity

Calculation of piston velocity does present some problems, because it depends on an accurate estimate of the external loads on the piston. The following are the loads that might be present

- inertia loads caused by accelerating the mass attached to the piston,
- work load performed by the cylinder,
- friction,
- lifting or lowering a weight.

The sum of all these forces must be equated to the force across the piston.

Usually one of the design parameters is the cycle time, and this may mean the choice of a cylinder larger than that strictly necessary to apply the maximum load. The average force may not be the same as the maximum force. Once the pressure in the cylinder and the frequency of operation have been determined, the pressure drop across the valve can be found and a proper choice of valve made, using the valve data as explained above. It

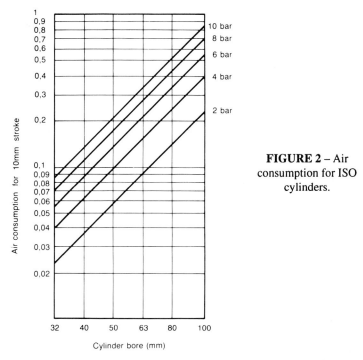

FIGURE 2 – Air consumption for ISO cylinders.

may not be necessary to use valves and piping equal to the port size of the cylinder to which it is connected. Too large a pipe size will be wasteful of air and may increase the start-up time (Figure 3).

The maximum piston velocity which can be expected for various pipe sizes and cylinder

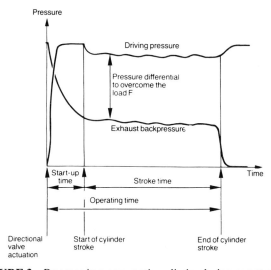

FIGURE 3 – Pressure in a pneumatic cylinder during one stroke.

TABLE 1 – Maxiumum piston velocity

Cylinder bore (mm)	Expected velocity (metres/second) Pipe size (BSP)					
	$^1/_8$	$^1/_4$	$^3/_8$	$^1/_2$	$^3/_4$	1
25	0.86	-	-	-	-	-
32	0.54	-	-	-	-	-
40	0.35	0.78	-	-	-	-
50	0.22	0.50	-	-	-	-
63	0.14	0.31	0.67	-	-	-
80	0.087	0.19	0.43	-	-	-
100	0.050	0.12	0.28	0.52	-	-
125	0.036	0.080	0.18	0.33	-	-
160	0.022	0.047	0.108	0.20	0.34	-
200	0.014	0.031	0.067	0.131	0.22	-
250	0.009	0.020	0.044	0.084	0.14	0.26
320	0.006	0.012	0.027	0.051	0.085	0.016

Cylinder bore (mm)	Expected velocity (metres/second) Tube o.d.mm (normal BSP valve size)					
	$4(^1/_8)$	$5(^1/_8)$	$6(^1/_4)$	$8(^1/_4)$	$10(^3/_8)$	$12(^1/_2)$
10	1.875	-	-	-	-	-
16	0.732	-	-	-	-	-
20	0.469	0.817	-	-	-	-
25	0.3	0.523	-	-	-	-
32	0.183	0.319	-	-	-	-
40	0.117	0.204	0.300	0.567	-	-
50	0.075	0.130	0.192	0.363	-	-
63	0.047	0.0823	0.121	0.229	-	-
80	0.0293	0.0510	0.075	0.1418	0.230	-
100	0.0188	0.0327	0.048	0.0908	0.147	0.217
125	0.0120	0.0209	0.0307	0.0581	0.0941	0.139
160	0.0073	0.0127	0.0188	0.0354	0.0574	0.0847
200	0.0047	0.0082	0.012	0.0227	0.0368	0.0542
250	0.0030	0.0052	0.0077	0.0145	0.0235	0.0347
320	0.0018	0.0032	0.0047	0.0089	0.0144	0.0212

These tables show the maximum piston velocity to be expected for various pipe sizes and cylinder bore sizes.
The tables have been computed at 6 bar, and should be used as a general guide.
Actual velocities will depend on the length of pipe runs used, valves, fittings, *etc.*

bore sizes can be assessed from Table 1. This is a guide only and should be used only in the absence of a full analysis. The table has been computed for a pressure of 6 bar.

Improving operational speed

If it should prove that the operating time of a cylinder is too long, there are various measures that should be considered to reduce it. If the pressure in the cylinder is close to the line pressure and the exhaust pressure is close to atmospheric, there is little one can do other than increase the supply pressure or change the cylinder diameter. If the pressure on the power side of the cylinder is too low, consider:

- placing the main valve nearer to the cylinder port,
- introducing pilot operation if necessary,
- increase the size of the valve and pipework but no more than the port size of the cylinder.

If pressure on the exhaust side of the cylinder is too high, consider introducing a quick exhaust valve.

If the problem is too rapid an operating time, the preferred solution is the introduction of a restriction valve in the exhaust line.

Pressure limitations

The maximum safe working pressures for valves and cylinders are given by the manufacturers and should not be exceeded. The system operating pressure is set at the compressor and air receiver by a pressure regulating valve and safety valve. As the air is compressible and is rarely stored at pressures much higher than 10 bar, high peak pressures are seldom encountered as they are with hydraulic valves.

The development of higher working pressures requires an increasingly disproportionate expenditure of power for compression, as well as increased component stresses and explosion hazard in the event of failure. The difference between hydraulics and pneumatics is due to the compressibility of the air.

For general applications, the only advantage offered by higher compression ratios is in storage systems, where air at high pressure can be stored in a smaller vessel and reduced to conventional delivery pressure when required.

For specialised applications such as rock drilling, air blasting and metal forming, higher pressures are called for.

The following pressure ranges can be distinguished:

- low pressure for conventional industrial pneumatics,
- moderate pressure (17 to 35 bar) derived from a compressor or reservoir charged from a compressor,
- high pressure derived from a charged air bottle,
- very high pressure derived from a high pressure receiver or intensifiers for small volume applications,
- ultra-high pressures supplied from multi-stage reciprocating compressors.

CIRCUIT ANALYSIS

In this chapter, some basic circuits incorporating standard valves and actuators are presented. Every application is different so it is not possible to cover all the possible arrangements, but with the use of these basic simple elements, many more complex circuits can be developed.

It should not be thought that circuit design is always as straightforward as may seem from these circuits. Circuit design is very much an art, to be learnt through experience. One very good way of learning the fundamentals of design is to attend one of the many customer training courses given by most of the major pneumatic companies.

When developing complex circuits, difficulties may be caused by a failure to appreciate that certain circuit states may occur with unexpected results. A common fault is to keep on adding circuit elements to cover problems as they arise, with each addition bringing in its own difficulties. In such an instance, it is always better thoroughly to analyse the cause of the problem which may often be solved by simplification rather than by adding complexity. Pneumatic valves are readily obtainable and are comparatively cheap to buy, so one often sees unnecessarily complex circuits, which quite obviously have been developed by a process of continuous additions. It is worthwhile spending some time with even the simplest circuit to make sure that it cannot be improved.

In the circuits described here, the symbols used are the CETOP symbols, according to BS 2917. These are listed extensively in the Data Section. The convention used is that the valves and cylinders are shown in their un-operated or rest position. All the valves used are standard types, available from a number of suppliers. It is conventional to indicate the power connections by solid lines and the pilot connection by dotted lines, but it is not always easy to define which is which, and in some layouts, the distinction is not kept.

Single-acting cylinder operation

These cylinders are power operated in one direction only, the return stroke being achieved either by a spring, which may be internal to the cylinder, or more usually by an external spring or weight. Figure 1 shows such a cylinder, operated by a three way valve. By utilising the difference in area between the two sides of the piston, it is also possible to use a double acting cylinder operated by a three-way valve, with a constant supply to the return

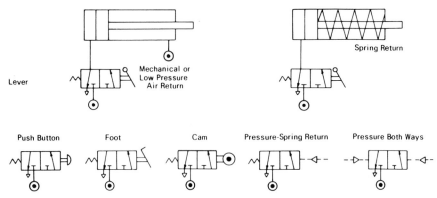

FIGURE 1 – Basic function single-acting cylinders.

side as shown. Valves may be operated by a variety of mechanical means, by air pressure acting on the spool ends, by electric solenoids or by a combination of these.

Double-acting cylinder operation

Double-acting cylinders are power operated in both directions and require a four-way valve as shown in Figure 2. The purpose of the four-way valve is to insure that, as one side of the piston is pressurised, the other can be vented. Note that a four-way valve is so called because there are four paths or ways through the valve. Such a valve usually has five ports – one inlet, two outlets to the cylinder and two exhaust ports. For pilot operated valves there will also be one or two pilot ports. The valve shown in the figure is a four-way, five-port, two-position valve. Five-port valves can also be used when different pressures are required for extension and retraction.

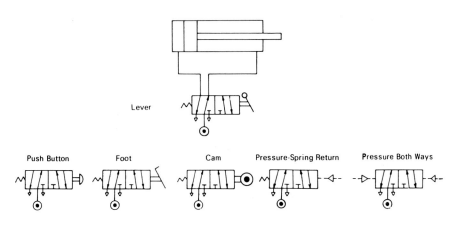

FIGURE 2 – Basic function double-acting cylinders.

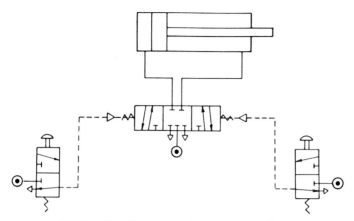

FIGURE 3 – Pilot control double-acting cylinders.

Remote control by pilot operation

It is often convenient to operate a cylinder from a remote position, as in Figure 3. The main valve, which is sized to take the full flow, is mounted close to the cylinder and is actuated by pilot signals from a pair of three-way valves. The main valve in this example is a three-position, four-way valve. A pilot signal shifts the main valve from the neutral or central position; removal of the pilot signal allows a spring to return the valve to neutral. In the central position, the valve traps the air on both sides of the piston; another form of three-way valve has a central position in which the air is vented from the cylinder. The former type is more economical in air consumption, the latter allows free movement of the piston in the central position

Speed control of cylinders

Usually it is desirable to control the speed of movement of a cylinder, at least in one direction. The most convenient way of doing this is to restrict the air flow in the exhaust

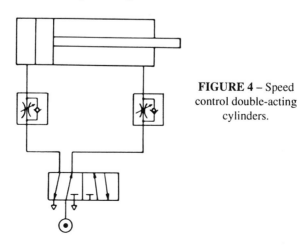

FIGURE 4 – Speed control double-acting cylinders.

side of the cylinder. This creates a back pressure on that side which slows down the piston movement. This is a much better way of speed control than restricting the inlet side which leads to instability.

In Figure 4 a typical circuit is given for the case where it is required to restrict the speed in both directions. Between the cylinder and main valve is a pair of restrictor valves. These valves control the flow rate in one direction through an adjustable, screw down restrictor; flow in the other direction is unrestricted through a non-return valve. A warning should be given about the use of restrictions in the exhaust: if there is an external force on the piston or if a high kinetic energy is generated, there may be excessive pressures produced in the exhaust circuit. This may require the incorporation of a pressure relief valve set at a safe pressure.

The other extreme of speed control is when maximum operating speed of actuation is desired, as for example when the piston is used for impact. In this case a quick exhaust valve, which allows unrestricted flow from one side of the piston to atmosphere, is used, as in Figure 5. This valve is symbolised as a pilot operated shuttle; typically it incorporates a resilient disc which moves from one seat to the other; on the exhaust side the passages are generously sized. To be effective, it must be mounted very close to the cylinder port which is to be exhausted. In the diagram, the valve forms part of the supply circuit when the cylinder is retracting, but it exhausts directly when pressurised from the cylinder.

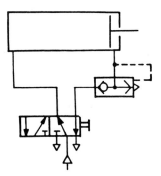

FIGURE 5 – Quick release valve.

Automatic cylinder return

A common requirement is for a single-stroke actuation upon operation of the main valve, followed by automatic return. This is achieved by incorporating a cam-operated pilot valve which sends a signal to the main valve reversing its position. In Figure 6a, the cycle is initiated by depressing the hand-operated pilot. A momentary supply of air is sufficient to move the main valve into the position where it causes the cylinder to extend; when the piston rod triggers the cam valve, it sends a signal to the main valve to return it to its former position, and the cylinder then retracts.

It is possible to actuate the cycle with a pilot valve which is hand-operated and pressure-returned, as in Figure 6b. Whilst this is functionally satisfactory, it is not good practice to allow pressure to move a valve handle, with the possibility of injury by trapping the hand of the operator.

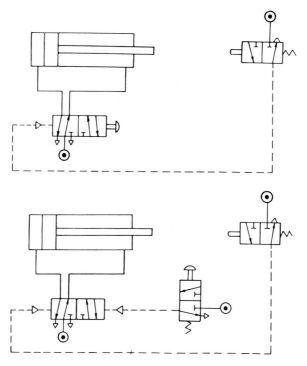

FIGURES 6a and b – Automatic return of cylinders.

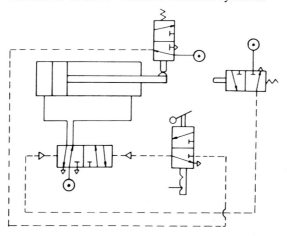

FIGURE 7 – Reciprocating cylinder control.

Automatic cylinder cycling

Figure 7 demonstrates how it is possible to achieve continuous reciprocation of the cylinder so long as the hand-operated valve remains in the selected position. Two cam-operated pilot valves define the extent of the piston travel which need not, indeed in most cases should not, be at the mechanical ends of the cylinder. The hand-operated pilot valve has a detent which allows the valve to be left continuously in either the actuated or the rest position. In this example, the cylinder always stops in the retracted position, but it could equally well be arranged to stop in the extended position.

It is important to design an automatic circuit so that it always stops with the components in the correct position for restarting. It would not, for example, be correct to initiate reciprocation by turning on an air supply to the main valve, because the cylinder might stop in mid-stroke when the supply was cut off and there would be uncontrolled movements when the supply was restored. Furthermore there would be no locked-in pressure and therefore no holding force in the stopped position.

Reciprocation without position valves

It is not always possible or convenient to arrange for the reciprocation to be triggered by mechanically operated pilot valves; an alternative way of initiating cylinder reversal is by sensing the cylinder pressures. A circuit which demonstrates the principle is shown in Figure 8. In this arrangement, the main supply between valve and cylinder is also passed via a restrictor valve to the pilot port of the main valve. Two reservoirs are incorporated in the pilot lines and the stroke length is governed by the build-up of pressure in the reservoirs, which are charged by restrictor valves. For fast cycling, the reservoirs may be unnecessary.

Another circuit is in Figure 9, which uses a differential-area, pressure-operated pilot valve. This valve switches with a lower pressure on one pilot; the lower pressure side is usually diaphragm actuated, commonly requiring only about 0.2 bar for operation. When the cycle is started by operation of the hand-operated pilot, the cylinder extends. Because

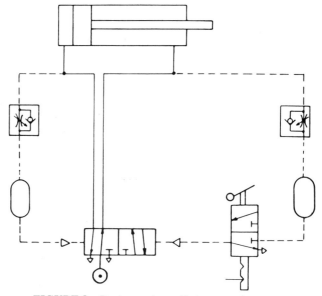

FIGURE 8 – Reciprocating cylinder control.

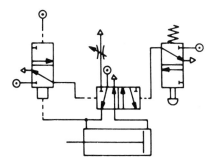

FIGURE 9 – Alternative reciprocation circuit.

the flow from the return volume is restricted by the restrictor valve, pressure builds up in that line and triggers the pilot.

The positional accuracy of the systems which rely on pressure is not as good as those with mechanically operated valves. If it is not easy to mount these valves, the designer should consider the use of cylinders with integral control valves or one of the piston sensing devices described in the chapter on Construction of Pneumatic Cylinders.

Controlling the speed of operation

The frequency of reciprocation can be controlled by incorporating restrictor valves in the exhaust lines as described above, but this may not always be adequate. In some applications it is desirable to arrange for a time delay at one end or both ends of the travel. A purely pneumatic time delay can be arranged by the use of a reservoir, which is merely a chamber which can be charged up by a controlled air supply from a restrictor valve. One such circuit is illustrated in Figure 10. Towards the end of the piston stroke a cam-operated pilot valve is actuated; but before the main valve is switched, the air from this pilot has to pass through the variable restrictor and charge up the reservoir. It is only when the pressure in the reservoir reaches the pressure required to operate the main valve that it switches, allowing the piston to return. This circuit is suitable for fairly small time delays (in practice up to about 10 s). Longer delays require the use of very fine bleed holes or large reservoirs.

A reservoir is usually made from the same section tubing as a standard cylinder, with a port in the end instead of a piston. For adjusting the time delay, it is possible to use a variable volume reservoir, but these are not readily available and a much cheaper way of

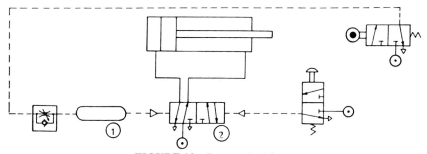

FIGURE 10 – Pneumatic delay.

achieving the same end is by the use of a variable restrictor.

The circuit of Figure 10 can be considered to be the "classical" method of obtaining a time delay, but there are now available special time-delay valves which are able to effect the same result with just the one device. The main advantage of these valves is that the time delay can be set by a graduated control knob; they do not rely on trial and error. Valves with time delay settings between 0.2 seconds and 100 minutes are available. These valves can be looked at as three-way pilot valves with a time delay on the pilot circuit and are available in normally closed or normally open types.

These techniques of controlling actuation times can equally well be replaced by electrical methods.

Dual control of a cylinder

It may be necessary to control a cylinder from more than one position. Two ways of doing this are shown in Figure 11. The *in* actuation uses a pair of valves connected by a shuttle, the *out* actuation uses a number of pilot valves in series.

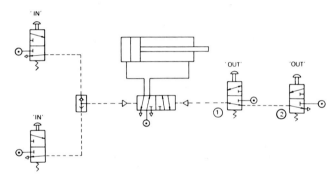

FIGURE 11 – Dual control of a cylinder.

Safety circuits

It is sometimes necessary to ensure that both an operator's hands or his feet are occupied during a cycle, for example in the operation of a press or while a cutter is rotating. One way of doing this is by the circuit of Figure 12, which has two palm-operated valves in series. While this has the merit of simplicity, it is not to be recommended for two reasons: the operator can by-pass the safety system by tying down one of the valves or a valve may stick in a dangerous position. Such a circuit may not meet the safety code that applies to the use of the machine being operated. There are safer circuits that are less susceptible to operator misuse; they require not only the operation of both valves but also that they be actuated simultaneously.

One example is shown in Figure 13. If only valve A is operated, the air supply from A passes through B and D, which causes C to switch and vent the supply line; if only B is operated, the air supply from B passes through D and also switches C. It is only when A and B are operated simultaneously that the left pilot line of C is pressurised, allowing air

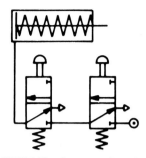

FIGURE 12 – Sample safety circuit.

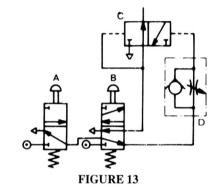

FIGURE 13

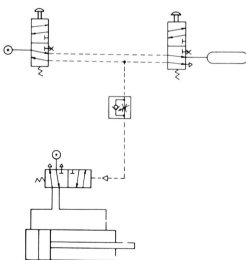

FIGURE 14 – Safety control.

to pass to the cylinder. The air to the right hand pilot is delayed by the restrictor valve, so the left hand pilot signal dominates.

An alternative safety circuit is shown in Figure 14. In the un-operated position, air is

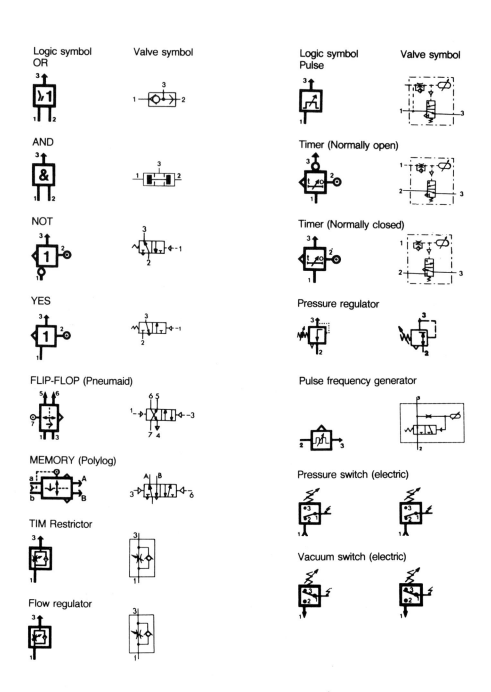

FIGURE 17 – Logic symbols and their pneumatic equivalents. *(CompAir Maxam)*

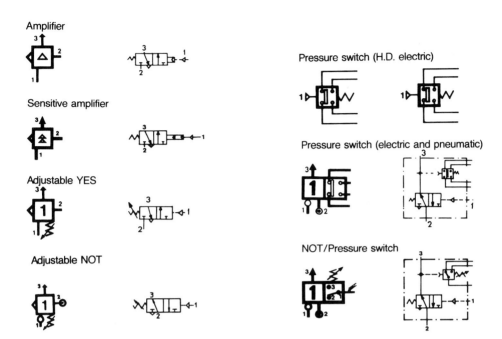

FIGURE 17 (continued) – Logic symbols and their pneumatic equivalents. *(CompAir Maxam)*

circuit. There are other logic functions which can be achieved by use of combinations of two or more basic elements, these are NOR, NAND and INHIBITION.

Where A and B are the two inputs, and S the result, the logic functions can be described algebraically as:

NOT	S = not A	S is present only if input signal is not present.
AND	S = A and B	S is present only when A and B are simultaneously present.
OR	S = A or B	S is present when A or B is present.
YES	S = A	This can be looked upon as an amplifier, or an inverted NOT with a second input. B is the supply pressure input and A is the signal input.
MEMORY		A flip-flop device. A signal A will give output state S1, and will remain in that state with A removed. A signal B changes the output to S2, which remains in that state until receipt of next A.
NOR	S = not A or not B	S is present when there is not a signal at A or not a signal at B.
NAND	S = not A and not B	Output is present when there is not a signal at A and not a signal at B.
INHIBITION	S = B and not A	Output is present when there is an input at B and not an input at A. When a signal is present at A it inhibits the signal path of B.

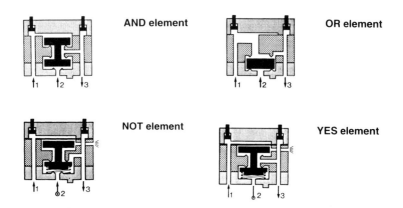

FIGURE 18 – Diagrammatic construction of basic logic elements. *(Crouzet)*

Most manufacturers can supply MPLs of the first five kinds. The others may be available either as valves in their own right or as a combination of two or more of the others. For example NOR is realised by an OR with its output connected to a NOT, and NAND by an AND with its output connected to a NOT.

Figure 18 shows how the various MPLs can be realised by simple poppet valves.

The logic circuit has to interface with sensing units at the input stage and the power system at the output stage. The sensing units may themselves be pneumatic or electronic, so it is necessary to have available a range of interface devices which can be assembled onto the same sub-base system.

Parts of the circuit may be required to operate at pressures different from that of the line. To accommodate that feature there is an adjustable YES and an adjustable NOT as well as adjustable timers and pressure regulators.

A sensitive amplifier is necessary if one wishes to take a signal from a proximity sensor, for example, which may work on a pressure measured in millibars and convert it into line pressure in bars.

Other devices include pulse generators, pulse frequency generators, various pressure switches, all of which have their place in a complete control circuit.

Usually the logic devices have a visual indicator on them to show their state, which is useful when developing the system or trouble-shooting.

AUTOMATION AND ROBOTICS

Compressed air for robotic applications

The versatility and economy of compressed air as a power medium leads to its use in automation and in robots. It is, of course, important that its particular characteristics are thoroughly understood, but once that is done, there are many applications and proprietary products on the market which can be used in robotic applications, more specifically in 3-D Manipulators or "Pick and Place" Modular robots. This subject was briefly covered in the section on pneumatic cylinders, but the subject is of such growing importance that a more extended treatment will be of assistance.

The advantages that compressed air possesses in this application are:

- speed of operation;
- precision (an accuracy of 0.1 mm is possible);
- safety in hazardous environments;
- cleanliness for food product processing;
- flexibility of application;
- stroke adjustments are easily made to meet a variety of applications;
- modular kits are available from a number of suppliers, making do-it-yourself robot assemblies a practical possibility;
- integration with electronic control units and with electric power drives.

Available modules

Basically a robotic system comprises some or all of the following modules:

1. Precision linear units for horizontal or vertical movements in a straight line. Some suppliers differentiate between long and short stroke units; the long stroke units are designed to be stiff to carry unsupported loads and include hydraulic damping for end of stroke damping; the short stroke units are attached to the end of the long stroke units for the final precise positioning movement. Although for some applications it is possible to use a conventional pneumatic cylinder or a rodless cylinder with position sensors, in most cases higher precision is required and so special units are used. Figure 1 shows one form of construction. Another type, shown in Figure 2, incorporates telescopic feed tubes in the unit to supply the air

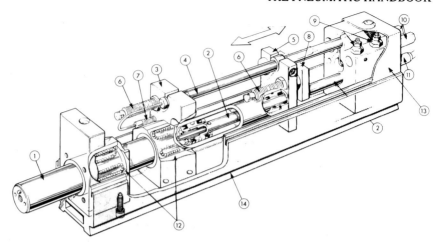

1.Cylinder. Outer tube moves to provide movement.   8.Strikers to set stroke length.
2.Fixed piston and air feed.   9.Speed control valves.
3.Inboard bearing block. Incorporates recirculating ball bearing,   10.Air connection.
  hydraulic damper and inductive sensor.   11.Electrical connection.
4.Guide tube, used with bridge 5 to prevent rotation.   12.Front bearing block.
5.Bridge carries inward stroke damper.   13.Cover.
6.Hydraulic damper.   14.Dovetail slide.
7.Induction sensors.

FIGURE 1 – Linear unit. *(Parker)*

for the further modules that are attached to it; up to 16 connections are possible.

Because of the overhang where these modules are set up for horizontal movement, the permissible load is severely restricted for long strokes; so reference should be made to the suppliers to ascertain the maximum allowable load for the envisaged application.

2. Base rotate units on which the linear units are mounted. These are similar to rotary indexing tables using either cylinders or a rotary actuator. Multiple angles in intervals usually of about 15° are available. It is likely that a base unit will have to support an arm which, including its attachments, will have a high rotary inertia; so care should be taken to ensure that the speed is controlled and end-of-stroke cushioning is provided.

3. Wrist rotate units, which are usually mounted on the end of the linear units. Angular travel of 180° or in some cases 270° is possible.

4. Grippers for holding components. They may be single acting with spring return or double acting. Different gripping hands to cater for variously shaped components as well as multiple grips for handling several components at a time are also available. Pneumatic chucks may be used for turned components; for glass or metallic sheets, a suction head can also be used. For components made, or partly made, of magnetic materials, a permanent or electro-magnet may be used.

With these four modules it is possible to secure the most complex of transfer motions. Each modular element is provided with induction sensors to detect the end of the stroke

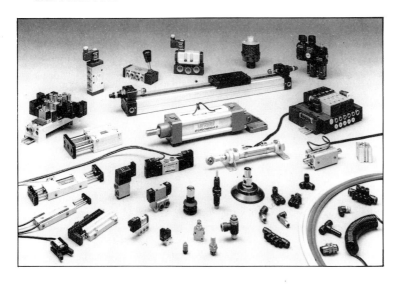

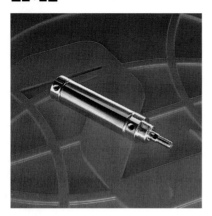

Linear module   Wrist rotate unit.

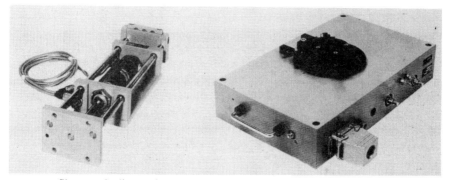

Short stroke liner unit.   Base rotate unit.

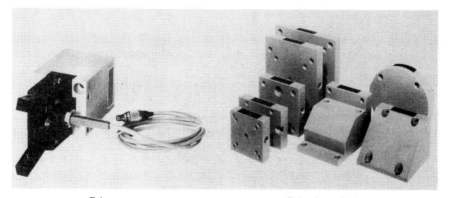

Grippers.   Selection of adaptor plates.

FIGURE 2 – Robot modules. *(Norgren)*

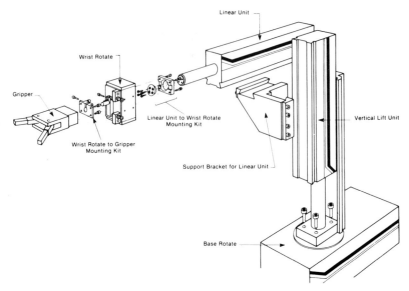

FIGURE 3 – Typical assembly of the various modules showing the use
of adaptor plates. *(Parker)*

and, in some types, intermediate stroke positions. In addition, it will usually incorporate
speed control valves, which are set during development.

If the user keeps to one supplier, it is possible to use a range of standard mounting kits
to link the modules together, but provided one is prepared to makes one's own connections, this is not a necessary restriction. Figure 3 shows how the modules can be connected
by the appropriate adaptor plates.

If the velocity of one of the units needs to be governed either to a maximum value or
to ensure that the travel is smooth, without end-of-travel bounce, an air/oil tandem cylinder
may be incorporated. These are discussed in the chapter on Actuators.

Two further modules, which are modifications of the linear unit, are the precision
positioning unit and the gantry unit. The positioning unit allows very precise, infinitely
variable stroke length. One type is shown in Figure 4. Power for movement is supplied by
the internal cylinder and positioning is achieved by a ball screw linked to an incremental
encoder and a magnetic brake. When the encoder signals that the correct position has been

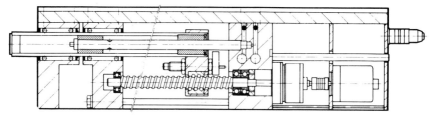

FIGURE 4 – Precision positioning unit employing electromagnetic brake
and ball screws.

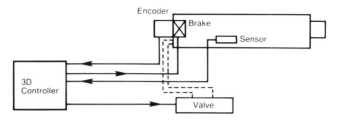

FIGURE 5 – 3-D controller for precision positioning unit. *(Parker)*

reached, the brake is applied, locking the unit in position. Variable positioning is possible with an accuracy of 0.1 mm. A positioning unit is controlled by a 3-D controller as in Figure 5.

The gantry unit is for applications where the component weight requires a rigid support or where the distances to be traversed are too long for a basic linear unit. Using a rodless cylinder allows a maximum stroke of up to 3 metres, without the load restrictions of an overhung unit. It is more suitable for permanent installations, for it lacks the versatility of the basic linear module. A gantry unit is one example of where a combination of electrical and pneumatic operation is advantageous. As an alternative to a pneumatic cylinder, a d.c. servo-motor can be used to drive the carriage along the gantry to stop at a number of pre-determined positions. Figure 6 shows a gantry unit for charging a milling machine. The cylinder is designed to have four stop positions.

FIGURE 6 – Heavy-duty portal frame feeding a milling machine.
Horizontal motion is by means of a rodless pneumatic cylinder. *(Parker)*

Use of multi-motion actuators

Fairly simple factory automation can be introduced with multi-motion actuators. These provide linear and rotary motion in the same unit and are described in the chapter on Actuators. They can be used, in conjunction with a gripper to position parts for machining or further assemblies. Some possibilities are shown in Figure 7.

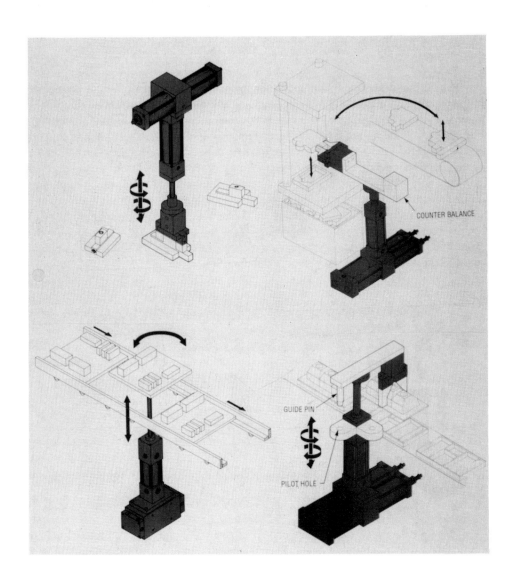

FIGURE 7. *(PHD inc)*

Control systems

The versatility of robotic operation can only be realised with a well-designed programmable controller (PLC), of which there are several on the market. A 3-D controller can be linked directly to a PLC. Manufacturers of pneumatic components can also offer PLCs which interface with their own equipment, and which can be programmed either with the help of a personal computer or a hand-held programmer, using a standardised form of programming.

There are also pneumatic logic components which allow the control to be done purely pneumatically. These are discussed in the chapter on Circuit Analysis.

Fieldbus systems

There has been much interest in recent years in the development of controls based on a two wire cable (a fieldbus), rather than the conventional cabling normally associated with solenoid valves and read-out devices. The great advantage of a fieldbus is the reduced wiring requirements, with space saving and low maintenance. The digital signal and processing allows factory control with a much simpler system than hitherto. Figure 8 compares a fieldbus with a conventional system. Baud rates of up to 12 Mbits/s are possible with a standard fieldbus. RS 485 is the standard communication technology.

To take advantage of this form of control, pneumatic solenoid valves with the appropriate digital signalling have to be available, as does the appropriate interface with the PLC. Fortunately a number of manufacturers can supply these. One has to choose a suitable protocol, which is common across the various components in use. For a low-cost local system, one might choose Pneubus; one of the most widely promoted international

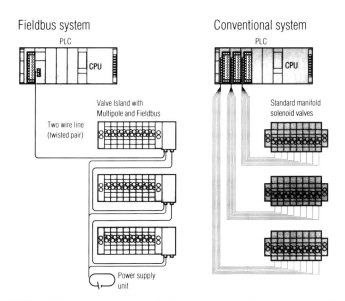

FIGURE 8 – Comparison between fieldbus and conventional systems. *(Norgren)*

FIGURE 9 – Valve island with fieldbus connection. *(Norgren)*

systems is PROFIBUS; other proprietary systems specific to a single manufacturer, are available.

Valve islands with up to 24 valves on a single island can be pre-wired and ready for use, see Figure 9.

Semi-robotic devices

In addition to these fully automatic systems, there are also available pneumatic manipulators which remain always under manual control but which take the effort out of the

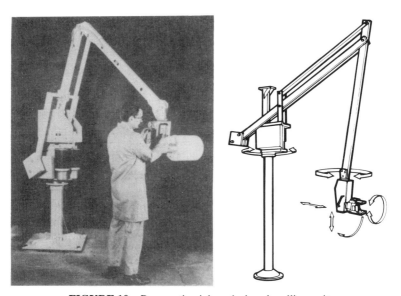

FIGURE 10 – Pneumatic pick-and-place handling unit.

handling. They incorporate some of the features discussed above, so it is convenient to include them here, but it should be emphasised that they are intended solely to assist the operator and not to replace him. Often, because the movement is always under human control, the operation can be faster than in a pneumatic device.

Figure 10 shows a typical pick-and-place handling unit of this kind. The weight of the component (in this example a roll of glass fibre) is reacted through the pantograph mechanism to pneumatic cylinders in the base of the unit. The grippers are pneumatically actuated, but apart from this the placing of the component is performed by the operator. A number of special optional features can be built into this system; the payload can be either completely balanced or, in cases where varying weights have to be moved, by a lever under the operator's control. Safety features are built in, so that if the air pressure is suddenly lost or if the payload becomes detached, the system remains fail-safe.

SECTION 10

Vacuum and Low Pressure

VACUUM PUMPS

VACUUM TECHNIQUES

AIR BLOWERS

VACUUM PUMPS

Theory

Mechanical vacuum pumps can be looked on as special kinds of compressors and so the theoretical analysis of their behaviour is based on the same principles. A vacuum pump is a compressor in which the inlet pressure is that of the chamber to be evacuated and the delivery pressure is that of the atmosphere (at least in those pumps which are designed to evacuate air back to the atmosphere).

The analysis of compressor behaviour as described in earlier chapters should be referred to for background information on the theory. There are several vacuum regimes that can be studied. One of the common situations is where the pump is used to evacuate a chamber which is initially at atmospheric pressure, reducing gradually until the final depression is reached. It will be apparent that, in a pump which can generate a depression of 0.01 torr, the effective compression ratio varies from unity to 76 000. It would be impossible to attain optimum efficiency over such a range, so the design of vacuum pumps is usually a compromise.

The characteristics of the vacuum circuit can affect considerably the behaviour and efficiency of the pump and, when designing an installation, equal weight should be given to the system as to the pump.

Units

Vacuum pressure is quoted in torr (1 torr = 1 mm Hg) or in kPa. Alternatively, depression is quoted as a percentage of atmospheric pressure, in which case the percentage is numerically equal to the vacuum pressure in kPa. Thus 85% vacuum is equal to a pressure of -85 kPa.

The quantity of a gas in a partial vacuum is expressed as a product of pressure and volume (commonly called PV units) because this is a measure of the total number of molecules in a given volume. A variety of PV units are in use: torr litres is the traditional unit; micron litres (*ie* micrometres of Hg) is also used; in SI units, the preference is for Newton metres (1 torr litre = 0.1333 N m).

Leakage rate is often expressed as torr litre/s, lusec (micron litre/s), clusec (0.01 lusec) or in SI units N m/s.

Work of depression

The equation governing the work of depression, related to unit volume of free air, is:

$$W = \frac{\gamma}{\gamma - 1} P_1 \left[\left(\frac{P_2}{P_1} \right)^{\frac{\gamma - 1}{\gamma}} - 1 \right]$$

P_1 is the pressure in the vacuum chamber and P_2 is the atmospheric pressure (or the pressure against which the chamber is evacuated).

Expressed in practical units, this equation becomes

$$W = 0.1 \frac{\gamma}{\gamma - 1} P_1 \left[\left(\frac{P_2}{P_1} \right)^{\frac{\gamma - 1}{\gamma}} - 1 \right]$$

where W is the work rate in kW per l/s of the pump displacement, and the pressures are in bar.

It will observed that the work is zero for $P_1 = P_2$ and for $P_1 = 0$. It is a maximum when $dW/dP_1 = 0$, which occurs when

$$\frac{P_2}{P_1} = \gamma^{\frac{\gamma}{\gamma - 1}}$$

For diatomic gases ($\gamma = 1.4$),

$$\frac{P_2}{P_1} = 3.25$$

If the vacuum required is lower than is represented by this value of pressure ratio, the work increases to a maximum and then decreases; the drive to the pump must be capable of supplying this maximum.

With $\gamma = 1.4$, the maximum value for the work rate is 0.043 kW per l/s of pump flow.

Conductance

An important concept in vacuum technology is that of conductance in a pipe or an aperture. It is usually given the symbol U and is defined as:

$$\frac{\text{flow of gas in PV units}}{\text{pressure difference across the device}}$$

The units are litre/s.

For the determination of conductance in various types of flow restriction, reference should be made to specialist texts, but two of the most common restrictions are long pipes and apertures, for which the following approximations may be used. For a long pipe of constant diameter and air at 25°C,

$$U = 0.125 \, d^3/L \qquad l/s$$

d is the internal diameter and L the length in mm.

For an aperture under the same conditions,

$$U = 0.0935 \, d^2 \qquad l/s$$

Note these equations are empirical so are correct only with the units as stated.

For parallel connection of conductances, the compound conductance is the sum of the individual conductances. For series connection, the compound conductance is given by

$$1/U = 1/U_1 + 1/U_2 + 1/U_3 \, ...$$

Vacuum pump capacity

A common problem is the calculation of the time taken to reduce the pressure in a chamber from atmospheric down to a prescribed value.

Two different formula are applicable, one for reciprocating pumps and the other for continuous flow pumps such as ejectors and diffusion pumps.

In the following formulae, C is the volume of the chamber from which gas is being withdrawn, P_0 is the initial and P_n the final pressure in the chamber.

For reciprocating pumps:

$$\left(1 + \frac{V}{C}\right)^n = \frac{P_0 - q/S}{P_n - q/S}$$

V is the swept volume per revolution of the pump.

If the pump speed is N r/min and the time interval is t, then

$$n = Nt$$

For continuous flow pumps acting isothermally:

$$\frac{S}{C} = \frac{1}{t} \log_e \left(\frac{P_0 - q/S}{P_n - q/S}\right)$$

For continuous flow pumps acting adiabatically:

$$\frac{S}{C} = \frac{1}{\gamma t} \log_e \left(\frac{P_0 - q/S}{P_n - q/S}\right)$$

S is the effective suction flow rate of the pump in l/s and q is the leakage flow in Pa l/s.

S is more correctly defined as:

$$\frac{\text{net flow entering the pump in PV units}}{\text{pressure at the entry port}}$$

The maximum possible flow rate of a pump is the conductance of its entry port, *ie* (as given above) $0.0935 \, d^2$ l/s.

Usually an inlet passage is placed between the chamber to be evacuated and the pump entry flange, in which case the conductance of the passage will have a marked effect on the pump flow. The effective flow becomes:

$$S_e = S \ U/(U+S)$$

S_e replaces S in the equations. When designing a vacuum system, this factor should be taken into account. A common fault is the use of a large pump connected to a working chamber through a pipe possessing a conductance so small that the potential capacity of the pump is wasted.

Leakage in vacuum installations is often significant and is conveniently taken into account, as in the above equations, by replacing P by (P – q/S). Since S appears on both sides of the equations, iteration may be needed for a solution; usually the leakage rate is small, so the solution quickly converges. The leakage includes the gases given off by the process and back flow through the pump . In a practical system, leakage can often be determined by test or from a knowledge of similar circuits.

It should be noted that the above theory is valid only down to medium vacuum (about 0.01 torr). For lower pressures, other factors such as "out-gassing" (the desorption of gas from the surface of the container) come into play. In this regime, experience and test are the best guides for performance estimation.

The two major parameters in the design of a vacuum system are the rate of evacuation required and the ultimate pressure to be realised and held. These parameters determine the number and size of pumps required. This may involve the use of backing pumps capable of pumping from atmospheric pressure downwards, followed by further pumps which only begin to operate below a certain limiting pressure level.

The majority of backing pumps are of the mechanical type although fluid jet pumps, condensers, liquid air traps and cryogenic pumps will work from atmospheric pressure

TABLE 1 – Typical performances of mechanical vacuum pumps

Type	Design	Minimum partial pressure (approx)
Reciprocating	Dry — single-stage	10 torr
	Dry — double-stage	1 torr
	Wet	20 torr
Rotary vane	Single-blade — single-stage	5×10^{-1} torr
	Single-blade — double-stage	2×10^{-5} torr
	Double-blade — single-stage	5×10^{-5} torr
	Double-blade — double-stage	2×10^{-5} torr
	Multi-blade	5×10^{-1} torr
Rotary plunger	Single-stage	5×10^{-3} torr
	Double-stage	2×10^{-5} torr
Twin rotor	Roots — single-stage	100 torr
	Roots — double-stage	10 torr
	Roots — type booster	10 torr down to 10^{-3} torr
	Lysholm	2×10^{-1} torr
	Northey	2 torr

down. Mechanical pumps are widely used for producing coarse vacua and can extend into the medium vacuum range by including a second stage. The first stage then acts as a fore-pump for the second stage.

Selection of pumps

The production of vacuum pressure implies exhaustion of a given vessel by suitable pumping action. Certain types of compressor can also work as exhausters, but in general, the machines used for producing vacuum pressures are known as vacuum pumps.

The pump or pump combination is selected on the capacity needed for the range to be covered, with consideration to pump down-time and duration of the cycle, which determine the most suitable pump size. Determination of pump size is straightforward in the case of coarse vacua, for it is related solely to the size of the container. In the medium to medium-high vacuum range, the shape as well as the volume must be taken into account because the lower the final pressure, the greater the effect of the surface areas. Selection of pump capacity in this range is best based on empirical data. Figure 1 gives a broad indication of the range covered by various types of pump.

In the high and ultra-high vacuum range, the pump size required is related almost entirely to the size and state of the surface of the container. Thus at pressures below 10^{-3} torr there are more adsorbed or absorbed molecules in the system than free gas molecules. The performance of the pump must ensure that at the required working pressure, a rate of evacuation is maintained equivalent to the rate at which gas is emitted from the surfaces of the system. This varies widely with different surfaces.

This condition may be further complicated by the processes which take place in the container being evacuated: drying, degassing and distilling. In such cases it is necessary to form an estimate of the amount of vapour likely to be released during the process so that the pump can be sized accordingly. Where large quantities of vapour have to be dealt with it may be necessary to install an efficient condenser system before the pump.

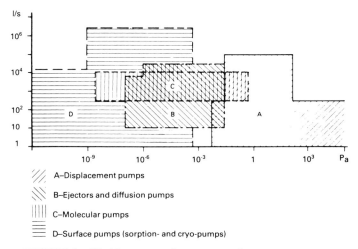

A–Displacement pumps

B–Ejectors and diffusion pumps

C–Molecular pumps

D–Surface pumps (sorption- and cryo-pumps)

FIGURE 1 – Working range of some types of vacuum pumps.

TABLE 2 – Vacuum pumps – types and characteristics

Type	Normal or typical operating range	Remarks
Reciprocating, dry type	Single-stage — down to 10 torr Two-stage — down to 1 torr	Vertical or horizontal Single-, twin- or multi-cylinder
Reciprocating, wet type	Single-stage — down to 5 torr	Slow running speeds normal
Vane, dry type	Down to 5×10^{-1} torr	With or without gas ballast
Vane, oil sealed	Single-stage — down to 5×10^{-2} torr Two-stage — down to 5×10^{-5} torr	
Meshing rotor	Single-stage, down to 1 torr, depending on type Two-stage, down to 10^{-1} torr	With or without gas ballast
Rotary plunger	Down to 10^{-2} torr	With or without gas ballast
Molecular drag, finned rotor	10^{-1} to 10^{-5} torr	Require fore-pump
Molecular drag, plain rotor	10^{-1} to 10^{-6} torr	
Steam ejector	Single-stage — down to 26 in Hg Two-stage — down to $29^{1}/_{2}$ in Hg Three-stage — down to $29^{3}/_{4}$ in Hg Four-stage — down to 1 torr Five-stage — down to 10^{-1} torr	Normally employed with intermediate condensers
Liquid ring	Down to 50 torr	Single- or two-stage Liquid used is generally water. Remove condensable vapours.
Water ejector	Down to 50 torr	Larger than steam ejector for similar duty.
Vapour booster	From 10 to 10^{-2} torr	Combines diffusion pump backed by ejector pump
Diffusion, oil	Non-fractionating — from 10^{-1} to 10^{-6} torr Semi-fractionating — from 10^{-3} to 10^{-7} torr Fractionating — from 10^{-2} to 5×10^{-8} torr	Require fore-pump for backing pressure
Diffusion, mercury	From 10^{-2} to 10^{-8} torr	Regular freezing traps may be preferred for oil-free systems
Cryogenic	Down to 10^{-10} torr	Condensation pumps employing liquefied gas refrigerant
Chemical (absorption)	Down to 2×10^{-2} torr	Alumino-silicate chemical compound at −200°C (liquid nitrogen coolant)
Sorption	Down to 10^{-3} torr	Molecular sieves at *circa* −200°C (liquid nitrogen coolant)
Getter–ion (evaporative ion)	5×10^{-4} to 10^{-11} torr	Usually titanium or zirconium getter. Outgassing reduced by uniform baking prior to pumping (*eg* at 450°C).
Cold cathode (penning pump)	5×10^{-2} to 10^{-11} torr	Outgassing reduced by uniform baking prior to pumping (*eg* at 500°C).

The working pressure to be produced or maintained in the system thus determines the type of pump required (or pump combination where a fore-pump is necessary), and indicates whether the pump size has to be chosen on the volume or surface of the container.

The pump-down time has to be considered when deciding on a suitable pump size and balanced against other operational factors such as the work cycle and whether the process is a continuous one with the pumps continuing to operate after pumping down to the working pressure.

Blank-off pressure

It is recommended that in selecting vacuum pumps, the minimum 'blank-off' pressure capability of a pump should be of the order of a half to one decade lower than the ultimate operating pressure to be attained by the pump. Thus a rotary piston pump with a blank-off capability of 0.01 torr would be suitable for use on a system requiring an ultimate operating pressure above 0.05 to 0.2 torr. Increasing the margin between blank-off and operating pressure is advantageous, but decreasing it below the recommended value will result in a reduction of system reliability.

Gas ballast

If water or other vapour is present in the system it will cause trouble by condensing during the compression cycle of mechanical pumps, contaminating the oil circulating in them. On entering the high vacuum side, this condensate will re-evaporate, reducing the ultimate pressure and pumping speed. This can be overcome by using the technique of gas ballasting on the backing pump in the system.

Gas ballasting, Figure 2, involves the ingestion of a small quantity of gas (usually air) at a constant rate of mixture, preventing condensation during compression, provided the sealing oil temperature is above the saturation temperature of water vapour at the reduced partial pressure. Normally, for effective gas ballast, the oil temperature should be maintained at 60°C to 70°C, where its lubricating and sealing properties are not impaired due to a reduction in viscosity.

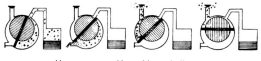

Vane pump working with gas ballast
(water vapour ejected with gas ballast).

Vane pump working without gas ballast
(condensed water vapour remaining over
a certain level if partial evacuation is
needed).

FIGURE 2

The use of gas ballast limits the blank-off and operating pressure capability of the pump. Depending on the particular design and characteristics of the pump, there is an upper limit to the pressure at which gas ballast is effective. This is normally in the region of 25 to 30 torr, which limits the quantity of water vapour which can be handled by a particular capacity pump. Where large quantities of water are present, the mechanical pumps should be protected by a water-cooled or refrigerated condenser or trap, as condensing is the most efficient means of pumping water vapour.

Condensers

Condensers usually consist of a vacuum envelope containing tubes or parallel plates, which may be either water-cooled or refrigerated. The most efficient vapour trapping path is achieved by correct positioning of the vapour inlet and outlet.

Condenser performance is independent of the capacity of the backing pump, which merely maintains the required low pressure. However, high permanent gas pressures must be avoided as they will blanket the cooled surface and reduce condenser efficiency.

If very high water vapour pumping speeds are required, it is necessary to use a water cooled surface condenser, consisting of a tube unit which has a high cooling water velocity. Pumping speeds as high as 2.5 kg of water vapour per m^2 of condenser area per °C of mean temperature difference can be achieved. For lower partial pressures, refrigerated condensers can be used, but with a risk that the vapour may freeze.

Space must be left between the condenser tubes to prevent ice bridging across them and causing too great a temperature difference. The coefficient of heat transfer depends mainly on the thickness of the frozen layer; for water vapour which has formed an ice layer 6 mm thick, the value is about 320 kcal/m^2/°C. The temperature difference between the evaporating refrigerant and the ice also affects the coefficient and hence the rate of heat transfer.

A disadvantage of refrigerated condensers is that they must be defrosted periodically. Therefore, if continuous operation is required, two units must be installed in parallel, one condensing, the other defrosting.

An ideal pump/condenser combination which has excellent vapour handling characteristics, and can operate efficiently in a variety of conditions is a two-stage mechanical system consisting of a rotary pump and a booster pump with an interstage condenser. Very fast vapour pumping is possible due to the high interstage pressure and vapour can be condensed and collected in a suitable receptacle.

Contamination of the primary low pressure pump is avoided by maintaining an interstage vapour pressure below that at which condensation takes place in the pump. The backing pump which has a much smaller capacity, is protected by gas ballasting. Using this system, it is possible to cope with water vapour with a partial pressure of up to 100 torr. Solvents, particularly those with a high vapour pressure can be easily pumped, trapped and recovered.

Diffusion pumps

These require a fore-pump for backing pressure, when they are capable of working from 10^{-8} down to 10^{-9} torr, depending on the type. They are basically condensation pumps using mercury or oil as the working fluid. Oil-diffusion pumps are preferred for industrial applications and may be a fractionating or non-fractionating type.

Diffusion pumps are manufactured both as individual pumps and pump sets, incorporating a suitable fore-pump.

Oil-diffusion pumps are capable of much greater pumping speeds than mercury pumps of the same size, because of the difference in molecular weight of the operating fluid. Oils have a low vapour pressure compared with mercury and thus the pumping speed is relatively unrestricted. Also the use of a freezing trap can be avoided (usually necessary with a mercury diffusion pump to prevent mercury vapour from diffusing back into the vessel being exhausted), except in certain applications where it is necessary to ensure complete absence of oil vapour in the work-piece, as for example in the evacuation of envelopes of thermionic valves and similar devices.

The typical modern oil-diffusion pump is capable of attaining pressures of the order of 10^{-6} torr without the aid of cold traps or baffle systems and maintains a high running speed over the majority of the working range. Apiezon oils (produced by vacuum distillation of petroleum products) are commonly employed as the working fluid, having a low vapour pressure and being extremely reliable in use. Construction is usually of metal with water- or air-cooling. The jet system may be fractionating, semi-fractionating or non-fractionating according to the application.

Some back-streaming is inevitable with a diffusion pump, although this may be reduced by fitting a baffle plate directly above the pump inlet. Some baffle plates incorporate a valve as well as a baffle so that the pump can be completely isolated from the vacuum system when necessary. The use of a baffle plate means a reduction in pumping speed so where one is employed, the design should be a compromise between the complete reduction of back-streaming and pump efficiency. Where it is necessary to reduce back-streaming to a minimum, refrigerated baffles or cold traps have to be used.

For applications where the slightest trace of oil vapour in the vacuum system must be avoided, mercury diffusion pumps may be used. The presence of mercury vapour may be acceptable in some cases when oil is not; where it has to be eliminated, very low temperature traps can be included using liquified gases as coolants.

Vapour booster pumps

The performance characteristics of these are between conventional diffuser pumps and rotary pumps. They comprise a diffuser pump having one or more stages, backed by an ejector pump or ejector stages.

Fluid jet pumps

These work on the principle of introducing a high velocity jet of liquid or vapour into a mixing section where it meets the gas to be evacuated and entrains the gas by turbulent mixing. Some of the momentum of the jet is thus imparted to the gas which is carried down into a diffuser section and thence to a condenser, if necessary, for separation. When the operating fluid is a liquid, separation can be achieved in the diffuser; where it is a gas, separation is usually in a condenser. The complete system may comprise two or more stages, with or without intercooling, depending on the performance required. The

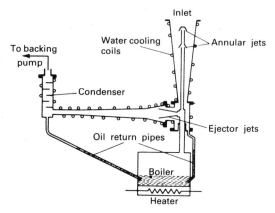

FIGURE 3 – Vapour booster pump.

available types are water jet pumps (fluid), steam ejector pumps (vapour) and vapour-booster pumps. The first two are similar in principle and application, whilst the vapour-booster pump comprises a two-stage system with a condenser, see Figure 3.

The steam ejector pump has widespread industrial application and has largely replaced other types of vacuum equipment for such applications as power plant condensers, because of its simplicity, reliability and low maintenance. It can also handle corrosive or hot gases, since many materials can be adopted for its construction: the major pump elements can be made from carbon or refractory materials, glazed ceramics to resist corrosive attack. It may be termed a jet pump, ejector, augmenter (or vacuum augmenter)

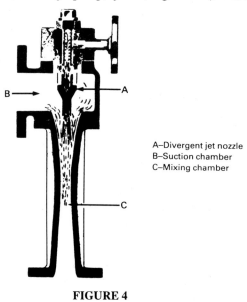

A–Divergent jet nozzle
B–Suction chamber
C–Mixing chamber

FIGURE 4

or thermo-compressor, depending on the duty to be performed. When used to produce and maintain a condition of vacuum by the removal of incondensable gases it is correctly termed an ejector. If used to compress vapours from a condition of high vacuum to a lower vacuum at which condensation can take place, it is termed a vacuum augmenter.

The design of all steam jet pumps is similar and follows the form shown in Figure 4. Steam is introduced via a divergent nozzle into a converging section of the body. The body section also incorporates a suction branch connected to the vessel or system to be evacuated and gases drawn in through this opening are entrained with the steam jet and carried down with it into the diffuser section or compression tube. During the process of entrainment, the steam jet loses a proportion of its energy which is imparted to the aspired gases. In the diffuser section, the kinetic energy of the mixture of steam and gases is re-converted into pressure energy. The compression which can be achieved is proportional to the amount of kinetic energy each unit mass of mixture has when it enters the diffuser and thus to the quantity of steam passing through the nozzle. Single-stage ejectors have a steam consumption from about 25 to 400 kg per hour.

The compression ratio of the ejector is defined as the ratio of the absolute pressure of the steam–gas mixture at the discharge end to that of the absolute pressure of the aspired gas at the suction branch. Compression ratios of up to about 15:1 are possible with simple ejectors, although it is more usual to work with a lower value in order to achieve a better economy with steam consumption. The single-stage ejector produces economic operation down to vacua of about 85% For higher vacua a two- or multi-stage unit is to be preferred.

When two or more stages are used, it is usual to incorporate a condenser between stages, so that the steam of the first stage is condensed out and withdrawn before reaching the next stage; each stage has then only to deal with the incondensable vapours remaining. The usual method of cooling is to employ a cold water spray when the steam is condensed by direct contact with the cooling water, whilst at the same time the incondensable gases are cooled and reduced in volume before being drawn into the suction branch of the following stage.

Three-stage ejectors are provided with an intermediate condenser for the first stage and a further condenser following the second stage. Four-stage units seldom require a condenser after the first stage, since the amount of steam consumed from this stage is small and can be effectively condensed by the following stages. Five- and six-stage units follow on the lines of four-stage units with the addition of one or two non-condensing stages. The bulk of industrial applications are covered by one-, two- or three-stage ejectors.

Water-operated air injector pumps operate on the same principle except that the velocity of the water jet is due to pressure alone rather than expansion and is therefore much lower than the typical steam jet velocity of 1200 m/s. For a similar duty and gas volume, the water ejector pump is considerably larger than the steam pump. Initially the air is in contact with the outer surface of the water jet only and subsequently is entrained in the cone.

Air driven ejectors

For "coarse" vacuum applications, such as those illustrated in the chapter on vacuum techniques, ejectors driven by compressed air are a compact and economical solution. By careful design it is possible to produce a depression of 85%. A pump of this kind can be

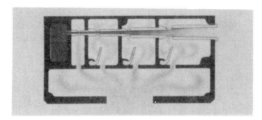

FIGURE 5 – Multi-ejector vacuum pump. Three-stage pump with non-return valves.

based on single or multi-ejectors. One multi-ejector type is illustrated in Figure 5; the compressed air expands in stages with non-return valves on each stage. Typical characteristics of this ejector are as follows:

Air consumption at optimum feed pressure of 4 bar = 0.06 m³/min
Max vacuum = –85 kPa
Weight = 195 g
Sound level: unloaded 65 dB(A); loaded 50 dB(A)

TABLE 3

Vacuum pressure (-kPa)	Induced air flow (m³/min free air)	Time to evacuate 1 litre to the desired vacuum (s)
5	0.088	0.027
10	0.06	0.075
20	0.04	0.017
30	0.028	0.50
40	0.016	0.85
50	0.012	1.25
60	0.008	1.95
70	0.004	3.0
80	0.002	6.0

From this table, it can be seen that, at low values of suction, it is possible to induce a volume flow (in terms of free air) greater than that of the supply volume. Single ejectors are usually worse than this.

VACUUM TECHNIQUES

Vacuum technology includes all systems utilizing a pressure less than atmospheric, taken as 1.013 bar. Vacuum is usually measured in the unit torr (after Torricelli who invented the barometer) which is equal to a pressure of 1 mm of mercury. Torr is not an SI unit, so for consistency in calculations it is preferable to quote vacuum in pascal (Pa), kilopascal (kPa) or bar. The conversion factors are:

$$1 \text{ torr} = 0.133322 \text{ kPa} = 13.5951 \text{ mm } H_2O = 1.33322 \text{ mbar} = 0.01934 \text{ lbf/in}^2$$

The reason for preferring torr is that vacuum is most often measured by a barometer or a manometer which gives a direct reading of vacuum as a measurement of a liquid column.

The complete vacuum range can be divided into 5 classes:

- Low vacuum – pressures down to 25 torr.
- Medium vacuum – 25 torr down to 10^{-3} torr.
- High vacuum – 10^{-3} down to 10^{-6} torr.
- Very high vacuum – 10^{-6} down to 10^{-10}.
- Ultra-high vacuum – less than 10^{-10}.

The most common use of vacuum in industrial processes is to reduce the boiling point of water contained in a product in order to accomplish one of the following:

- to prevent thermal damage to the product when heat is applied to vaporise and effect mass transfer to the water and hence dry the product.
- to speed up the process.
- to ensure very low residual moisture content difficult to dry by other means. examples are distillation, evaporation, vacuum drying of liquids and solids and freeze drying-processes.

Vacuum cooling is closely allied to vacuum drying and is a simple mass transfer process which results in cooling of the product by the removal of a controlled quantity of moisture.

Depression of the vaporisation point is also the objective of vacuum coating processes in which metals and other materials are melted, evaporated and eventually sublimated on a prepared substrate. The high vacuum condition prevents oxidation of the material evaporated at high temperature.

Simple removal of atmospheric air and moisture so that they can be replaced by another gas or liquid is the reason for the use of vacuum in impregnation; in filling of capillaries and containers with small orifices; in some vacuum investment and die casting processes and in refrigeration, cryogenic and gas filled cooling system evacuation; electric lamp and switch gear evacuation and inert gas welding.

In packaging of food, vacuum is applied to remove atmospheric air from the pack so that the contained product will have an increased shelf life, once it is sealed with an impervious pack. An important aspect of vacuum packaging for some products is the differential pressure between interior and exterior which causes the flexible pack material to collapse and cling to the product, providing increased structural strength.

Vacuum coating

This technique is used for applying a thin (0.1 μm) aluminium coating on metals and plastics, which appears as a reflective film. The aluminium is evaporated by tungsten filaments at a vacuum of 0.2 x 10^{-3}.

Ion plating is a further development which allows the deposition of various metals, titanium and carbides. Rather thicker coatings are possible. The air in the vacuum chamber is replaced by argon gas. A high potential produces a gas stream and evaporates a metal source which is then deposited on the surface of the component.

Mechanical processes

There are numerous ways in which low vacuum can be used for mechanical handling and low cost automation. Some examples are:

- High speed paper feeding for printing and processing.
- Vacuum forming of thermoplastics.
- Liquid filling.
- Dry powder filling.
- Continuous liquid/solid separation.
- Air sampling.
- Vacuum feeding of metal timber and glass sheets.
- Machine tool chucks and vices.
- For gripping components during automation.

Applications of low or coarse vacuum

Some typical applications using low vacuum are illustrated in Figure 1. It will be noted that in each application, a filter, vacuum gauge and relief valve are required in the circuit. In addition, according to the application, a liquid trap and a vacuum release valve may be required. When using a suction cup for material handling, a vacuum check valve is fitted for security in the event of pump failure; this ensures that vacuum is retained in the cup.

Laminating Press

Vacuum is used to eliminate bubbles in laminating and veneering hard board, plastics, and wood.

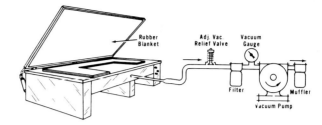

Mobile Material Handling with Vacuum

Vacuum equipped lift trucks eliminate fork or clamp damage. Rolls can be rotated to either stack or load in machinery.

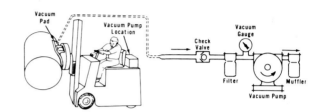

Blood Taking Equipment

Plastic blood collection bag inserted in rigid flask. Vacuum applied to flask draws blood from patient into plastic bag.

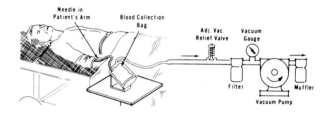

Vacuum Egg Lifters

Feather-light egg handling with vacuum uses a special accordion pleated suction cup which lifts and holds each egg.

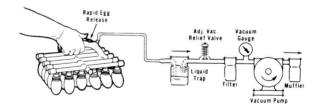

Centrifugal Pump Priming

Vacuum priming of centrifugal pumps assures quick start pumping. Prevents pump burnout from running dry.

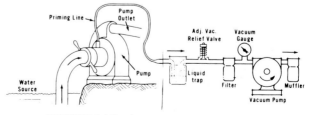

FIGURE 1 *(Gast)*

High Speed Labeling Machine

Vacuum handling of pressure sensitive labels is jam proof, accurate, and ideal for automatic labeling of odd shaped, soft and flexible items.

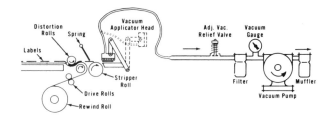

Vacuum Feeding Sheet Steel, Glass, etc.

Vacuum pads will lift any non-porous material: glass, aluminum, brass, wood, etc., and transfer to conveyor or processing machinery. No scuffing or scratching.

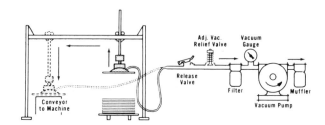

Vacuum Solder Gobbler/Iron

Excess solder on printed circuits is removed with vacuum tube and glass holder mounted on soldering iron.

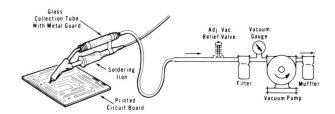

Vacuum Column to Hold Slack in Computer Tape

High speed computer tape requires slack to protect tape from snapping apart during fast starts and stops. Slack in tape is held and controlled by vacuum.

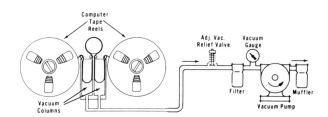

Dry Powder Vacuum Filler

Dry powders can be rapidly vacuum-bottled by drawing vacuum on container causing dry material to flow from hopper into conatiner.

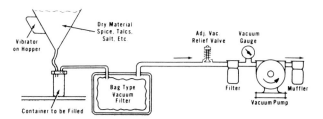

FIGURE 1 (continued) *(Gast)*

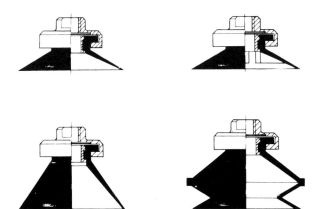

FIGURE 2 – Suction cups. Top left: for plane and slightly curved surfaces.
Top right: flat with cleats for plane surfaces. Bottom left: deep when handling curved
surfaces. Bottom right: bellow for use when level compensation is required.

Suction cups and pads

A variety of designs of suction cups are available, for use according to the shape of the surface to which they are to be attached. Figure 2 shows some typical shapes.

In calculating the holding force of a cup it is customary to take the maximum force as the product of the area of the inner sealing lip and a pressure corresponding to 75% vacuum. In addition it is recommended that minimum safety factors of 2 for horizontal surfaces and 4 for vertical surfaces should be used.

When using vacuum pads, the response time (the time taken for the vacuum to reach the design level) should be calculated. When using suction for component handling, it is important the next stage of the operation should not start until the correct depression is reached. The total internal volume of the pipework and fittings up to the pad should first be calculated and the formulas in the next chapter used to determine the response time.

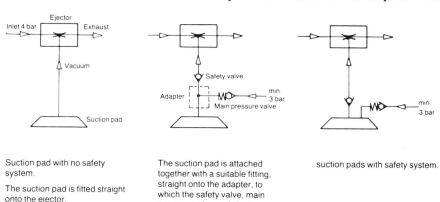

Suction pad with no safety system.

The suction pad is fitted straight onto the ejector.

The suction pad is attached together with a suitable fitting, straight onto the adapter, to which the safety valve, main pressure valve and ejector are also connected.

suction pads with safety system.

FIGURE 3

Cups are made of oil resistant rubber or, if a wide range of temperatures is required, of silicone rubber. Nitrile, urethane and fluoro-rubbers may be used. They are commercially available in diameters from 6 mm to 300 mm. It is usually preferable to use several small cups rather than one large one.

Figure 3 shows some typical circuits using ejectors with suction cups; note the various attachment methods. These venturi ejectors are frequently used in this application. Although they consume mains air and overall are not as efficient as vacuum pumps, their simplicity and reliability makes them the favoured method for many automation applications. See Figure 4.

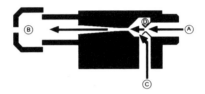

FIGURE 4 – Venturi ejector. Compressed air A is blown through nozzle D, which causes a depression at C. Air flows from C to the atmosphere at B. *(Vickers)*

Combined ejectors and solenoid- (or pilot-operated) valves are available for building into an application, Figure 5.

It may be necessary, in some applications, to apply a positive pressure to release the cup. Special combined valves are available for this purpose.

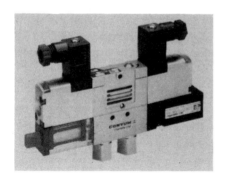
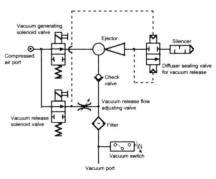

FIGURE 5 – Vacuum ejector incorporating solenoid-operated air valve, adjustable vacuum switch, vacuum release and silencer. *(Vickers)*

AIR BLOWERS

The description air blower is applied to types of air pump designed to deliver small volumes of air at low to moderate pressure. Some types can also work in reverse as exhausters, the Roots pump being a typical example. Other types have fan type impellers capable of moving larger volumes of air at low pressures and are called fan blowers or turbo-blowers. They differ from positive displacement pumps which are designed to work at higher pressure ratios.

The Roots blower, Figures 1 and 2, employs two identical rotors turning in opposite directions inside a cylindrical casing. The rotors intermesh but maintain positive clearance by timing gears. Internal lubrication is unnecessary, so the air is delivered free of oil. Compression takes place by back-flow every time a rotor tip uncovers the discharge port. High volumes of air can be displaced since the rotors can be driven at high speeds by synchronous or directly-coupled electric motors. The pressure generated is usually less than 1 bar in a single stage (see the chapter on Compressor Performance).

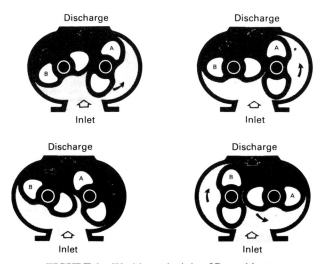

FIGURE 1 – Working principle of Roots blower.

FIGURE 2 – Roots type air blower. *(Dresser Roots)*

Each revolution of the rotors causes four volumes of air to be conveyed through the blower. The cross-sectional area of the rotor is, to a close approximation, equal to one-half of the cross-section of the rotor diameter, so the total theoretical delivery is given by:

$$q = \frac{\pi}{4} D^2 L n$$

where L is the length of the rotors, D the diameter and n the rotational speed.

The actual delivery is less than this, primarily through leakage losses. These losses result from the inter-lobe clearances and the gap between the rotors and casing. The actual leakage is almost independent of speed and is given by a relationship of the form:

$$q_v = K A_L \sqrt{(\Delta P/\rho)}$$

q_v is the leakage volume, K is a constant of proportionality, A_L is the leakage area, ΔP is the pressure rise through the blower and ρ is the density at inlet conditions.

The volumetric efficiency is:

$$\eta_v = 1 - q_v/q$$

The rotor clearances are kept very small in modern Roots blowers; the rotational speed is kept high (up to 5000 rev/min in the smaller sizes) and so the volumetric efficiency is also high.

The theoretical power needed to compress the air is

$$W = 0.1 q \Delta P \qquad kW$$

where q is in l/s and ΔP is in bar.

As explained in the chapter on Compressor Performance, the overall efficiency is not as good as if there were a built-in volume compression ratio. This inefficiency is tolerated for the sake of simplicity of production. Because the pressure rise is small the actual difference in power is usually not significant.

In addition to the power needed to compress the air, there is a further power loss to overcome friction in the timing gears, bearings and seals as well as dynamic losses in the compression chamber; the total is of the order of 5%. An advantage of this type of blower is that, because there is no internal compression, the off-load power is approximately equal to this small amount.

Positive-displacement blowers

The boundary line between a compressor and a blower is not always clear. One machine which can be considered as either a blower or a compressor is the helical screw machine illustrated in Figure 3.

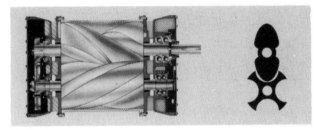

FIGURE 3 – Helical screw blower. *(Gardner Denver)*

This lies between the helical screw compressor and the Roots blower. It is more efficient than the Roots (up to 10%, at maximum pressure), but is more expensive to manufacture because of the complexity of the screw manufacture. It requires timing gears to maintain correct meshing. This machine can generate a pressure of 1.2 bar or a vacuum of 50 %.

Fan blowers

A fan impeller consists of a number of blades mounted at a pitch angle and assembled on, or integral with, a hub mounted on a driven shaft. Rotation generates a slip-stream in a similar manner to an airscrew, although the efficiency is low compared with a true airscrew for several reasons: usually the geometric pitch along the length of the blade is not constant; a large number of blades are provided, which increases the volumetric flow at the expense of efficiency; the cross-section of the blades is not of aerofoil form, indeed they are often made from flat steel plate, curved to an approximate aerodynamic form. Fans are used to move large volumes of air at very low pressures and are to be found in ventilation systems, engine and compressor cooling, conveying, vacuum cleaning and supercharging.

Fan concepts and definitions

The effect of a fan is to change both the static pressure and the air velocity from those measured upstream of the fan (the inlet) to those measured downstream (the outlet). There will also be changes in temperature and density but these are small and are frequently ignored in the analysis. If v_1 and v_2 are the inlet and outlet air velocities and p_1 and p_2 the inlet and outlet static pressures, the following definitions apply:

$$\text{Inlet velocity pressure, } p_{1d} = \tfrac{1}{2} \rho \, v_1^2$$
$$\text{Outlet velocity pressure, } p_{2d} = \tfrac{1}{2} \rho \, v_2^2$$
$$\text{Total inlet pressure, } p_{1t} = p_{1d} + p_1$$
$$\text{Total outlet pressure, } p_{2t} = p_{2d} + p_2$$

Fan total pressure, p_t, is the difference between the total pressures at the fan outlet and inlet.

Fan velocity pressure, p_d, is the velocity pressure corresponding to the average velocity at the fan outlet (*ie* the volume flow of air divided by the area of the discharge orifice).

Fan static pressure, p_s, is the fan total pressure minus the fan velocity pressure.

Note that the fan static pressure is not the rise in static pressure through the fan but is defined in terms of the velocity pressures.

Total fan air power $= p_t \, q$

Static air power $= p_s \, q$

Fan total efficiency,

$$\eta_{tA} = \frac{p_t q}{W_A}$$

Fan static efficiency,

$$\eta_{sA} = \frac{p_s q}{W_A}$$

W_A is the fan shaft power.

If both the fan inlet and outlet are open, the fan total pressure is equal to the fan velocity pressure, and the fan static pressure is zero. As the volume flow decreases, the fan total pressure increases, reaching a maximum value at a volume flow which depends on the particular fan design. Such characteristics may be seen in Figure 4. When interpreting manufacturer's catalogues, this factor should be borne in mind.

Fan performance characteristics

The subject of fan design is outside the scope of this chapter, but it may be helpful to examine some of the design parameters which can be used to evaluate fan performance. Dimensional analysis leads to the following pairs of relationships (each pair is a representation of the same relationship expressed in a different form). The first expression of each pair is to be preferred because the coefficient is independent of the units used (provided of course that the units are consistent), but both are to be found in technical literature, so are included here. When using the second form, care must be taken to use the proper units.

Volume coefficients

$$q = \phi(\pi d^2/4) \, u$$
$$q = K_q d^3 \, (N/1000)$$

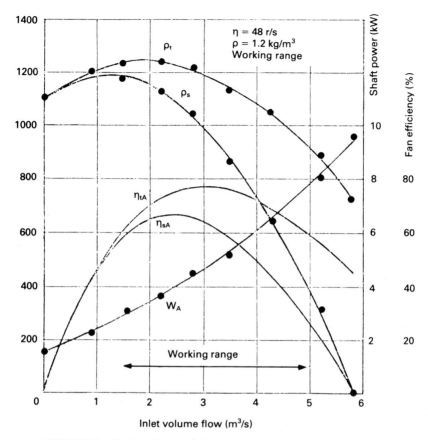

FIGURE 4 – Typical characteristics of a fan as measured in accordance with BS 848 : Part 1.

Pressure coefficients

$$p_t = \psi \, \frac{1}{2} \, \rho u^2$$

$$p_t = K_p d^2 \, (N/1000)^2$$

Fan power coefficient

$$W_A = \lambda \, (\pi \, \frac{d^2}{4}) \, \frac{1}{2} \, \rho u^3$$

$$W_A = K_w d^5 \, (N/1000)^3$$

Note that

$$\lambda = \frac{\phi}{\eta_{tA}}$$

ϕ, ψ and λ are empirical coefficients corresponding to K_q, K_p and K_w.

TABLE 1 – Coefficients for Different Fan Designs
r = ratio form diameter/hub diameter

r	ϕ	ψ	$\lambda\eta$	$\eta_s x10^{-3}$
	1	2 to 4	2 to 4	40 to 100
0.9	1	2 to 3	2 to 3	70 to 90
0.75	0.3	0.75	0.225	100
0.7	0.2	0.6	0.12	100
0.5	0.13	1.0	0.13	60
0.3	0.03	1.1	0.033	25
0.15	0.0018	1.1	0.002	6
	0.1 to 0.2	0.05 to 0.01	0.005 to 0.02	250 to 600

In these relationships, u is the peripheral fan speed, d is the overall fan diameter and N is the rotational speed in rev/min.

Specific speed

A further concept, which is also often found when comparing the performance of fans is that of specific speed. It is defined as the rotational speed of a unit fan, *ie* the speed at which a fan geometrically similar to the given fan would need to run to give unit volume flow at unit fan pressure. Specific speed is a theoretical concept only and its real value lies in use in comparing the performance of fans in an homologous series (*ie* geometrically similar). Its value depends on the chosen units of volumetric flow and pressure. The preferred units are m^3/s and Pa which are employed here. The specific speed, n$_S$, of a fan is calculated from:

$$n_s = nq^{\frac{1}{2}}/p_t^{\frac{3}{4}}$$

where n is the speed of the given fan in rev/s. Alternative expressions for specific speed are also to be found.

Factors to be considered in design

The following factors need to be considered when designing or choosing a fan

- Maximum efficiency is satisfied by correct choice of n$_S$.
- Low noise is obtained with a low peripheral velocity and a high value of ψ.
- Large capacity is achieved by a high value of ρ.
- Large capacity in a minimum size, leading to the cheapest design, implies a high value of λ.

Table 1 gives some typical values of the coefficients discussed above.

Fans should be tested in accordance with the recommendations of BS 848, to which reference should be made for further information on the various kinds of fan.

Fan noise

Noise level and frequency content are influenced by the fan power, volume flow and the number of blades. General characteristics are summarised in Table 2.

The noise spectra of fans vary with the type of fan but in general they are typical of all aerodynamic noise in that the acoustic power is generated over a wide spectrum, rising to a shallow peak at a frequency determined by the type, size and duty of the fan. Discrete tones will be generated at a blade passing frequency of nx/60 Hz, where n is the fan speed in r/min and x is the number of blades.

Centrifugal fans having forward or backward curved blades tend to produce a spectrum which falls at approximately 5 dB per octave with increasing frequency, although there is a tendency for backward curved impellers to generate more energy at high frequencies and correspondingly less at low frequencies.

TABLE 2 – Fan noise parameters

Type of fan	Air velocity	Turbulence	Speed	Remarks
Propeller	Low	Low	Low	Inherently low noise level.
Axial flow	Low*	Low*		*Low pressure types
	High†	High†	High	†High pressure types
Centrifugal				
Forward-curved	High	High	High	Large number of blades.
Backward-curved	Low	Low	Low	
Straight radial blade	High	High	Medium	

Axial flow fans have a much flatter spectrum and contain a much greater proportion of high frequency energy.

Not all noise generated by fans and transmitted along the ductwork comes from aerodynamic sources at the impeller. Some of the noise may be attributed to fan unbalance, bearing noise, structural resonance of the casing or to noise generated at the motor, its coupling or belts. The fan noise may also be affected by its installation: the sound power of a centrifugal fan does not seem to be greatly affected by variations in inlet and discharge design. Care should be taken to ensure that axial flow fans are installed in accordance with the manufacturer's recommendations. Quoted sound power levels are generally applicable to the case when the axial flow is coupled to inlet and discharge ducts having the same diameter as the fan casing, and a minimum straight duct run upstream and downstream of the fan equal in length to two duct diameters. The rated sound power levels are applicable to fans having coned inlets and inlet transformation sections provided that the included angle does not exceed 60°.

The octave band sound power level and the spectrum of the fan noise vary not only with the speed and duty of the fan, but also with the type and the manufacturer. A standardised procedure for the measurement of fan sound power is described in BS 848:Part 2. Most manufacturers publish details of the noise generation in accordance with this Standard. Refer also to BS 848:Part 5 (Guide to mechanical and technical safety)

Ducted fans

The performance of a simple propeller type fan can be considerably improved by enclosing it within a close-fitting open-ended cylindrical casing, Figure 5. The design is critical. It may be parallel sided or convergent–divergent with the fan at the throat. Diffuser vanes may be added. Much depends on the individual installation as to how efficient the fan is, although quite high pressures can be realised with suitably proportioned ducted fans.

With peripheral velocities of the order of 100 m/s a pressure rise of 0.07 bar is possible, but it is more common, with industrial fans, to limit the peripheral speed to 50 m/s and a pressure of 0.007 bar so as to limit the noise generation. Multi-stage units are also possible.

Advantages of propeller type fans are simple construction, low cost and an ability to handle contaminated air. Their short axial length, even including the motor, makes them suitable for installation directly in walls or ceilings or for compact installation elsewhere. They are widely used for ventilation, forced draught and cooling purposes and are

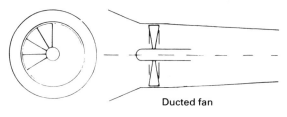

Ducted fan

FIGURE 5

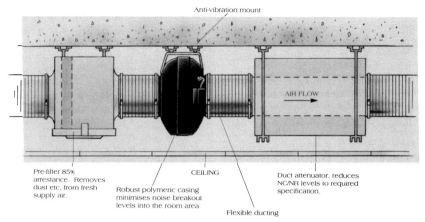

Anti-vibration mount

AIR FLOW

Pre-filter 85%
arrestance. Removes
dust etc, from fresh
supply air.

CEILING

Robust polymeric casing
minimises noise breakout
levels into the room area.

Flexible ducting

Duct attenuator, reduces
NC/NR levels to required
specification.

FIGURE 6 – Ducted fan installation. *(Ventaxia)*

normally quiet running at low or moderate speeds, being directly coupled to an electric motor, with a maximum speed of the order of 2700 r/min.

They may generate noise and vibration if run at higher speeds, although if care is taken with the installation, these can be reduced by incorporating duct attenuators and anti-vibration mountings (see Figure 6).

Axial flow fans are a development of the simple propeller type. They are ducted with a small clearance between fan blades and casing (consistent with safety and noise). The smaller the clearance, the more effective the casing is in preventing recirculation at the blade tips and consequent loss of efficiency. The sides of the casing are normally parallel, although entry may be convergent and long enough completely to enclose both fan and motor. An axial fan can generate a higher pressure than a simple propeller type fan (up to two and a half times) with a high efficiency.

The critical parameters of a fan design are the aerofoil blade form, the number of blades, angle and area, diameter, hub ratio and tip clearance. For low pressure working, the hub ratio is small (small diameter with long blades). For high pressure working, hub ratio is large and the aerofoil section chosen differently. Guide or diffuser vanes may be fitted downstream to straighten the flow and improve pressure development.

Noise generation limits the practical maximum speed of a fan, the critical factor being the entry velocity which should be limited to 50 m/s, above which the noise can be intolerable. The presence of guide vanes can also cause noise problems by setting up local oscillatory flow, which can generate high pitched noise.

The pressure available from an axial flow fan can be increased by two-staging. The two stages may be driven off the same shaft (by a motor with a shaft at each end) or may be contra-rotating with two motors. In the former case, straightening vanes are necessary between the two impellers to ensure efficient working of the second stage. With contra-rotating fans, no straightening vanes are necessary, but the drive system is more complicated. In practice the two fans are driven by separate motors.

A further development is the bifurcated co-axial fan where the motor is isolated from

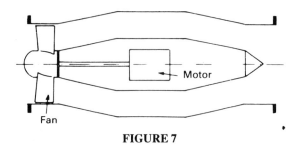

FIGURE 7

the downstream flow by taking it through separate ducts as in Figure 7. This can be useful where the air is contaminated or hot, with the possibility of damage being caused to the motor. It also improves access to the motor for maintenance purposes. The drive motor can also be placed outside the duct and the fan driven by a V-belt or the fan can be placed at a bend in the ducting allowing the drive shaft to be taken through the wall to an external motor.

Radial flow fans

Radial flow impellers have similar characteristics to centrifugal compressors. The inducted air is turned through a right angle and ejected radially from the casing. This provides a greater pressure generation. The volume of air moved depends on the form of the impeller; three basic blade forms are employed on blowers:

- forward-curved blades;
- backward-curved blades;
- straight radial blades.

The forward curved impeller, Figure 8, has a pressure generating capacity about five times that of an axial flow fan and a high volume capacity. It is suitable for handling a large

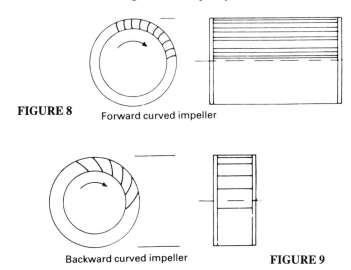

FIGURE 8   Forward curved impeller

Backward curved impeller   **FIGURE 9**

quantity of air where higher pressures are required than could be supplied by an axial fan. However for the same capacity it will be larger in diameter and more expensive. Its efficiency is lower and the speed may have to be limited to keep the noise level to an acceptable value, which reduces its usable capacity. They are used with short discharge ducts, since the high discharge velocity should not be converted into pressure.

The backward-curved impeller, Figure 9, is also a high volume machine, capable of generating high flow at a lower speed than the forward curved fan. For the same size its capacity is higher or it can have the same capacity at a lower speed, with a consequently lower noise level. The flow is smoother, which also serves to reduce the noise level. The characteristics depend on the shape of the blades; aerofoil blades produce the smoothest flow but are more expensive to manufacture. Flat plate blades are used on the simpler designs at a loss of efficiency but still with an adequate performance.

The pressure generated by either type of impeller is generally sufficient for all blower duties, so more than one stage is not usually required. Multi-stage units are used for low pressure compressors.

The radial bladed impeller, Figure 10, is a high pressure low capacity fan (the opposite of an axial flow fan). It has a low efficiency, its chief advantage being its simple construction and the open form of the impeller, which makes it suitable for working with dirty air – hence its common application as an exhaust fan in contaminated atmospheres. If necessary, the impeller can be made non-clogging by fitting a back plate to prevent

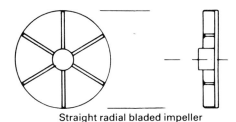

Straight radial bladed impeller

FIGURE 10

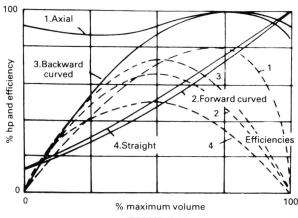

FIGURE 11

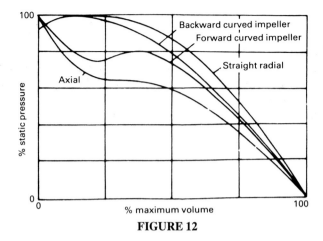

FIGURE 12

fibrous material collecting at the back of the hub. It can be protected in a corrosive atmosphere by a suitable coating treatment.

Typical performance characteristics of axial flow fans and the three basic forms of centrifugal blowers are given in Figures 11 and 12. Figure 11 shows the power and efficiency curves and Figure 12 the static pressure curves. It should be noted that these are related to 100% performance and no quantitative comparison between the different types should be drawn for pressure and volume.

Regenerative blowers

One type of blower which is capable of generating a pressure of 0.5 bar in a single stage is the regenerative blower. In this design, as the impeller rotates, centrifugal action moves the air from the root of the blade to the tip. Leaving the blade tip, the air flows round the

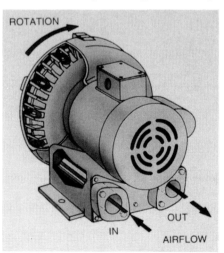

FIGURE 13 – Principle of regenerative blowers. *(EG and G Rotron)*

round the annular housing back down to the root of the succeeding blade where the flow pattern repeats. This action provides a staging effect to increase the pressure difference. At the outlet the annular housing is reduced in size to produce a close fit with the impeller blades and the air is then discharged. The advantage of the regenerative principle is that the pressures generated are equal to those obtained by multi-stage or positive displacement blowers. The principle is illustrated in Figure 13.

The same machine can be used to produce a vacuum of 0.3 bar.

SECTION 11

Engineering Data

STANDARDS AND PUBLICATIONS

GRAPHICAL SYMBOLS FOR PNEUMATIC SYSTEMS
AND COMPONENTS

STANDARDS AND PUBLICATIONS

British Standards

Standard	Subject	Cross Reference
BS 21:1985	Specification for pipe threads for tubes and fittings where pressure-tight joints are made on the threads (metric dimensions)	ISO 7/1 ISO 7/2
BS 143:1986 and BS 1256	Specification for malleable cast iron and cast copper alloy threaded pipe fittings	ISO 49
BS 673:1984	Specification for shanks for pneumatic tools and fitting dimensions of chuck bushings	ISO 1180
BS 848	Fans for general purpose	
Part 1	Methods of testing performance	
Part 2	Methods of noise testing	ISO 5136
Part 5	Guide for mechanical and electrical safety	
Part 6	Method of measurement of fan vibration	
BS 1042	Methods of measurement of fluid flow in closed conduits	
BS 1123	Safety valves, gauges and fusible plugs for compressed air or inert gas installations	
Part 1:1987	Code of practice for installation	ISO 4126
BS 1134	Assessment of surface texture	ISO 468 ISO 4287/1
BS 1256	see BS 143	
BS 1306:1975	Specification for copper and copper alloy pressure piping systems	
BS 1387:1985	Specification for screwed and socketed steel tubes and tubulars and for plain end steel tubes suitable for welding or for screwing to BS 21 pipe threads	ISO 65

BS 5543:1978	Vacuum technology – graphical symbols	ISO 3753
BS 5555:1993	Specification for SI units and recommendations for the use of their multiples and of certain other units	ISO 1000
BS 5755	Specification for dimensions of basic features of fluid power cylinders	ISO 3320, 3322, 4393, 4395, 7181
BS 5791	Glossary of terms for compressors, pneumatic tools and machines	
Part 1:1979	General	ISO 3857/1
Part 2:1979	Compressors	ISO 3857/2
Part 3:1990	Pneumatic tools and machines	ISO 3857/3
BS 5793	Industrial process control valves	IEC 534
BS 6005:1993	Specification for moulded transparent polycarbonate bowls used in compressed air filters and lubricators	
BS 6244:1982	Code of practice for stationary air compressors	ISO 5388
BS 6413	Lubricants, industrial oils and related products (class L)	
Part 3:1992	Classification for family D (compressors)	
Section 3.1	Air compressor	ISO 6743
BS 6759	Safety valves	
Part 2:1984	Specification for safety valves for compressed air or inert gases	ISO 4126
BS 6842:1987	Guide to the measurement and evaluation of human exposure to vibration transmitted to the hand	ISO 5349
BS 7086:1989	Glossary of terms for pneumatic tools	ISO 5391
BS 7129:1989	Recommendations for mechanical mounting of accelerometers for measuring mechanical vibration and shock	ISO 5348
BS 7389	Pneumatic fluid power valve mounting surfaces	
Part 1:1990	Specification for five-port directional control valves (without electrical connector)	ISO 5599-1
BS 7714:1993	Guide for the care and handling of seals for fluid power applications	
BSEN 10028	Specification of flat products made of steel for pressure purposes	
BSEN 10207: 1992	Steels for simple pressure vessels. Technical delivery requirements for plates, strips and bars	
BS ISO 6431	Pneumatic fluid power. Single rod cylinder,	

	1000 kPa (10 bar) series with detachable mountings, bores from 32 mm to 320 mm. Mounting dimensions
BSEN 139:1995	Respiratory protective devices. Compressed air line breathing apparatus for use with a full face mask, half mask or a mouthpiece assembly. Requirements, testing and marking
BSEN 270:1995	Respiratory protective devices. Compressed air line breathing apparatus incorporating a hood. Requirements, testing and marking
BSEN 271	Respiratory protective devices. Compressed air line or powered fresh air hose breathing apparatus incorporating a hood for use in abrasive blasting operations. Requirements, testing, marking
BSEN 286	Simple unfired pressure vessels designed to contain air or nitrogen
BSEN 28662	Hand-held portable power tools. Measurement of vibration at the handle
BSDDENV 25349	Mechanical vibration. Guidance for the measurement and assessment of human exposure to hand-transmitted vibration

CETOP publications *(available in English, French and German)*

RP4P	Pneumatic cylinders – suggested data for inclusion as a minimum in manufacturers, sales and literature
RP 19P to 30P, and RP 57P	Recommended data for inclusion in manufacturers, sales literature
RP19P	Pneumatic directional control valves
RP20P	Pneumatic flow control valves
RP22P	Pneumatic shuttle, non-return and quick exhaust valves
RP23P	Pneumatic pressure intensifiers
RP24P	Pneumatic rectilinear piston type cylinders
RP28P	Connections
RP29P	Pneumatic quick-action couplings
RP30P	Pneumatic rotating and telescopic joints
RP34P	Couplings for fluid power hoses
RP38P	Guidance on relation between port threads and pipe hose diameters
RP40P	Hose couplings claw type
RP41P	Hydraulic and pneumatic circuits – circuit diagrams

RP49P	Technological symbols for fluid logic and related devices with and without moving parts
RP54P	Specification for polyamide tubing 11 and 12 for pneumatic transmissions
RP57P	Pressure relief valves
RP68P	Rules for the identification of ports and operators of pneumatic control valves and other pneumatic components
RP76	Outside diameters for tubes in fluid power applications
RP80	Cone type connections – 24° – for fluid power tubes and hoses
RP85P	Characteristics of pneumatic components
RP105P	Acceptance tests for pneumatic cylinders
RP107P	Fixed pivot bracket for pneumatic cylinders
RP111P	Switching characteristics of pneumatic directional control valves and moving part logic components
RP114P	Pneumatic valves, subplates for five-port valves with mounting surfaces in accordance with ISO 5599/1

BFPA (British Fluid Power Association) publications

BFPA/P3	Guidelines for the safe application of hydraulic and pneumatic fluid power equipment
BFPA/P27	Guidelines on understanding the electrical characteristics of solenoid for fluid power control valves and their application in potentially explosive atmospheres
BFPA/P28	Guidelines for errors and accuracy of measurements in the testing of hydraulic and pneumatic fluid power components

BCAS publications

Guide to the selection and installation of compressed air services

Buyers guide to compressed air plant, equipment and services

BCAS Recommendations for filters for compressed air – methods of test

BCAS Guidance and interpretation for the pressure systems and transportable gas containers Regulations, SI 2169:1989

Machinery Safety legislation (covers Directives 89/392/EEC and 91/368/EEC)

Compressed air condensate

Tools guide to the selection and application of pneumatic tools

Pneurop publications *(mostly available in English, French and German)*

PN2CPTC1	Acceptance test code for bare displacement air compressors
PN2CPTC2	Acceptance test code for electrically driven packaged displacement air compressors
PN2CPTC3	Acceptance test code for IC engine-driven packaged displacement air compressors
5617	Graphical symbols for control and instrument panels on rock drilling rigs

Test procedures for the measurement of dust emissions from hand-held (portable) power driven tools:

01/1987	General measuring regulations
02/1987	Percussive and chipping tools
03/1987	Saws, shapes and planers
04/1987	Grinders
05/1987	CAGI/Pneurop – Standard for electronic interfaces for pneumatic tools

5607(1972)	Vacuum pumps, rules of acceptance
	    Part II Vapour pumps
	    Part III Turbomolecular pumps
	    Part IV Sputter ion pumps

6601(1978)	Application of national standards for acceptance and capacity measurements of steam jet vacuum pumps and steam jet compressors
6602(1979)	Vacuum pumps, rules of acceptance: Part I Positive displacement pumps – Roots pumps
PN5ASRCC/5	Pneurop Acceptance Specification for refrigerator cooled cryo-pumps – Part 5
6606(1981)	Vacuum flanges and connections – dimensions
6612(1984)	Acceptance specification for liquid ring vacuum pumps

Note: CETOP, Pneurop and BCAS publications are available from:
BCAS, 33/34 Devonshire Street, London W1N 1RF

BFPA and CETOP publications are available from:
BFPA, Cheriton House, Cromwell Business Park, Chipping Norton, OX7 5SR

GRAPHICAL SYMBOLS FOR PNEUMATIC SYSTEMS AND COMPONENTS

The following symbols are specified in BS 2917 (ISO 1219)

General symbols (basic and functional)

The symbols for hydraulic and pneumatic equipment and accessories are functional and consist of one or more basic symbols and one or more functional symbols. They are neither to scale nor oriented in any particular direction.

BASIC SYMBOLS

Description	Application	Symbol
Line: – continuous – long dashes – short dashes – double – long chain thin (optional use).	Flow lines. Mechanical connections (shafts, levers, piston-rods). Enclosure for several components assembled in one unit.	$L > 10E$ $L < 5E$ $D < 5E$ L = length of dash E = thickness of line D = space between lines
Circle, semi-circle.	As a rule, energy conversion units (pump, compressor, motor). Measuring instruments.	

BASIC SYMBOLS (continued)

Description	Application	Symbol
	Non-return valve, rotary connection, *etc.*	○
	Mechanical link, roller, *etc.*	○
	Semi-rotary actuator.	D
Square, rectangle.	As a rule, control valve(s), except for non-return valves.	□ ▢ ▢▢
Diamond.	Conditioning apparatus (filter, separator, lubricator, heat exchanger).	◇
Miscellaneous symbols.	Flow line connection.	d $d \approx 5E$ E = thickness of line
	Spring.	⋀⋀
	Restriction: – affected by viscosity	≍
	– unaffected by viscosity.	∨ ∧

FUNCTIONAL SYMBOLS

Description	Application	Symbol
Triangle:	The direction of flow and the nature of the fluid.	
– solid	Hydraulic flow.	▼
– in outline only.	Pneumatic flow or exhaust to atmosphere.	▽

FUNCTIONAL SYMBOLS *(continued)*

Description	Application	Symbol
Arrow	Indication of: – direction	
	– direction of rotation	
	– path and direction of flow through valves.	
	For regulating apparatus as in Pressure Control Valves both representations with or without a tail to the end of the arrow are used without distinction.	
	As a general rule the line perpendicular to the head of the arrow indicates that when the arrow moves the interior path always remains connected to the corresponding exterior path.	
Sloping arrow.	Indication of the possibility of a regulation or of a progressive variability.	

PUMPS AND COMPRESSORS

Description	Remarks	Symbol
Fixed capacity compressor. (Always one direction of flow)		
Fixed capacity pneumatic motor: – with one direction of flow		
– with two directions of flow.		

PUMPS AND COMPRESSORS *(continued)*

Description	Remarks	Symbol
Variable capacity pneumatic motor: – with one direction of flow – with two directions of flow.		
Oscillating motor: – hydraulic – pneumatic.		
Pump/motor units.	Unit with two functions, either as pump or as rotary motor.	
Fixed capacity pump/motor unit: – with reversal of the direction of flow – with one single direction of flow – with two directions of flow.	Functioning as pump or motor according to direction of flow. Functioning as pump or motor without change of direction of flow. Functioning as pump or motor with either direction of flow.	
Variable capacity pump/motor unit: – with reversal of the direction of flow – with one single direction of flow – with two directions of flow.		

CYLINDERS

Description	Remarks	Symbol	
		Detailed	Simplified
Single-acting cylinder: – returned by an unspecified force – returned by spring.	Cylinder in which the fluid pressure always acts in one and the same direction (on the forward stroke). General symbol when the method of return is not specified.		
Double-acting cylinder. – with single piston rod – with double-ended piston rod.	Cylinder in which the fluid pressure operates alternately in both directions (forward and backward strokes).		
Differential cylinder.	The action is dependent on the difference between the effective areas on each side of the piston.		
Cylinder with cushion: – with single fixed cushion – with double fixed cushion – with single adjustable cushion – with double adjustable cushion.	Cylinder incorporating fixed cushion acting in one direction only. Cylinder with fixed cushion acting in both directions.		
Telescopic cylinder: – single-acting – double-acting.	The fluid pressure always acts in one and the same direction (on the forward stroke). The fluid pressure operates alternately in both directions (forward and backward strokes)		

PRESSURE INTENSIFIERS

Description	Remarks	Symbol	
		Detailed	Simplified
For one type of fluid.	*eg* a pneumatic pressure *x* is transformed into a higher pneumatic pressure *y*.		
For two types of fluid.	*eg* a pneumatic pressure *x* is transformed into a higher hydraylic pressure *y*.		
Air–oil actuator.	Equipment transforming a pneumatic pressure into a substantially equal hydraulic pressure or vice versa.		

CONTROL VALVES

Description	Remarks	Symbol
Method of representation of valves.	Made up of one or more squares and arrows. *(In circuit diagrams hydraulic and pneumatic units are normally shown in the unoperated conditions).*	
One single square.	Indicates unit for controlling flow or pressure, having in operation an infinite number of possible positions between its end positions so as to vary the conditions of flow across one or more of its ports, thus ensuring the chosen pressure and/or flow with regard to the operating conditions of the circuit.	
Two or more squares.	Indicate a directional control valve having as many distinct positions as there are squares. The pipe connections are normally represented as connected to the box representing the unoperated condition. The operating positions are deduced by imagining the boxes to be displaced so that the pipe connections correspond with the ports of the box in question.	

CONTROL VALVES *(continued)*

Description	Remarks	Symbol
Simplified symbol for valves in cases of multiple repetition.	The number refers to a note on the diagram in which the symbol for the valve is given in full.	

DIRECTIONAL CONTROL VALVES

Description	Remarks	Symbol
Flow paths: – one flow path – two closed ports – two flow paths – two flow paths and one closed port – two flow paths with cross connection – one flow path in a by-pass position, two closed ports.	Square containing interior lines.	
Non-throttling directional control valve.	The unit provides distinct circuit conditions each depicted by a square. Basic symbol for two-position directional control valve. Basic symbol for three-position directional control valve. A transitory but significant condition between two distinct positions is optionally represented by a square with dashed ends. A basic symbol for a directional control valve with two distinct positions and one transitory intermediate condition.	

CONTROL VALVES *(continued)*

Description	Remarks	Symbol
Designation: The first figure in the *designation* shows the number of ports (excluding pilot ports) and the second figure the number of distinct positions.		
Directional control valve 2/2: – with manual control – controlled by pressure operating against a return spring (*eg* on air unloading valve).	Directional control valve with two ports and two distinct positions.	
Directional control valve 3/2: – controlled by pressure in both directions – controlled by solenoid with return spring.	Directional control valve with three ports and two distinct positions.	
Directional control valve 4/2: – controlled by pressure in both directions by means of a pilot valve (with a single solenoid and spring return).	Directional control valve with four ports and two distinct positions.	Detailed Simplified

CONTROL VALVES *(continued)*

Description	Remarks	Symbol
Directional control valve 5/2: – controlled by pressure in both directions.	Directional control valve with five ports and two distinct positions.	
Throttling directional control:	The unit has two extreme positions and in infinite number of intermediate conditions with varying degrees of throttling. All the symbols have parallel lines along the length of the boxes. Showing the extreme positions and a central (neutral) positions.	
– with two ports (one throttling orifice)	*eg* tracer valve plunger operated against a return spring.	
– with three ports (two throttling orifices)	*eg* directional control valve controlled by pressure against a return spring.	
– with four ports (four throttling orifices).	*eg* tracer valve, plunger operated against a return spring.	
Electro-hydraulic servo-valve: Electro-pneumatic servo-valve: – single-stage	A unit which accepts an analogue electrical signal and provides a similar analogue fluid power output. – with direct operation.	
– two-stage with mechanical feedback	– with indirect pilot-operation.	

CONTROL VALVES *(continued)*

Description	Remarks	Symbol
– two-stage with hydraulic feedback.	– with indirect pilot-operation.	

NON-RETURN VALVES, SHUTTLE VALVE, RAPID EXHAUST VALVE

Description	Remarks	Symbol
Non-return valve:		
– free	Opens if the inlet pressure is higher than the outlet pressure.	
– spring-loaded	Opens if the inlet pressure is greater than the outlet pressure plus the spring pressure.	
– pilot-controlled	With pilot control it is possible to prevent: – closing of the valve	
	– opening of the valve.	
– with restriction.	Unit allowing free flow in one direction but restricted flow in the other.	
Shuttle valve.	The inlet port connected to the higher pressure is automatically connected to the outlet port while the other inlet port is closed.	
Rapid exhaust valve.	When the inlet port is unloaded the outlet port is freely exhausted.	

PRESSURE CONTROL VALVES

Description	Remarks	Symbol

Pressure control valve:
– one throttling orifice
  normally closed
 	
– one throttling orifice
  normally open
 	
– two throttling orifices,
  normally closed.		

Pressure relief valve (safety valve) – with remote pilot control.	Inlet pressure is controlled by opening the exhaust port to the reservoir or to atmosphere against an opposing force (for example a spring). The pressure at the inlet port is limited, or to that corresponding to the setting of a pilot control.	
Proportional pressure relief.	Inlet pressure is limited to a value proportional to the pilot pressure.	
Sequence valve.	When the inlet pressure overcomes the opposing force of the spring, the valve opens permitting flow from the outlet port.	
Pressure regulator or reducing valve (reducer of pressure) – without relief port – without relief port with remote control	A unit which, with a variable inlet pressure, gives substantially constant output pressure provided that the inlet pressure remains higher than the required outlet pressure. Outlet pressure is dependent on the control pressure.	

SOURCES OF ENERGY *(continued)*.

Description	Remarks	Symbol
– flexible pipe – electric line.	Flexible hose, usually connecting moving parts.	
Pipeline junction.		
Crossed pipelines.	Not connected.	
Air bleed.		
Exhaust port: – plain with no provision for connection – threaded for connection.		
Power take-off: – plugged – with take-off line.	On equipment or lines, for energy take-off or measurement.	
Quick-acting coupling: – connected, without mechanically opened non-return valve – connected, with mechanically opened non-return valves – uncoupled, with open end – uncoupled, closed by free non-return valve.		
Rotary connection: – one-way – three-way.	Line junction allowing angular movement in service.	
Silencer.		

RESERVOIRS

Description	Remarks	Symbol
Reservoir open to atmosphere: – with inlet pipe above fluid level – with inlet pipe below fluid level – with a header line.		
Pressurized reservoir.		
Accumulators.	The fluid is maintained under pressure by a spring, weight or compressed gas (air, nitrogen, *etc*).	
Filter or strainer.		
Water trap: – with manual control – automatically drained.		
Filter with water trap: – with manual control – automatically drained.		
Air dryer.	A unit drying air (*eg* chemical means).	
Lubricator.	Small quantities of oil are added to the air passing through the unit, in order to lubricate equipment receiving the air.	

RESERVOIRS *(continued).*

Description	Remarks	Symbol
Conditioning unit.	Consisting of filter, pressure regulator, pressure gauge and lubricator.	*Detailed symbol* *Simplified symbol*
Heat exchangers.	Apparatus for heating or cooling the circulating fluid.	
Temperature controller.	The fluid temperature is maintained between two-pre-determined values. The arrows indicate that heat may be either introduced or dissipated.	
Cooler.	The arrows in the diamond indicate the extraction of heat. – without representation of the flow lines of the coolant – indicating the flow lines of the coolant.	
Heater.	The arrows in the diamond indicate the introduction of heat.	

CONTROL MECHANISMS

Description	Remarks	Symbol
Rotating shaft: – in one direction – in either direction.	The arrow indicates rotation.	
Detent.	A device for maintaining a given position.	
Locking device.	The symbol * for unlocking control is inserted in the square.	

CONTROL MECHANISMS (continued)

Description	Remarks	Symbol
Over-centre device.	Prevents the mechanism stopping in a dead centre position.	
Pivoting devices: – simple – with traversing lever – with fixed fulcrum.		

CONTROL METHODS

Description	Remarks	Symbol
Muscular control:	General symbol (without indication of control type).	
– by push button		
– by lever		
– by pedal.		
Mechanical control: – by plunger or tracer		
– by spring		
– by roller		
– by roller, operating in one direction only.		
Electrical control: – by solenoid	– with one winding	

CONTROL METHODS (continued)

Description	Remarks	Symbol
– by electric motor.	– with two windings operating in opposite directions – with two windings operating in a variable way progressively operating in opposite directions.	
Control by application or release of pressure Direct-acting control: – by application of pressure.		
– by release of pressure – by different control areas.	In the symbol the larger rectangle represents the larger control area, *ie* the priority phase.	
Indirect control, pilot actuated: – by application of pressure – by release of pressure.	General symbol for pilot directional control valve.	
Interior control paths.	The control paths are inside the unit.	
Combined control: – by solenoid and pilot directional valve	The pilot directional valve is actuated by the solenoid.	

CONTROL METHODS *(continued)*

Description	Remarks	Symbol
– by solenoid or pilot-directional valve.	Either may actuate the control independently.	1)
Mechanical feedback	The mechanical connection of a control apparatus moving part to a controlled apparatus moving part is represented by the symbol which joins the two parts connected.	2) 1) Controlled apparatus. 2) Control apparatus.

SUPPLEMENTARY EQUIPMENT

Description	Remarks	Symbol
Pressure measurement: – pressure gauge.	The point on the circle at which the connection joins the symbol is immaterial.	
Temperature measurement: – thermometer.	The point on the circle at which the connection joins the symbol is immaterial.	
Measurement of flow: – flow meter – integrating flow meter.		
Other apparatus Pressure electric switch.		

LOGIC SYMBOLS AND THEIR PNEUMATIC EQUIVALENTS.

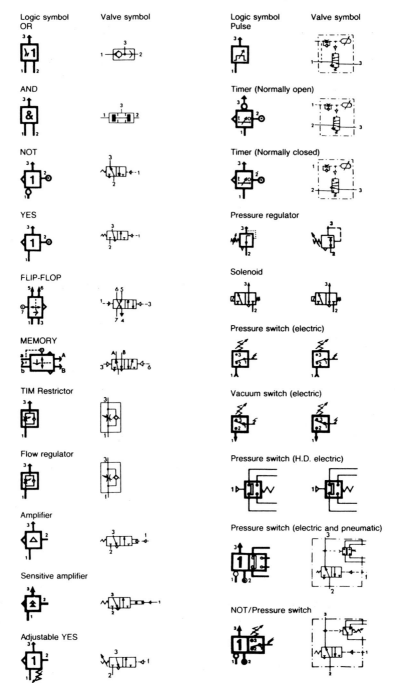

Logic symbol | Valve symbol
Logic symbol | Valve symbol

OR

AND

NOT

YES

FLIP-FLOP

MEMORY

TIM Restrictor

Flow regulator

Amplifier

Sensitive amplifier

Adjustable YES

Pulse

Timer (Normally open)

Timer (Normally closed)

Pressure regulator

Solenoid

Pressure switch (electric)

Vacuum switch (electric)

Pressure switch (H.D. electric)

Pressure switch (electric and pneumatic)

NOT/Pressure switch

CETOP recommendations for circuit diagrams

The following rules and recommendations are extracted from CETOP RP41 – Hydraulic and Pneumatic Systems Circuit Diagrams. The purpose of this document is to facilitate the design, construction, description and maintenance of fluid power systems by aiding communication through uniform presentation. The following recommendations are based on the appropriate CETOP documents and applicable ISO standards (mostly adopted as British Standards). It is not necessary for a circuit diagram to incorporate all the requirements listed but anything which is shown must be in accordance with the appropriate section. They apply generally to all fluid power systems.

Design of circuit diagrams

The circuit diagram must show fluid circulation for all the control and motion functions. All hydraulic and pneumatic components and flow line junctions in the plant will be represented functionally. In addition all control components that affect the operational processes of the fluid power system will be shown.

The circuit diagram need not be constructed with reference to the physical arrangement of the system. Interdependent circuits should be shown on one diagram. For electro-hydraulic or electro-pneumatic controls, the circuit diagram will however be divided into one hydraulic and one pneumatic circuit diagram and if necessary a separate electrical circuit.

The following circuit diagram formats should be used:

- ISO A format.
  Length preferably 297 mm, exceptionally 420 or 594 mm.
  Width up to 1189 mm.

- Diagrams will be folded to ISO A format with a binding margin (except for diagrams for reproduction).

- The layout of the circuit diagram must be clear. Care should be taken to see that the circuit paths are easily followed. Pneumatic circuits should be drawn if possible in the sequence of the operational process of functions. The components of the individual control chains and groups should be shown wherever possible in the direction of energy flow.

- Cylinders and directional control valves should preferably be shown horizontally. Fluid lines should be represented wherever possible by straight lines without intersections.

- The location of control components such as sequence switch cams and simple cams should be indicated. If a signalling element is to be actuated by a one-way trip, then the actuating direction will be indicated by an arrow. In the case of electro-hydraulic and electro-pneumatic control systems, signalling and final control elements such as limit switches and electro-magnetic valves will be shown in both circuit diagrams.

- Individual sub-circuits may be identified. The function of each operating element should be indicated beside it (for example clamping or lifting). Where necessary for a better understanding, a function chart can be drawn up. Components should be drawn up in accordance with ISO 1219 (BS 2917) for fluid power equipment. Where no symbol exists, the operation of the part should be clearly shown by means of a simplified sectional representation, or the like.

• Symbols that are repeated in the diagram can be represented by a numbered rectangle and the corresponding symbols shown separately.

Data to go on circuit diagram

Every component must be clearly identified by an item number. Components in a given sub-circuit may be given a common prime identification plus a running sub-number, for example sub-circuit 2, component 3 = 2.3.

Component operating positions

In the case of valves with discrete switching positions, these positions may be identified by letter, the letter 'O' being used only for spring return positions (spring-loaded valves).

Identification of ports

Ports will be identified on the circuit diagram by the characters marked on the components of the connection plate. This applies to individual components and sub-assemblies.

Identification of oil or air lines

The pipelines are to be drawn according to their function in line with BS 2917.

In cases where hydraulic and pneumatic circuits are represented on one diagram, they should be identified in accordance with BS 2917.

Should it be necessary to identify the pipelines more precisely, the following code applies:

full line	pressure line.
dashed line	pilot control line.
short dotted line	drain or bleed line.
full line with short cross lines	return or exhaust line.
full line with crosses	replenishing line.
full line with arrows	pump inlet line.

If pipelines are to be identified by colour, the following code should be used:

red line	pressure line.
red dashed line	pilot control line.
blue line	return or exhaust line.
blue dashed line	drain or bleed line.
green line	replenishing line.
yellow line	pump inlet line.

Advertisers Buyer's Guide

ADVERTISERS INDEX

BUYER'S GUIDE

TRADE NAMES INDEX

ADVERTISERS INDEX

Elsevier Science, Regional Sales Office
Customer Support Department
9-15 Higashi-Azabu, 1-chome, Minato-ku
Tokyo, 106 Japan
Tel: +81 3 5561 5033   Fax: +81 3 5561 5047
E-mail: info@elsevier.co.jp

Elsevier Science, Regional Sales Office
Customer Support Department
PO Box 211, 1000 AE Amsterdam
The Netherlands
Tel: +31 20 485 3757   Fax: +31 20 485 3432
E-mail: info-f@elsevier.nl

Flair International Ltd, Hazelton Interchange, Lakesmere Road,   Facing page 203
Horndean, Hants, PO8 9JU, United Kingdom   Facing page 209
Tel: +44 (0)1705 591021   Fax: +44 (0)1705 596799

IMI Norgren Ltd, P.O. Box 22, Eastern Avenue, Lichfield,   Facing page 553
Staffordshire, WS13 6SB, United Kingdom
Tel: +44 (0)1543 414333   Fax: +44 (0)1543 268052

Ingersoll-Rand European Sales Ltd, P.O. Box 2,   Facing page 66
Chorley New Road, Horwich, Bolton, BL6 6JM, United Kingdom
Tel: +44 (0)1204 690690   Fax: +44 (0)1204 690388

Numatics Limited, Cherry Court Way, Leighton Buzzard,   Facing page 424
LU7 8UH, United Kingdom
Tel: +44 (0)1525 370735   Fax: +44 (0)1525 382567

Parker Hannifin Plc, Pneumatic Division, Walkmill Lane,   Facing Contents
Bridgtown, Cannock, Staffordshire, WS11 3LR, United Kingdom
Tel: +44 (0)1543 456000   Fax: +44 (0)1543 456161

PSI (Precision Scientific Innovations Ltd),   Facing page 167
Industrial Estate South, Bowburn, Durham, DH6 5AD,
United Kingdom
Tel: +44 (0)191 377 0550   Fax: +44 (0)191 377 0769

RGS Electro Pneumatics Ltd, West End Business Park,   Facing page 424
Blackburn Road, Oswaldtwistle, Near Accrington,
Lancashire, BB5 4W, United Kingdom
Tel: +44 (0)1254 872277   Fax: +44 (0)1254 390133

BUYER'S GUIDE

AIR COMPRESSORS
Screw
Atlas Copco Compressors Ltd
CFP
CompAir BroomWade Ltd
Ingersoll-Rand European Sales Ltd
Thomas Wright/Thorite Group Ltd

Reciprocating
Atlas Copco Compressors Ltd
CompAir BroomWade Ltd
Ingersoll-Rand European Sales Ltd
Thomas Wright/Thorite Group Ltd

Diaphragm
Thomas Wright/Thorite Group Ltd

Lubricated
Atlas Copco Compressors Ltd
CFP
CompAir BroomWade Ltd
Ingersoll-Rand European Sales Ltd
Thomas Wright/Thorite Group Ltd

Vane
Thomas Wright/Thorite Group Ltd

Rotary tooth
Atlas Copco Compressors Ltd

Dynamic (centrifugal and axial)
Atlas Copco Compressors Ltd
Ingersoll-Rand European Sales Ltd

Non-lubricated
Atlas Copco Compressors Ltd

CFP
CompAir BroomWade Ltd
Ingersoll-Rand European Sales Ltd
Thomas Wright/Thorite Group Ltd

Mobile
Atlas Copco Compressors Ltd
CFP
Ingersoll-Rand European Sales Ltd

Stationary
Atlas Copco Compressors Ltd
CFP
CompAir BroomWade Ltd
Ingersoll-Rand European Sales Ltd
Thomas Wright/Thorite Group Ltd

Roots-type blowers
Ingersoll-Rand European Sales Ltd

Compressor installations
Atlas Copco Compressors Ltd
CFP
Ingersoll-Rand European Sales Ltd
Thomas Wright/Thorite Group Ltd

Noise reduction techniques for compressors
Atlas Copco Compressors Ltd
Ingersoll-Rand European Sales Ltd

Noise reduction techniques for tools
Scroll
Atlas Copco Compressors Ltd

AIR RECEIVERS
Construction
Atlas Copco Compressors Ltd
CFP
Cool Technology Ltd

Legislation
Atlas Copco Compressors Ltd
Cool Technology Ltd

Radiators
Atlas Copco Compressors Ltd
Cool Technology Ltd

Heat exchangers
Atlas Copco Compressors Ltd
Cool Technology Ltd

Coolers
Atlas Copco Compressors Ltd
CFP
Cool Technology Ltd

AIR DRYERS
Breathing air
Atlas Copco Compressors Ltd
CFP
Flair International Ltd
ultrafilter international
Zander (UK) Ltd

Process air
Atlas Copco Compressors Ltd
CFP
Flair International Ltd
Ingersoll-Rand European Sales Ltd
Thomas Wright/Thorite Group Ltd
ultrafilter international
Zander (UK) Ltd

Adsorption
Atlas Copco Compressors Ltd
CFP
CompAir BroomWade Ltd
Flair International Ltd
Ingersoll-Rand European Sales Ltd
ultrafilter international
Zander (UK) Ltd

Absorption
Atlas Copco Compressors Ltd
CFP
Flair International Ltd
ultrafilter international

Heatless
Atlas Copco Compressors Ltd
CFP
CompAir BroomWade Ltd
Flair International Ltd
Ingersoll-Rand European Sales Ltd
ultrafilter international
Vickers System Division
Zander (UK) Ltd

Refrigerant
Atlas Copco Compressors Ltd
CFP
CompAir BroomWade Ltd
Flair International Ltd
Ingersoll-Rand European Sales Ltd
Thomas Wright/Thorite Group Ltd
ultrafilter international
Vickers System Division
Zander (UK) Ltd

Air filters
Atlas Copco Compressors Ltd
CFP
CompAir BroomWade Ltd
Flair International Ltd
IMI Norgren Ltd
Ingersoll-Rand European Sales Ltd
ultrafilter international
Vickers System Division
Zander (UK) Ltd

Pipe flow
Atlas Copco Compressors Ltd
CFP
Ultrafilter Ltd

PIPE MATERIALS AND FITTINGS
Aluminium
Ingersoll-Rand European Sales Ltd

Copper
IMI Norgren Ltd
Thomas Wright/Thorite Group Ltd

Nylon
IMI Norgren Ltd
Parker Hannifin Plc
SMC Pneumatics (UK) Ltd
Thomas Wright/Thorite Group Ltd
Vickers System Division

AIR FLOW MEASUREMENT
CFP

REGULATING VALVES
Pressure
Atlas Copco Compressors Ltd
CFP
IMI Norgren Ltd
Parker Hannifin Plc
RGS Electro Pneumatics Ltd
SMC Pneumatics (UK) Ltd
Thomas Wright/Thorite Group Ltd
Vickers System Division

Flow
Atlas Copco Compressors Ltd
CFP
Parker Hannifin Plc
RGS Electro Pneumatics Ltd
SMC Pneumatics (UK) Ltd
Thomas Wright/Thorite Group Ltd
Vickers System Division

Drain valves
Atlas Copco Compressors Ltd
CFP
IMI Norgren Ltd
Ingersoll-Rand European Sales Ltd
Parker Hannifin Plc
Thomas Wright/Thorite Group Ltd

Pressure Gauges
Atlas Copco Compressors Ltd
IMI Norgren Ltd
Ingersoll-Rand European Sales Ltd

Parker Hannifin Plc
RGS Electro Pneumatics Ltd
Thomas Wright/Thorite Group Ltd
Vickers System Division

Lubricators
Atlas Copco Compressors Ltd
IMI Norgren Ltd
Ingersoll-Rand European Sales Ltd
Parker Hannifin Plc
RGS Electro Pneumatics Ltd
SMC Pneumatics (UK) Ltd
Thomas Wright/Thorite Group Ltd
Vickers System Division

TOOLS
Industrial
Ingersoll-Rand European Sales Ltd
Thomas Wright/Thorite Group Ltd

Contractors
Ingersoll-Rand European Sales Ltd

Mining & quarrying
Ingersoll-Rand European Sales Ltd

Air motors
CFP
Ingersoll-Rand European Sales Ltd
Parker Hannifin Plc
Thomas Wright/Thorite Group Ltd

Mining & quarrying equipment
Ingersoll-Rand European Sales Ltd

Compressed air in martine applications
IMI Norgren Ltd
Ingersoll-Rand European Sales Ltd

Air bubble techniques
CFP
Ingersoll-Rand European Sales Ltd

Paint spraying
Thomas Wright/Thorite Group Ltd

Air industry standards
CFP

Two-pressure installations
CFP

Pneumatic conveying
CFP
Vickers System Division
Ingersoll-Rand European Sales Ltd

Air springs and actuators
IMI Norgren Ltd
RGS Electro Pneumatics Ltd
Thomas Wright/Thorite Group Ltd

Cylinders
CFP
IMI Norgren Ltd
Ingersoll-Rand European Sales Ltd
Parker Hannifin Plc
RGS Electro Pneumatics Ltd
SMC Pneumatics (UK) Ltd
Thomas Wright/Thorite Group Ltd
Vickers System Division

VALVES
Manual
IMI Norgren Ltd
Ingersoll-Rand European Sales Ltd
Parker Hannifin Plc
RGS Electro Pneumatics Ltd
SMC Pneumatics (UK) Ltd
Thomas Wright/Thorite Group Ltd
Vickers System Division

Pilot
CFP
IMI Norgren Ltd
Parker Hannifin Plc
RGS Electro Pneumatics Ltd
SMC Pnemuatics (UK) Ltd
Thomas Wright/Thorite Group Ltd
Vickers System Division

Electrical
CFP
IMI Norgren Ltd
Parker Hannifin Plc
RGS Electro Pneumatics Ltd

Thomas Wright/Thorite Group Ltd
Vickers System Division

Automation and Robotics
CFP
IMI Norgren Ltd
Parker Hannifin Plc
SMC Pneumatics (UK) Ltd
Vickers System Division

Lubricants
Ingersoll-Rand European Sales Ltd

Sensors and controllers
CFP
Parker Hannifin Plc
Vickers System Division

Sensors
CFP

Silencers
IMI Norgren Ltd
Parker Hannifin Plc
RGS Electro Pneumatics Ltd
Thomas Wright/Thorite Group Ltd
Vickers System Division

Vacuum pumps
CompAir BroomWade Ltd
IMI Norgren Ltd
Ingersoll-Rand European Sales Ltd
Thomas Wright/Thorite Group Ltd

Vacuum systems
IMI Norgren Ltd
Parker Hannifin Plc
Vickers System Division

Waste heat usage
Ingersoll-Rand European Sales Ltd
ultrafilter international

Vibration reduction in tools
IMI Norgren Ltd
ingersoll-Rand European Sales Ltd

Flow meters
IMI Norgren Ltd

Air preparation equipment
Parker Hannifin Plc

Air line filters
IMI Norgren Ltd

Rodless cylinders
IMI Norgren Ltd

Impact cylinders
IMI Norgren Ltd

Pneumatics training courses
IMI Norgren Ltd
CFP

Pneumatics equipment servicing
IMI Norgren Ltd

Quick release couplings
IMI Norgren Ltd

Tube fittings
IMI Norgren Ltd

Alarm units/systems
CFP

Electronic drains
Zander (UK) Ltd

Oil/water separations
Zander (UK) Ltd

Oil removal filters
Zander (UK) Ltd

Sterile filters/vent filters
Zander (UK) Ltd

Oil vapour absorbers
Zander (UK) Ltd

High pressure dryers (350 bar)
Zander (UK) Ltd

Exhaust silencers and reclassifiers
Zander (UK) Ltd

Membrane dryers
Zander (UK) Ltd

Pneumatic actuators/cylinders
SMC Pneumatics (UK) Ltd

Pneumatic fittings & tubing
SMC Pneumatics (UK) Ltd

Pneumatic control systems
SMC Pneumatics (UK) Ltd

Pneumatic instrumentation
SMC Pneumatics (UK) Ltd

Pneumatics training
SMC Pneumatics (UK) Ltd

Condensate management, drains & condensate separators
ultrafilter international

Air valves
Numatics Ltd

Air cylinders
Numatics Ltd

TRADE NAMES INDEX

AIR AUDIO – Pneumatic alarm units and systems – CFP

AQUAFIL-K – Emulsion separator – Zander (UK) Ltd

ARO – Tools, pumps, valves & cylinders – Ingersoll-Rand European Sales Ltd

ATLAS AUTOMATION – Pneumatic control and automation equipment – Parker Hannifin Plc

Beech – Glandless spool valves – IMI Norgren Ltd

BTD – Thermal refrigerant dryers – CompAir BroomWade Ltd

Cemtac – Centrifugal air compressor – Ingersoll-Rand European Sales Ltd

CIRCUITMASTER – Pneumatic sequential control system – CFP

COOLTECH – Air receivers, air/oil separators, filter bodies, paint containers, dryer vessels, intercooler/aftercoolers, air blast radiators, cooling towers, pressure vessels – Cool Technology Ltd

CUBIC TRAINER – Multi-dimensional training rigs – CFP

Cyclon – Lubricated rotary screw compressors – CompAir BroomWade Ltd

Cyclone – Air tools – Ingersoll-Rand European Sales Ltd

Delair – Refrigerant/dessicant dryers – Flair International Ltd

Deltech – Filtration – Flair International Ltd

Dollinger – Filtration – Flair International Ltd

Dryclon – Oil-free rotary screw compressors – CompAir BroomWade Ltd

DSD/DXD/DRD – Desiccant Air Dryers – CompAir BroomWade Ltd

ECODRAIN – Electronic condensate drain- Zander (UK) Ltd

ECOSEP-S – Oil/water separator – Zander (UK) Ltd

ECODRY – Heatless adsorption dryers – Zander (UK) Ltd

Enots – Tube fittings – IMI Norgren Ltd

EUROMIDAS – Pneumatic cylinders – Vickers System Division

Fleetfit – Push-in tube fittings for vehicles – IMI Norgren Ltd

Ingersoll-Rand – Compressors, air dryers, tools, valves – Ingersoll-Rand European Sales Ltd

LANG – Pneumatic control equipment – Vickers System Division

Lintron – Rodless cylinders – IMI Norgren Ltd

Mantonair – Cylinders and valves – IMI Norgren Ltd

MICROFILTER – Oil removal filters – Zander (UK) Ltd

MIDAS – Pneumatic cylinders – Vickers System Division

Norgren – Air line units – IMI Norgren Ltd

Nugget – Directional control valves – IMI Norgren Ltd

Numasizing (USA) – Accurate component sizing technique – Numatics Ltd

Numatics Actuator (USA) – Air cylinders – Numatics Ltd

Numatics Flexiblok (USA) – Valve manifolds, filters, regulators & lubricators – Numatics Ltd

Numatrol (USA) – Air logic controls – Numatics Ltd

Olympian/Olympian Plus – Air line units – IMI Norgren Ltd

PEGASUS – Air treatment and control equipment – Vickers System Division

Pneufit – Push-in tube fittings – IMI Norgren Ltd

PNEUMETHODS – Special pneumatic devices and systems – CFP

Puraine – Air line filter units – IMI Norgren Ltd

SCHRADER BELLOWS – Pneumatic control and automation equipment – Parker Hannifin Plc

Sierra – Oil-free screw compressor – Ingersoll-Rand European Sales Ltd

Simplair – Aluminium pipework system – Ingersoll-Rand European Sales Ltd

SSR – Rotary screw compressor – Ingersoll-Rand European Sales Ltd

Technolab – Filtration – Flair International Ltd

TELEPNEUMATIC – Pneumatic control and automatisation equipment – Parker Hannifin Plc

Thorite – Air compressors – Thomas Wright/Thorite Group

TRANSPORTER – Fluidic transfer and processing systems – CFP

Type 30 – Reciprocating compressor – Ingersoll-Rand European Sales Ltd

ultrac – High efficiency filters – ultrafilter international

ultraine – Air line filter units – IMI Norgren Ltd

ultradri – Water separator – ultrafilter international

ultradepth – Sterile filters for gases – ultrafilter international

ultradepth – Vent filters – ultrafilter international

ultradepth – Sterile air system – ultrafilter international

ultrapoly – Fine filter – ultrafilter international

ultraporex – Fine filter – ultrafilter international

ultrex – Steam filters – ultrafilter international

ultrapur – Pure gas filters – ultrafilter international

ultrapolyplea – Totally disposable filters – ultrafilter international

ultrapolymem – Totally disposable filters – ultrafilter international

ultranylomem – Totally disposable filters – ultrafilter international

ultradrain – Condensate drain – ultrafilter international

ultramat plus pneumo – Condensate drain – ultrafilter international

ultramat plus tronic – Condensate drain – ultrafilter international

ultramatic – Condensate drain – ultrafilter international

ultramatic sensor – Condensate drain – ultrafilter international

ultrasep superplus – Oil/water separator – ultrafilter international

ultraaqua – Oil/water separator – ultrafilter international

ultraaqua caramen autoclean – Oil/water separator – ultrafilter international

ultrapure – Breathing air system – ultrafilter international

ultracheck vario – Test device – ultrafilter international

ultrapac – High efficiency dryers – ultrafilter international

ultrastrockner – High efficiency dryers – ultrafilter international

ultrasorp – High efficiency dryers – ultrafilter international

ultracool watercooled – Aftercoolers – ultrafilter international

ultracool aircooled – Aftercoolers – ultrafilter international

oilfreepac – Purification packages – ultrafilter international

pharmapac – Purification packages – ultrafilter international

ultrapurepac – Purification packages – ultrafilter international

ultratroc – Refrigerant dryers – ultrafilter international

V Compact – Reciprocating compressors – CompAir BroomWade Ltd

Veratec – Air tools – Ingersoll-Rand European Sales Ltd

V Major – Reciprocating compressors – CompAir BroomWade Ltd

Walter – Glandless spool valves – IMI Norgren Ltd

WS – Oil-free rotary screw compressors – CompAir BroomWade Ltd

Editorial Index

EDITORIAL INDEX